LATTICE QUANTUM FIELD THEORY
of the **DIRAC** and **GAUGE FIELDS**
Selected Topics

LATTICE QUANTUM FIELD THEORY
of the **DIRAC** and **GAUGE FIELDS**
Selected Topics

Belal Ehsan Baaquie

INCEIF, The Global University of Islamic Finance, Malaysia

World Scientific

NEW JERSEY · LONDON · SINGAPORE · BEIJING · SHANGHAI · HONG KONG · TAIPEI · CHENNAI · TOKYO

Published by

World Scientific Publishing Co. Pte. Ltd.

5 Toh Tuck Link, Singapore 596224

USA office: 27 Warren Street, Suite 401-402, Hackensack, NJ 07601

UK office: 57 Shelton Street, Covent Garden, London WC2H 9HE

Library of Congress Cataloging-in-Publication Data

Names: Baaquie, B. E., author.

Title: Lattice quantum field theory of the Dirac and gauge fields : selected topics / Belal Ehsan Baaquie.

Description: New Jersey : World Scientific Publishing, [2020] | Includes bibliographical references and index.

Identifiers: LCCN 2019047773 | ISBN 9789811209697 (hardcover) |

    ISBN 9789811209703 (ebook for institutions) | ISBN 9789811209710 (ebook for individuals)

Subjects: LCSH: Lattice gauge theories. | Quantum field theory. | Lattice

  field theory. | Dirac equation. | Quantum chromodynamics.

Classification: LCC QC793.3.G38 B33 2020 | DDC 530.14/3--dc23

LC record available at https://lccn.loc.gov/2019047773

British Library Cataloguing-in-Publication Data

A catalogue record for this book is available from the British Library.

For any available supplementary material, please visit

https://www.worldscientific.com/worldscibooks/10.1142/11536#t=suppl

This book is dedicated to the memory of my lifelong friends and companions, Masud Sadique Chullu (1943-2017), Yamin Chowdhury (1951-2018), Syed Muhammad Hashem Raja (1957-2018) and Shahadat Hossain (1953-2020). Their courage, integrity, sincerity, loyalty, spiritual insight and prescient vision continues to inspire all those who knew them – and their memories live on.

Acknowledgments

I would like to acknowledge and express my thanks to many outstanding scholars and researchers whose work motivated me to study quantum field theory and to grapple with its mathematical formalism. Two peerless scholars who greatly influenced me are Kenneth G. Wilson and Richard P. Feynman. I did my PhD thesis under the guidance of Kenneth G. Wilson at a time when he first introduced the concept of lattice gauge theory. Ken Wilson's visionary conception of quantum field theory greatly enlightened and inspired me, and continues to do so till today. As an undergraduate I had the privilege of meeting and conversing a number of times with Richard P. Feynman, and which left a lasting impression on me.

I thank Du Xin, Muhammad Mahmudul Karim and Nazmi Haskanbancha for valuable input in the preparation of the book. I thank Ali Namazie, Frederick H. Willeboordse, Spenta Wadia, Avinash Dhar, Michael Spalinski, H. R. Krishnamurthy, Michael Peskin, Steve Shenker, Daniel Rosenhouse, Don Lewis, Ivan Todorov and Waseem A. Sayed for many useful discussions.

I thank my precious family members Najma, Arzish, Farah and Tazkiah for their love, affection, delightful company and warm encouragement. They have made this book possible.

Preface

Quantum field theory is undoubtedly one of the most accurate and important scientific theory in the history of science. Relativistic quantum fields are the theoretical backbone of the Standard Model of particles and interactions. Relativistic and non-relativistic quantum fields are extensively used in a myriad branches of theoretical physics, from superstring theory, high energy physics and solid state physics to condensed matter, quantum optics, nuclear physics, astrophysics and so on.

Non-Abelian Yang-Mills gauge fields coupled to fermions are the workhorse of all advanced studies of quantum field theory. The Standard Model, and in particular, QCD (quantum chromodynamics) – which is the theory of quarks coupled to colored gluons – are exemplars of such quantum field theories.

Almost all books on quantum field theory study the continuum formulation of the Dirac and Yang-Mills fields [Polyakov (1986), Peskin and Schroeder (1995), Ryder (2001)] and there are many books that offer a comprehensive perspective on quantum field theory (for example, Weinberg (2010), Zinn-Justin (1993)).

To provide a different perspective to gauge fields, K. G. Wilson (1974) introduced lattice gauge theory: the formulation of quantum field theory of quarks and gluons defined on a lattice spacetime. The study of lattice Dirac and Yang-Mills fields provides a fresh perspective to the study of non-Abelian gauge field theories. The lattice gauge field has stood the test of time – continuing to provide new results – and is one of Wilson's lasting legacy to the edifice of theoretical physics [Baaquie et al. (2015)].

Quantum fields, and in particular fermions and gauge fields, defined on a lattice seems to be the only formulation that provides a mathematically rigorous definition of a quantum field theory. The lattice, in addition to providing an ultra-violet cut-off, also allows for the use of computational tools quite distinct from the continuum formulation. It is expected that with the increase in computational power, the lattice formulation will allow for an in-depth study of nonlinear quantum field theories and reveal features and secrets of quantum fields that are not possible to deduce using approximations, such as the weak or strong coupling perturbation theory.

There are many books on lattice field theory (e.g. Creutz (1983); Roth (1997); Montvay and Munster (1994)). Given the vast and increasingly complex mathematics of lattice quantum fields, it is virtually impossible for any one book to cover the entire terrain of lattice quantum field theory. The purpose of this book is to provide *a primer* and *an introduction*, to the *mathematical foundations* of the vast and ever changing landscape of lattice quantum fields.

A primer with a limited selection of topics can in itself be very interesting and useful. A primer focusing on a few selected aspects is easier for the reader to comprehend as compared with of an exhaustive book that serves as a reference tome meant for the specialists. The topics and calculations in this primer have been chosen to provide such an introduction.

Contents

Acknowledgments   vii

Preface   ix

1.   Synopsis   1

    1.1   Mathematical Background   2

    1.2   Lattice Gauge Field   3

    1.3   Lattice Dirac Field   4

    1.4   Lattice Gauge Theory Hamiltonian   4

Mathematical Background   7

2. $SU(\mathcal{N})$ Compact Lie Groups   9

    2.1   Introduction   9

    2.2   Lie Algebras 10

    2.3   Vielbein of $SU(\mathcal{N})$ 12

    2.4   Metric on $SU(\mathcal{N})$ Group Space 13

         2.4.1 $SU(2)$ metric 14

    2.5   The Invariant Haar Measure and Delta Function 15

         2.5.1   Delta function on group space 16

         2.5.2 $SU(2)$ measure 16

         2.5.3 $SU(2)$ measure: invariant 17

    2.6   Campbell-Baker-Hausdorff (CBH) Formula 17

    2.7   Irreducible Representations of $SU(\mathcal{N})$ 18

    2.8   Peter-Weyl Theorem 20

    2.9   Lie Algebra Generators: Differential Operators 21

    2.10 Vielbein e_{ab} and f_{ab} for $\mathcal{SU}(2)$ 23

    2.11 Left and Right Invariant Generators 24

         2.11.1 Differential realization of $SO(3)$ generators on state space . 26

    2.12 $SU(\mathcal{N})$ Character Function Expansion 26

xi

2.13 Summary   28

3. $SU(\mathcal{N})$ Kac-Moody Algebra   29

    3.1   Introduction   29
    3.2   Kac-Moody Algebras   30
    3.3   Functional Differentiation   31
        3.3.1   Chain rule   32
    3.4   Kac-Moody Generators   33
        3.4.1   Kac-Moody generator and the 2-cocycle   36
    3.5   Chiral Field   38
    3.6   The WZW Lagrangian   40
    3.7 $SU(2)$ Loop Group   43
    3.8 $SU(2)$ Kac-Moody Algebra   44
    3.9   Kac-Moody Commutation Equations   46
    3.10 Virasoro Generator: Point-Split Regularization   49
    3.11 Summary   53
    3.12 Appendix   54

4. $SU(\mathcal{N})$ Path Integrals   55

    4.1   Introduction   55
    4.2   Hamiltonian Operator   56
        4.2.1 $U(1)$ and $SU(2)$ evolution kernel   57
    4.3   Lattice Action for $SU(\mathcal{N})$ 58
    4.4   Classical Paths and Winding Number   59
    4.5 $U(1)$ Path Integral   63
    4.6 $U(1)$ Path Integral: Classical Paths   64
    4.7 $SU(\mathcal{N})$ Continuum Action   65
    4.8 $SU(2)$ Path Integration   66
        4.8.1 $SU(3)$ path integration   70
        4.8.2 $SU(\mathcal{N})$ path integration   72
    4.9   Summary   73
    4.10 Appendix: Continuum Limit of $SU(\mathcal{N})$ Path Integral   74

5. $SU(3)$ Character Functions   81

    5.1   Casimir Operator for $SU(3)$ 81
    5.2   Evolution Kernel for $SU(3)$ 83
    5.3   Character Functions of $SU(3)$ 84
    5.4   Summary   86

6. Fermion Calculus   87

    6.1   Fermionic Variables   88
    6.2   Fermion Integration   89

6.3   Fermion Hilbert Space 90
        6.3.1   Fermionic completeness equation 92
        6.3.2   Fermionic momentum operator 93
6.4   Antifermion State Space 94
6.5   Fermion and Antifermion Hilbert Space 95
6.6   Real and Complex Fermions: Gaussian Integration 96
        6.6.1   Complex Gaussian fermions 98
6.7   Fermionic Path Integral and Hamiltonian 100
6.8   Fermionic Operators 102
6.9   Fermion-Antifermion Hamiltonians 103
6.10 A Quadratic Hamiltonian 104
        6.10.1 Orthogonality and completeness 104
6.11 Fermion-Antifermion Lagrangian 105
6.12 Fermionic Transition Probability Amplitude 107
6.13 Quark Confinement 108
6.14 Summary 109

Lattice Gauge Field   **111**

7.   Non-Abelian Lattice Gauge Field   113

7.1   Introduction 113
7.2   The Weak Coupling Approximation 116
7.3   Gauge-Fixing the Lagrangian 118
7.4   Zero Mode 118
7.5   Gauge-Fixed Path Integral 119
7.6   The Faddeev-Popov Counter-Term 120
7.7   Abelian Gauge-Fixed Path Integral 121
7.8   Lattice Faddeev-Popov Non-Abelian Ghost Action 123
7.9   Lattice BRST Symmetry 126
7.10 Summary 130

8.   Abelian Lattice Gauge Field in $d = 3$ 131

8.1   Introduction 131
8.2   Strong Coupling Representation 132
8.3   Weak Coupling Representation 134
8.4   Gauge Invariance 135
8.5   Wilson Loop 137
8.6   Phase Transition 137
        8.6.1   Mean field approximation 139
8.7   Summary 140
8.8   Appendix A 141

8.9   Appendix B 142

9. Lattice Gauge Field Mass Renormalization   143

    9.1   Introduction 143
    9.2   The Propagator and Mass Renormalization 144
    9.3   The Computational Scheme 146
    9.4   Determination of m_α and m_μ 147
    9.5   Determination of m_c for $\mathcal{S}_c[\theta + B]$ 148
    9.6   Expansion of the Gauge Field Action 149
    9.7   Determination of m_A for $\mathcal{S}[\theta + B] + \mathcal{S}_\alpha[B]$ 151
    9.8   One-Loop Mass Renormalization 153
    9.9   Slavnov-Taylor Identity 154
    9.10   Summary 155

10. Gauge Field Block-Spin Renormalization   157

    10.1   Introduction 157
    10.2   Two-Dimensional Lattice Gauge Field 158
    10.3   The Renormalization Group Transformation 159
           10.3.1   Renormalization group and fixed points 165
    10.4   The Recursion Equation for d Dimensions 165
    10.5   Strong Coupling Approximation for $SU(2)$ 169
    10.6   Weak Coupling Expansion for $SU(2)$ 170
    10.7   Weak Coupling Approximation for $d = 4 + \epsilon$ 174
    10.8   Weak Coupling $SU(2)$ Gauge Field: β-Function 175
    10.9   Numerical Solution of the Recursion Equation 176
           10.9.1   Change of integration variable 176
           10.9.2   Numerical algorithm 178
           10.9.3   Total grid size $S(I)$ 179
    10.10 Numerical Results 182
    10.11 Summary: Confinement and Asymptotic Freedom 184
    10.12 Appendix 185

11. Lattice Gauge Field Hamiltonian   187

    11.1   Lattice Gauge Field Hamiltonian 188
    11.2   Gauge-Fixed Chromoelectric Operator 191
    11.3   Gauss's Law 194
    11.4   Gauge-Fixed Lattice Gauge Field Hamiltonian 197
    11.5   Hamiltonian and Covariant Gauge: Faddeev-Popov Quantization . 198
    11.6   Ghost State Space and Hamiltonian 199
           11.6.1   BRST cohomology: state space 201
    11.7   BRST Charge Q_B 202
    11.8 Q_B and State Space 204

11.8.1 Gupta-Bleuler condition 206

11.9 Summary 208

Lattice Dirac Field 209

12. Dirac Lattice Path Integral 211

12.1 Introduction 211

12.2 Dirac Field Coordinates 212

12.3 Dirac Lattice Lagrangian 213

12.4 Lattice Fermions and Chiral Symmetry 216

12.5 Dirac Field: Boundary Conditions 219

12.6 Dirac Fermionic State Space 220

12.7 Hilbert Space Metric and Transfer Matrix 223

12.8 Dirac Lattice Hamiltonian 227

12.9 Lattice Path Integral 229

12.9.1 Normalization constant 232

12.10 Evolution Kernel 233

12.11 Energy Eigenfunctions 235

12.12 Propagator 238

12.13 Summary 241

13. Dirac Hamiltonian 243

13.1 Fermionic Operators 244

13.2 Lattice Dirac Hamiltonian 245

13.3 Continuum Hilbert Space 247

13.4 Continuum Hamiltonian 248

13.5 Dirac Field's Energy Eigenfunctionals 254

13.6 Dirac Charge Operator 256

13.6.1 Momentum and spin operators 258

13.7 Finite Time Dirac Action 259

13.8 Continuum Evolution Kernel 261

13.9 Evolution Kernel: General Quadratic Case 263

13.10 Evolution Kernel: Dirac Hamiltonian 265

13.10.1 Chiral charge operator 267

13.11 Summary 267

Lattice Gauge Theory Hamiltonian 269

14. Lattice Gauge Theory Hamiltonian 271

14.1 Introduction 271

14.2 Finite Time Action and Transfer Matrix 272

14.3   Axial Gauge: $U_{n0} = \mathbb{I}$ 275
14.4   Noncanonical Fermion Anticommutation Equations 276
14.5   Lattice Gauge Theory Hamiltonian 277
14.6   Canonical Fermions 279
14.7   Color Charge Operator and Gauss's Law 280
14.8   Lattice Action from Lattice Hamiltonian 284
14.9   Summary 286
14.10 Appendix A: Fermion Calculus with Gauge Field 286
14.11 Appendix B: Matrix M 288
14.12 Appendix C: Lagrangian for an Asymmetric Lattice 289
14.13 Appendix D: Classical Continuum Limit 291

Bibliography   293

Index   297

Chapter 1

Synopsis

Lattice quantum field theory is quantum field theory defined on a lattice space-time. Quantum field theory based on the spacetime continuum is a vast subject that has been extensively studied since the 1950's – and many good books have been written on it. Almost all books on quantum field theory study the continuum formulation of the scalar, Dirac and Yang-Mills gauge fields [Polyakov (1986), Zinn-Justin (1993), Peskin and Schroeder (1995), Ryder (2001)], and there is not much that is gained by re-deriving well known results.

Lattice quantum field theory has come to the forefront of theoretical physics since the 1970's, largely due to the work of Ken Wilson. It is natural to start one's study of lattice quantum field theory with scalar quantum fields. This path has been eschewed since the primary result of the study of scalar lattice quantum fields is in developing numerical algorithms and applications based on these algorithms. Instead, the more complicated Dirac and gauge fields are studied to examine and study the mathematical structures and novel features that emerge from their formulation on a lattice spacetime.

Lattice gauge theory is the quantum field theory of the Dirac and gauge fields on a lattice spacetime. Lattice gauge theory was introduced by Wilson (1974) to study the theory of strong interactions, and in particular the confinement of quarks. Both the lattice Dirac and lattice gauge fields have features that are quite distinct from their continuum formulation and many new ideas have to be introduced to study their generalization to a lattice spacetime.

The book is grouped into the following Four Parts.

- Mathematical Background: Chapter 2 to Chapter 6.
- Lattice Gauge Field: Chapter 7 to Chapter 11.
- Lattice Dirac Field: Chapter 12 to Chapter 13.
- Lattice Gauge Theory Hamiltonian: Chapter 14.

```
┌ ─ ─ ─ ─ ─ ─ ─ ─ ─ ─ ─ ─ ─ ─ ┐                          ┌ ─ ─ ─ ─ ─ ─ ─ ─ ─ ─ ─ ┐
╎ 2.SU(N) Compact Lie Group   ╎────┐                      ╎ 6. Fermion Calculus   ╎
└ ─ ─ ─ ─ ─ ─ ─ ─ ─ ─ ─ ─ ─ ─ ┘    │                      └ ─ ─ ─ ─ ─ ─ ─ ─ ─ ─ ─ ┘
            │                       │
            ↓                       │
┌ ─ ─ ─ ─ ─ ─ ─ ─ ─ ─ ─ ─ ─ ┐      │
╎ 3.SU(N) Kac-Moody Algebra  ╎      │
└ ─ ─ ─ ─ ─ ─ ─ ─ ─ ─ ─ ─ ─ ┘      ↓
```

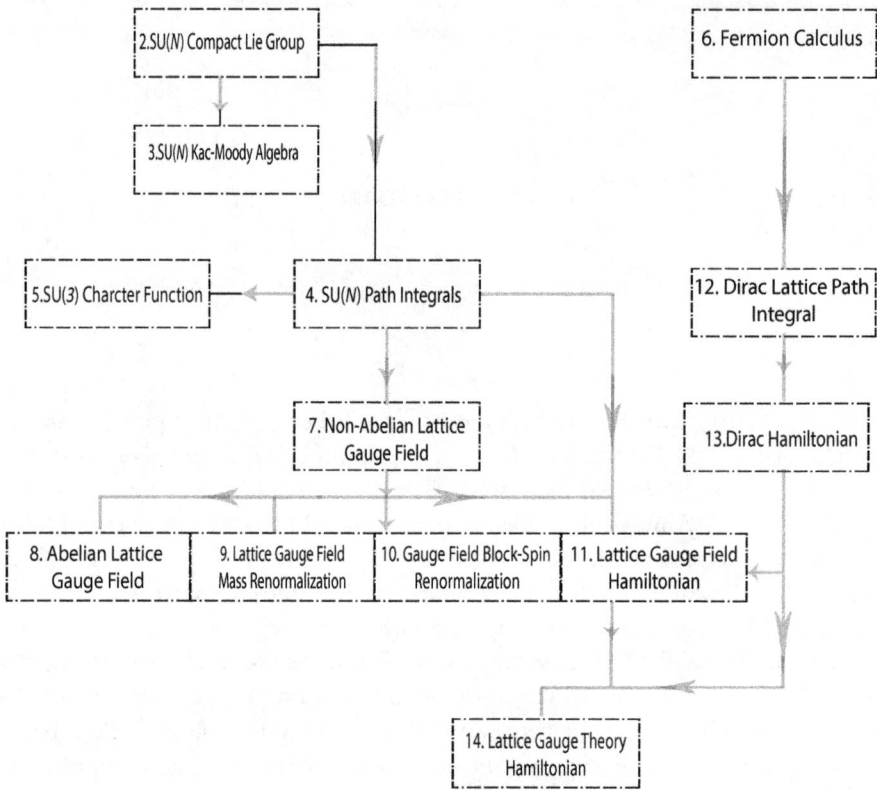

Fig. 1.1 Flow of the chapters.

1.1 Mathematical Background

This part is focused on the necessary mathematical background to the study of lattice gauge theory.

The fundamental degrees of freedom of lattice gauge theory are the finite elements of the $SU(\mathcal{N})$ Lie group and fermionic variables; the lattice gauge field is based on the calculus on compact group manifolds and the Dirac field requires fermionic variables.

Chapter 2 is a brief review of compact Lie groups. Finite Lie group elements are nonlinear variables. Integration over the compact group manifold is studied as the Feynman path integral is based on this theory of integration. The differential realization of the Lie group's generators is analyzed as it is required in defining the lattice gauge field's Hamiltonian.

Chapter 3 is on Kac-Moody algebras and is not required for any of the other Chapters. It has been included as it shows how Lie groups have an infinite dimensional generalization by enlarging the Lie algebra to include the Heisenberg algebra. Kac-Moody algebras can be used for defining new varieties of (lattice) quantum fields [Baaquie and Parwani (1996)] and also play an important role in superstring theory.

Chapter 4 studies the quantum mechanics of the compact $SU(\mathcal{N})$ degree of freedom and applies the calculus of compact Lie groups to define a path integral for a chiral Lagrangian. The quantum mechanics of the Lie group's nonlinear variables is a precursor to the more complex lattice quantum gauge field. The Hamiltonian studied in this Chapter will be seen to be the kinetic operator of the lattice gauge field.

Chapter 5 applies the result of Chapter 4 to derive the exact expression for the character functions of $SU(3)$ to illustrate the utility of the path integral. The character functions of $SU(3)$ are obtained by techniques that are independent of algebraic techniques.

Chapter 6 is on fermion calculus. Fermion calculus is unlike the calculus of real variables and this Chapter provides a brief introduction to fermion calculus from first principles. In particular, the distinctive feature of fermions is the concept of fermions and antifermions and this is discussed for the simplest case. The Dirac field is based on infinitely many independent complex fermionic variables, and this Chapter is a precursor to the discussion on the Dirac field.

1.2 Lattice Gauge Field

The chapters on lattice gauge field study the nonlinear Yang-Mills field defined on a Euclidean lattice spacetime. The lattice quantum gauge field is studied using both the path integral and Hamiltonian formulations. The lattice provides an ultra-violet cutoff for the quantum field and many interesting features of the lattice quantum fields are developed using the concept of the lattice cutoff. The Lagrangian is defined for a four-dimensional lattice embedded in Euclidean spacetime and analyzed following the treatment given by Baaquie (1977b).

In Chapter 7 the lattice formulation of Yang-Mills gauge fields is reviewed and is shown to have many features quite different from the continuum formulation. The key idea of gauge-fixing is discussed and it is shown that, since Lie group variables are compact, the lattice gauge field does not need gauge-fixing. The lattice gauge field is a generalization of the continuum formulation; to illustrate that the lattice theory has all the not-so-apparent symmetries, BRST symmetry is derived for the lattice theory.

In Chapter 8 the three-dimensional Abelian gauge field is studied as a warm-up to the more complex non-Abelian gauge field. A number of exact results are obtained, including the equivalence of the Abelian gauge field to the discrete Gaussian quantum field.

In Chapter 9 mass renormalization for the lattice gauge field is studied to one-loop using the technique of gauge-fixing and Feynman perturbation theory. Quadratic mass divergence occurs for many diagrams, which are expressed in terms of lattice constants. Nontrivial lattice constant identities lead to an exact cancellation of the quadratic divergence. Zero mass renormalization is essential for the consistency and renormalizability of the lattice gauge field, and the one-loop calculation verifies this.

The block-spin renormalization of the lattice gauge field, discussed for the Migdal model in Chapter 10, is a procedure that encodes Wilson's formulation of renormalization – and brings out very clearly the infinitely many (coupled) length scales that are inherent in every quantum field theory.

A simplified model proposed by Migdal for numerically renormalizing the gauge field is discussed for the $SU(2)$ lattice gauge field. The Migdal recursion equation is solved numerically and one can study the cross-over from weak to strong coupling. The Migdal model shows how the coupling constant of an asymptotically free non-Abelian gauge field continuously becomes larger and larger and smoothly goes over to the strong coupling confining phase. The numerical solution of Migdal's model provided the first evidence of the confinement of quarks [Baaquie (1977c)].

The lattice gauge field Hamiltonian is studied in Chapter 11. Gauge-fixing for the Hamiltonian requires a formalism based on constraining differential field operators and is quite distinct from the path integral approach based on the action. The exact gauge field Hamiltonian is obtained.

1.3   Lattice Dirac Field

The Dirac field is studied using fermionic variables. The well-known 'doubling' of the fermionic states is solved using Wilson fermions – which are shown to arise from the state space interpretation of the path integral. The lattice formulation allows one to study the finite time path integral with boundary conditions fixed by the lattice theory [Baaquie (1983b)]. The evolution kernel is exactly solved for the lattice using the classical solution of the Dirac field. The Ginsparg-Wilson result for lattice fermions is briefly reviewed.

The continuum limit is taken and greatly simplifies the evolution kernel. Using the properties of fermionic variables, an explicit representation is obtained for all the eigenfunctions of the Dirac field. Given the importance of the Dirac field, the Hamiltonian of the Dirac field is defined using fermionic variables, directly for the continuum theory. All the eigenfunctions are then given an independent derivation based on the fermionic coordinates of the Dirac state space.

1.4   Lattice Gauge Theory Hamiltonian

The lattice Dirac and gauge fields have been discussed separately, without the two being coupled. In continuum gauge theories, the coupling of the gauge field to the charge carrying fermion field is done via *minimal coupling*. Minimal coupling leads to a gauge-invariant Lagrangian.

Chapter 14 discusses the coupling of the gauge field to the Dirac field. In lattice gauge theory, the coupling of the gauge field to the Dirac field is determined by the necessity of preserving local lattice gauge invariance. One starts with the free Dirac field's Lagrangian: the coupling the Dirac field to the gauge field is realized by placing appropriate products of gauge field links between any two fermion and antifermion field variables that are defined at distinct spacetime points.

To obtain the lattice Hamiltonian for the coupled theory, one needs to define the Lagrangian on an asymmetric lattice. The time lattice ϵ is taken to zero while the space lattice spacing a is held fixed. The asymmetric lattice Lagrangian needs to be gauge-invariant; there is also the additional constraint that one obtains the continuum Dirac-Yang-Mills Lagrangian by taking the limit of $\epsilon \rightarrow 0$, followed by the limit of lattice spacing $a \rightarrow 0$.

The result obtained for the lattice Hamiltonian is fairly intuitive: the fermion sector consists of the free Dirac lattice Hamiltonian rendered gauge invariant with the appropriate links between nearest neighbor fermionic field variables. The lattice gauge field Hamiltonian is added to the fermion sector. A nontrivial gauge-invariant metric on the state space of the lattice gauge theory results from the lattice Lagrangian and action functional, and is required to make the lattice Hamiltonian Hermitian.

PART 1
Mathematical Background

Chapter 2

$SU(\mathcal{N})$ Compact Lie Groups

2.1 Introduction

Group theory is a vast subject of mathematics and forms one of the mathematical cornerstones of theoretical physics. Lie group theory plays a major role in various quantum field theories. The representation theory of $SU(2)$ Lie algebra is well known to most physics undergraduates in the study of angular momentum. The quark model of particle physics is based on $SU(3)$ Lie algebra, and Yang-Mills gauge fields introduced $SU(\mathcal{N})$ Lie algebras to the study of gauge fields.

As mentioned earlier, lattice gauge theory is based on degrees of freedom that are the finite element of the $SU(\mathcal{N})$ group, which are the nonlinear variables required for defining the gauge field theories on a lattice spacetime. The review of Lie groups in this chapter is a necessary preparation for the study of lattice gauge fields. A few of the key concepts of Lie group are reviewed and the discussion is limited to only $SU(\mathcal{N})$. Good references for the material in this Chapter are the following: [Zelobenko (1973); Haber (2017); Fadin and Fiore (2005); Cutler and Sivers (1978)].

A group $\mathcal{G}$ is a collection of elements $g \in \mathcal{G}$ that obey the following four axioms.

- A group multiplication $*$ is defined such that, for $g_1, g_2 \in \mathcal{G}$, we have $g_1 * g_2 \in \mathcal{G}$.
- Group multiplication $*$ is associative so that, for $g_1, g_2, g_3 \in \mathcal{G}$, $(g_1 * g_2) * g_3 = g_1 * (g_2 * g_3)$.
- The identity element I exists such that for $g_1, I \in \mathcal{G}$, $I * g = g * I = g$
- The inverse exists, namely for $g \in \mathcal{G}$, there exists a $g^{-1} \in \mathcal{G}$ such that $g * g^{-1} = g^{-1} * g = I$

A Lie group is a group with the additional property that it is constituted by elements that form a continuous differentiable space, with group multiplication having continuity for elements that are continuously varied.

The most familiar Lie groups in physics are the unitary (matrix) groups, denoted by $U(N)$, which are the set of $N \times N$ matrices with complex elements that satisfy the condition that $U^\dagger U = I$, where $U^\dagger$ is Hermitian conjugation (defined as the complex conjugation and transposition of the matrix elements). Group multiplication $*$ is

9

the ordinary matrix multiplication. The group $SU(\mathcal{N})$ is the special unitary group that has the additional constraint that $\det U = 1$.

The Lie groups $U(1), SU(2), SU(3), SU(5), SO(10), E_6$ and $E_8 \times E_8$ appear in the study of the Standard Model of particle physics and in superstring theory.

2.2   Lie Algebras

Lie groups are a special case of groups that are generated by a *Lie algebra*. Let $U \in SU(\mathcal{N})$ be an element of the group; it can be represented by the *generalization of the exponential function* to a matrix function. *Repeated indices are always summed unless stated.* The Lie group element is given by the following

$$U = \exp\{iB^a X^a\} \quad ; \quad a = 1, 2, ..., \mathcal{N}^2 - 1 \tag{2.2.1}$$

where X^a are the generators of the Lie algebra that can be finite dimensional matrices. The variables B^a are the *canonical coordinates* of the Lie group element.

Note that the exponential function, which is given by $\exp(i\theta), \theta \in [-\pi, +\pi]$, is an element of the $U(1)$ group; the Abelian phase θ is generalized by the Lie group to *non-Abelian phases B^a*.

Let C^{abc} be the totally antisymmetric structure constants; for all compact groups C^{abc} are totally antisymmetric in the three indices.[1] Defining the commutator of two matrices by

$$[A, B] \equiv AB - BA$$

the Lie algebra is defined by the commutation equation

$$[X^a, X^b] = iC^{abc}X^c \quad ; \quad X^a = X_a = X_a^\dagger \ : \text{Hermitian}$$
$$\text{Tr}(X^a X^b) = \delta^{ab}/s^2(p)$$

The finite dimensional matrices $X^a = X^a(p)$ – called the generators of the Lie algebra – define the representation of the Lie algebra, are denoted by p, and discussed further in Section 2.7.

The structure constants C^{abc} for (finite dimensional) compact Lie groups can be completely classified and lead to the A, B, D and E series of Lie groups [Zelobenko (1973), Varadarajan (1974)].

Noteworthy 2.1: Notation

In all the derivations, no distinction is made between upper and lower indices; rather, the choice of upper or lower indices is based on notational convenience. The index for the generators and structure constants will be denoted by both subscripts and superscripts, which are equivalent for compact Lie groups. In particular

$$C^{abc} \equiv C_{abc} \quad ; \quad X^a \equiv X_a \quad ; \quad B^a \equiv B_a$$

---

[1] This is not the case for non-compact Lie groups such as the Lorentz group.

Repeated indices are to be summed over; the summation is sometimes explicitly indicated when required for greater clarity.

The commutators follow the rule for differentiation; similar to the chain rule of differentiation, one obtains the Jacobi identity for finite matrices A, B, C given by

$$[A, [B, C]] = [[A, B], C] + [B, [A, C]] \tag{2.2.2}$$

The Jacobi identity being obeyed by the generators of the Lie algebras is a consequence of group multiplication being associative. The Jacobi identity yields, for the structure constants

$$[X^A, [X^B, X^C]] = [[X^A, X^B], X^C] + [X^B, [X^A, X^C]]$$
$$\Rightarrow C^{A\alpha I} C^{BC\alpha} = C^{AB\alpha} C^{\alpha C I} + C^{AC\alpha} C^{B\alpha I} \tag{2.2.3}$$

The Jacobi identity leads to *constraints* on the allowed set of structure constants C^{abc}. The structure constants themselves provide the adjoint representation of the Lie algebra. Consider the following relabeling in Eq. 2.2.3

$$I = i \ ; \ \ C = j$$

Using the complete antisymmetry of C^{abc} under the interchange of any two indices, the Jacobi identity can be rewritten as follows

$$C^{iA\alpha} C^{\alpha Bj} - C^{B\alpha I} C^{\alpha Aj} = C^{AB\alpha} C^{i\alpha j} \tag{2.2.4}$$

Using the matrix notation for writing the ij components of the matrices yields, from Eq. 2.2.4

$$C^A_{ij} \equiv C^{iAj} \ \Rightarrow \ [C^a, C^b] = C^{abc} C^c \tag{2.2.5}$$

The generators of the adjoint representation are defined by

$$(F^a)_{ij} = i C_{iaj}$$

It follows from Eq. 2.2.5 that the structure constants provide the adjoint representation of the Lie algebra. The adjoint representation of the $SU(N)$ Lie algebra is defined by $N^2 - 1$ generators and is given by $N^2 - 1 \times N^2 - 1$ matrices. The generators of the adjoint representation yield the following representation of the group elements

$$(F^a)_{ij} = i C^{iaj} \ \Rightarrow \ \rho = e^{iF^a B_a} = e^{-\omega}$$

where the adjoint matrix ω is given by

$$\rho = e^{-\omega} \ ; \ \ \omega_{ab} = \sum_c C^{acb} B_c = -\omega_{ba} \ \Rightarrow \ \omega^T = -\omega \tag{2.2.6}$$

2.3   Vielbein of $SU(\mathcal{N})$

The $SU(\mathcal{N})$ group space is spanned by all the allowed values of the non-Abelian phases B^a; the space is a compact Riemann space, which is simply connected for $\mathcal{N} \geq 2$.

Define the vielbein of $SU(\mathcal{N})$ group space, denoted by e_{ab}, as follows

$$e_{ab}X_b = -iU^\dagger \partial_a U \;\; ; \;\; U = \exp\{iB_a X_a\} \tag{2.3.1}$$

Introduce the homotopy parameter τ by the following

$$U(\tau) = \exp\{i\tau B_\alpha X_\alpha\} \;\; ; \;\; e_{ab}(\tau)X_b = -iU^\dagger(\tau)\partial_a U(\tau)$$

$$\text{Boundary conditions} \;:\;\; e_{ab}(1) = e_{ab} \;\; ; \;\; e_{ab}(0) = 0 \tag{2.3.2}$$

Differentiating by τ yields

$$\frac{de_{ab}(\tau)}{d\tau}X_b = -iU^\dagger(\tau)(-iB_\alpha X_\alpha)\partial_a U(\tau) - iU^\dagger(\tau)\partial_a\{iB_\alpha X_\alpha U(\tau)\}$$

$$\Rightarrow \frac{de_{ab}(\tau)}{d\tau}X_b = U^\dagger(\tau)X_a U(\tau) \tag{2.3.3}$$

Using ρ_{ab} as the group element in the adjoint representation defined by

$$U^\dagger X_a U = \rho_{ab}X_b \tag{2.3.4}$$

yields from Eqs. 2.3.3

$$\frac{de_{ab}(\tau)}{d\tau}X_b = U^\dagger(\tau)X_a U(\tau) = \rho_{a\alpha}(\tau)X_\alpha \;\; \Rightarrow \;\; \frac{de_{ab}(\tau)}{d\tau} = \rho_{ab}(\tau) \tag{2.3.5}$$

For X_a in the fundamental representation

$$\rho_{ab} = 2\,\mathrm{Tr}(X_a U X_b U^\dagger) \;\; : \;\; \mathrm{Tr}(X_a X_b) = \frac{1}{2}\delta_{ab} \tag{2.3.6}$$

The generators of $SU(\mathcal{N})$ in the fundamental representation X_a obey the following identity [Haber (2017)]

$$\sum_a \mathrm{Tr}(AX_a)\,\mathrm{Tr}(BX_a) = \frac{1}{2}\left[\mathrm{Tr}(AB) - \frac{1}{\mathcal{N}}\mathrm{Tr}(A)\,\mathrm{Tr}(B)\right]$$

Hence

$$\sum_c \rho_{ac}(U)\rho_{cb}(V) = 4\sum_c \mathrm{Tr}(X_a U X_c U^\dagger)\,\mathrm{Tr}(X_c U V X_b V^\dagger) = \rho_{ab}(UV) \tag{2.3.7}$$

It is shown that definition of the adjoint representation given in Eq. 2.3.4 reproduces the result stated earlier in Eq. 2.2.6. Consider the following

$$U^\dagger(\tau)X_a U(\tau) = \rho_{a\alpha}(\tau)X_\alpha \;\; ; \;\; \rho_{a\alpha}(1) = \rho_{a\alpha} \;\; : \;\; \rho_{a\alpha}(0) = \delta_{a-\alpha} \tag{2.3.8}$$

This yields

$$\frac{d\rho_{a\alpha}(\tau)}{d\tau}X_\alpha = -iB_\beta U^\dagger(\tau)[X_\beta, X_a]U(\tau) = -i^2 B_\beta C^{\beta ac}U(\tau)X_c U^\dagger(\tau)$$

$$= -\omega_{ac}\rho_{cd}(\tau)X_d \;\; ; \;\; \omega_{ac} = B_\beta C^{a\beta c}$$

Hence, integrating the differential equation below yields, in matrix notation

$$\frac{d\rho_{aa}(\tau)}{d\tau} = -\omega_{ac}\rho_{ca}(\tau) \quad \Rightarrow \quad \rho(\tau) = e^{-\tau\omega} \quad \Rightarrow \quad \rho(1) = \rho = e^{-\omega} \qquad (2.3.9)$$

and yields the result given in Eq. 2.2.6.

From Eqs. 2.3.5 and 2.3.9, in matrix notation

$$\frac{de(\tau)}{d\tau} = \rho(\tau) = e^{-\tau\omega}$$

and hence, using boundary conditions Eq. 2.3.2, yields

$$e = \int_0^1 d\tau \frac{\partial e(\tau)}{\partial \tau} = \frac{1 - e^{-\omega}}{\omega} \qquad (2.3.10)$$

One has the following expansion

$$e = \frac{1 - e^{-\omega}}{\omega} \simeq 1 - \frac{1}{2}\omega + O(\omega^2) \qquad (2.3.11)$$

The definition of the vielbein given in Eq. 2.3.1 yields the following result

$$e_{ab}X_b = -iU^\dagger\partial_a U \quad ; \quad U = \exp\{iB_\alpha X_\alpha\}$$

$$\Rightarrow e_{ab}X_b = -i\frac{1}{\epsilon}U^\dagger(B)\Big\{U(B_\alpha + \delta_{a\alpha}\epsilon) - U(B)\Big\}$$

$$\Rightarrow U^\dagger(B)U(B_\alpha + \delta_{a\alpha}\epsilon) = 1 + i\epsilon e_{ab}X_b \simeq= \exp\{i\epsilon e_{ab}(B)X_b\} \qquad (2.3.12)$$

The result obtained above in Eq. 3.4.20 will prove to be useful in studying the differential representation of the generators.

2.4 Metric on *SU(N)* Group Space

The group manifold $SU(\mathcal{N})$ is a Riemann metric space with constant curvature. The metric is defined using the properties of the group elements.

Define the matrix differential of the group element U by dU

$$(dU)_{\alpha\beta} = d(\exp iB^a X^a)_{\alpha\beta} \qquad (2.4.1)$$

The metric $g^{ab}(B)$ is defined by [Zelobenko (1973)]

$$g^{ab}(B)dB^a dB^b = \frac{1}{s^2}\text{Tr}(dU\,dU^\dagger) \qquad (2.4.2)$$

g^{ab} is invariant under group multiplication; to see this, let $V, W \in SU(\mathcal{N})$ be numerical matrices. Then

$$U' = VUW \quad \Rightarrow \quad dU' = V(dU)W \quad \Rightarrow \quad \text{Tr}(dU'dU'^\dagger) = \text{Tr}(dU\,dU^\dagger) \qquad (2.4.3)$$

From Eqs. 2.4.2 and 2.3.1

$$g_{ab}(B)dB^a dB^b = \frac{1}{s^2}\text{Tr}(dUU^\dagger UdU^\dagger) = \frac{-i^2}{s^2}e_{ab}dB_b e_{a'b'}dB'_b\,\text{Tr}(X_a X'_a)$$

$$= e_{ab}dB_b e_{ab'}dB_{b'} = (e^T e)_{ab}dB^a dB^b$$

and the metric is given by

$$g_{ab} = (e^T e)_{ab} = e^T_{ac} e_{cb} = 2\left(\frac{\cosh(\omega) - 1}{\omega^2}\right)_{ab} \tag{2.4.4}$$

Note that for Lie groups the existence of the underlying generators leads to the factorization of the metric tensor. In matrix notation, Eqs. 2.3.11 and 2.4.4 yield the following

$$g = e^T e = -\frac{1 - e^\omega}{\omega} \cdot \frac{1 - e^{-\omega}}{\omega} = 2\frac{\cosh(\omega) - 1}{\omega^2} \simeq 1 + \frac{1}{12}\omega^2 + O(\omega^4) \tag{2.4.5}$$

The metric tensor g_{ab} defines a space with constant curvature. The space is an Einstein space with the Ricci tensor $\mathcal{R}_{ab}$ being proportional to the metric tensor and the Ricci curvature scalar $\mathcal{R}$ is a constant. More precisely, as shown by Marinov and Terentyev (1979)

$$\mathcal{R}_{ab} = 4g_{ab} \quad ; \quad \mathcal{R} = \frac{1}{4}\mathcal{N}$$

Note the remarkable fact that the constant curvature $\mathcal{R}$ of the Lie group depends only on the groups *dimensionality*.

The metric tensor g_{ab} yields the following geodesic, denoted by $U(t)$, between group elements $\mathbb{I}$ and U

$$U(t) = (U)^t \quad ; \quad U(0) = \mathbb{I} \quad ; \quad U(1) = U \quad : \quad t \in [0, 1]$$

The group elements $U(t)$ trace out a geodesic in the group manifold. Since $U(t)U(t') = U(t + t')$, the geodesic $U(t)$ forms an Abelian one parameter subgroup of G.

2.4.1 *SU(2) metric*

As an example, the metric for $SU(2)$ is calculated; the fundamental representation (discussed in Section 2.7) yields, for $\sigma^a =$Pauli matrices, the following

$$U = \exp\{iB^a \sigma^a / 2\}$$

$$= \cos\left(\frac{B}{2}\right)1 + in^a \sigma^a \sin\left(\frac{B}{2}\right) \tag{2.4.6}$$

where

$$B^2 = \sum_a B^a B^a, \quad n^a = \frac{B^a}{B} \tag{2.4.7}$$

Note

$$d(\vec{n}^2) = 2\vec{n} \cdot d\vec{n} = 0 \tag{2.4.8}$$

A straightforward calculation gives

$$\text{Tr}(dU dU^\dagger) = \frac{1}{2}(dB)^2 + 2\sin^2\left(\frac{B}{2}\right)(d\vec{n})^2$$

$$= g^{ab} dB^a dB^b \tag{2.4.9}$$

Change from Cartesian coordinates (B^1, B^2, B^3) to polar coordinates (B, θ, ϕ). Then

$$d(\overrightarrow{n})^2 = d\phi^2 + \sin^2\theta d\theta^2 \qquad (2.4.10)$$

Let $B = a_1, \theta = a_2, \phi = a_3$; then

$$g^{ab} dB^a dB^b = g^{ij} da^i da^j$$

where

$$g^{ij} = \begin{pmatrix} 1/2 & 0 & 0 \\ 0 & 2\sin^2(\frac{B}{2})\sin^2\theta & 0 \\ 0 & 0 & 2\sin^2(\frac{B}{2}) \end{pmatrix} \qquad (2.4.11)$$

$SU(2)$ is isomorphic to S^3, and the metric given by 2.4.11 is the metric on S^3.

2.5 The Invariant Haar Measure and Delta Function

The symbol dU denotes the invariant measure as well as the infinitesimal matrix. It will be clear from the context of its use what it means. In terms of the canonical coordinates, the measure is written as

$$dU = \mu(B) \prod_a dB^a \quad \Rightarrow \quad \int dU f(U) = \prod_a \int dB^a \mu(B) f(U) \qquad (2.5.1)$$

The Haar measure $\mu(B)$ is defined, as is the case for all curved manifolds, by the following

$$\mu(B) = \sqrt{g(B)} \;;\; g(B) = \det(g^{ab}(B)) \qquad (2.5.2)$$

The measure is invariant under group multiplication in that

$$\mu(U) = \mu(VUW) \;;\; U, V, W \in SU(N)$$

This follows from the invariance of the metric given in Eq. 2.4.3.

Since $\det(\exp A) = \exp(\mathrm{Tr}\, A)$, from Eq. 2.4.5, in matrix notation and using $C_{a\alpha\beta} C_{b\alpha\beta} = N\delta_{ab}$

$$\mu(B) = \sqrt{\det g} = \sqrt{\det\left\{\exp\left(\frac{1}{12}\omega^2\right)\right\}} + O(\omega^4)$$

$$= \exp\left\{\frac{1}{24}\mathrm{Tr}(\omega^2)\right\} = \exp\left(-\frac{N}{24}B^2\right) + 0(B^4) \qquad (2.5.3)$$

The result for the measure term given in Eq. 2.5.3 is required for the mass renormalization calculations in Chapters 4 and 9.

2.5.1   Delta function on group space

Following the procedure for a general curved manifold, the delta function on group space is defined by

$$\delta[U - V] = \frac{1}{\mu(B)} \prod_a \delta(B^a - \tilde{B}^a) \tag{2.5.4}$$

where

$$U = \exp\{iB^a X^a\} \quad ; \quad V = \exp\{i\tilde{B}^a X^a\}$$

$\delta[U - V]$ has the important property that, for σ in the group space,

$$\delta[\sigma(U - V)\sigma^\dagger] = \delta[U - V] \tag{2.5.5}$$

The δ-function for the group elements is similar to the usual δ-function; for V in the group space, we have

$$\int dU \delta[U - V] = 1 \quad ; \quad \int dU f(U)\delta[U - V] = f(V) \tag{2.5.6}$$

2.5.2   SU(2) measure

The measure for $SU(2)$, from 2.4.11, is given by

$$dU = \mu(B) \prod_a dB_a = c \sin^2\left(\frac{B}{2}\right) \sin\theta (dB d\phi d\theta) = c \sin^2(B/2) dB d\Omega$$

where

$$d\Omega = \sin\theta d\phi d\theta \quad : \text{solid angle}$$

Note that for $B \simeq 0$, the group space is isomorphic to the three-dimensional Euclidean space.

The range of the variables B, θ, ϕ are such that each group element occurs once and only once. We can fix this directly from Eq. 2.5.7; the solid angle is to be covered once (for $n > 2$, the same holds true for the higher dimensional solid angles) and the radial variable B take values such that $\sin^2(B/2) > 0$. The point where the radial factor is zero is taken to be the maximum of B. Hence

$$0 \le B \le 2\pi, \quad 0 \le \theta \le \pi, \quad 0 \le \phi \le 2\pi \tag{2.5.7}$$

Note the factorization of dU into a solid angle and a radial part is a general feature of $SU(\mathcal{N})$. For all values of the variables, $\mu(B) \ge 0$. The points where $\mu(B) = 0$ marks the limit of the variables. The constant c is fixed by the normalization condition

$$\int_G dU = 1 \tag{2.5.8}$$

which, in case of $SU(2)$ gives the final result of

$$dU = \frac{1}{4\pi^2} \sin^2(B/2) dB d\Omega \tag{2.5.9}$$

2.5.3 *SU*(2) *measure: invariant*

To illustrate the invariance of the Haar measure under right- and left-multiplication, the case of $SU(2)$ is studied. Choose a representation of $SU(2)$ group element[2]

$$U = \begin{pmatrix} a & b \\ -b^* & a^* \end{pmatrix} \;\Rightarrow\; U^\dagger U = \mathbb{I} \;;\; \det U = |a|^2 + |b|^2 = 1 \;;\; a, b \;:\; \text{complex numbers}$$

The left- and right-invariant Haar measure is defined by

$$\int dU = \int_{-\infty}^{+\infty} da_R da_I db_R db_I \,\delta(|a|^2 + |b|^2 - 1) \equiv \int da\,db\,\delta(|a|^2 + |b|^2 - 1) \quad (2.5.10)$$

The variables a, b with constraint $|a|^2 + |b|^2 = 1$ are coordinates for S^3; choosing polar coordinates for a, b yields the $SU(2)$ measure given in Section 2.5.2.

Consider another $SU(2)$ group element, with *constant elements*, given by

$$V = \begin{pmatrix} c & d \\ -d^* & c^* \end{pmatrix} \;\Rightarrow\; V^\dagger V = \mathbb{I} \;;\; \det V = |c|^2 + |d|^2 = 1 \;;\; c, d \;:\; \text{complex numbers}$$

Left-multiplication yields

$$VU = \begin{pmatrix} ac - b^*d & bc + a^*d \\ -(bc + a^*d)^* & (ac - b^*d)^* \end{pmatrix} \equiv \begin{pmatrix} \alpha & \beta \\ -\beta^* & \alpha^* \end{pmatrix}$$

The matrix V is equivalent to a rotation of a four-vector in four-dimensional Euclidean space and leads to

$$d\alpha\, d\beta = da\, db \;:\; \text{invariant}$$

Furthermore

$$|\alpha|^2 + |\beta|^2 = \det(VU) = \det(V)\det(U) = (|c|^2 + |d|^2)(|a|^2 + |b|^2) = |a|^2 + |b|^2$$

Hence, from Eq. 2.5.10 the measure is left-invariant and given by

$$\int d(VU) = \int d\alpha\, d\beta\, \delta(|\alpha|^2 + |\beta|^2 - 1) = \int da\, db\, \delta(|a|^2 + |b|^2 - 1) = \int dU$$

A similar derivation shows the measure is right-invariant as well.

2.6 Campbell-Baker-Hausdorff (CBH) Formula

Let e^x, e^y be finite group elements of Lie group G and consider the multiplication

$$e^x e^y = e^z$$

Then [Zelobenko (1973)]

$$z = x + y + [x, y] + \sum_{m=1}^{\infty} \prod_{i=1}^{m} \sum_{p_i, q_i = 0}^{m} \frac{(-1)^{m-1}}{m} \cdot \frac{[x^{p_1} y^{q_1} \cdots x^{p_m} y^{q_m}]}{\sum_i (p_i + q_i) \prod_i p_i! q_i!}$$

---

[2]The following matrix, used in quantum random walks, is an element of $U(2)$ but *not* of $SU(2)$

$$\begin{pmatrix} a & b \\ b^* & -a^* \end{pmatrix} \;\Rightarrow\; U^\dagger U = \mathbb{I} \;;\; \det U = -(|a|^2 + |b|^2) = -1 \;;\; a, b \;:\; \text{complex numbers}$$

with

$$p_i + q_i \neq 0 \quad \text{and} \quad [a_1 a_2 \cdots a_{n-1} a_n] = [a_1 [a_2 \cdots [a_{n-1}, a_n] \cdots]]$$

In particular, the leading terms in the expansion are the following [Varadarajan (1974)]

$$z = x + y + \frac{1}{2}[x,y] - \frac{1}{12}[[x,y],x] + \frac{1}{12}[[x,y],y]$$
$$- \frac{1}{48}[y,[x,y],x]] - \frac{1}{48}[x,[[x,y],y]] + ..$$
(2.6.1)

2.7   Irreducible Representations of $SU(\mathcal{N})$

A representation of the elements of $SU(\mathcal{N})$ consist of matrices that obey the multiplication law of the group, which in turn requires that the matrices embodying the generators X_a must obey the Lie algebra of the generators of $SU(\mathcal{N})$. It is noteworthy that matrices of higher and higher dimensions can all provide a representation of the underlying Lie group $SU(\mathcal{N})$.

A faithful representation is one for which each element of the Lie group corresponds to one element of the representation – in other words, the representation is an isomorphism of the group elements to elements of the representation. One can form block diagonal matrices, with each block corresponding to one representation of the Lie group. An irreducible representation *cannot* be further reduced to a block diagonal form. The fundamental and the adjoint representations are the two most important representations of the group elements of $SU(\mathcal{N})$ and are discussed below.

Let U be an element of the special unitary group $SU(\mathcal{N})$, and let $D_{ij}^{(p)}(U)$ be the matrix elements of the irreducible representation given by the composite representation label p. Note $1 < i,j < d_p$ where d_p is the dimension of the pth irreducible representation. The traceless Hermitian generators in the pth representation are given by $d_p \times d_p$ matrices $X^a(p), a = 1,2,\cdots,\mathcal{N}^2 - 1$, satisfying the Lie Algebra

$$[X^a(p), X^b(p)] = iC_{abc}X^c(p) \quad ; \quad \text{Tr}(X^a(p)X^b(p)) = \frac{1}{s^2(p)}\delta_{ab}$$
(2.7.1)

where C_{abc} are the antisymmetric structure constants of $SU(\mathcal{N})$.

In terms of dimensionless canonical coordinates B^a of $SU(\mathcal{N})$ an irreducible representation of $SU(\mathcal{N})$ is given by

$$D_{ij}^{(p)}(U) = (e^{iB^a X^a(p)})_{ij}$$
(2.7.2)

The coordinates B^a take value in the $SU(\mathcal{N})$ group space.

The simplest Lie Group is the Abelian group $U(1)$ with $C_{iaj} = 0$, $i,a,j = 1$. The fundamental representation given by $U = e^{i\theta} : \theta \in S^1$. The adjoint representation for $U(1)$ is trivial, equal to 1. All the irreducible representation of $U(1)$ are given by integers p where

$$p = \in [-\infty, +\infty] \quad ; \quad X(p) = p \quad ; \quad D_{ij}^{(p)}(U) = e^{ip\theta} \quad ; \quad d_p = 1$$

The quadratic Casimir operator for the pth irreducible representation for $SU(\mathcal{N})$ is defined by

$$X^2(p) \equiv \sum_{a=1}^{\mathcal{N}^2-1} X^a(p)X^a(p) \tag{2.7.3}$$

Note that the second Casimir operator commutes with all the generators since

$$\left[\sum_a X^a X^a, X^b\right] = \sum_a X^a[X^a, X^b] + \sum_a [X^a, X^b]X^a = \sum_{a,c} iC^{abc}(X^a X^c + X^c X^a) = 0$$

Using Schur's Lemma, it follows that the second Casimir operator is proportional to the identity matrix and yields

$$X^2(p) = c_2(p)\mathbb{I}_p \quad ; \quad \mathrm{Tr}(\mathbb{I}_p) = d_p \tag{2.7.4}$$

where $c_2(p)$ is the numerical value of the quadratic Casimir operator in the pth irreducible representation and $\mathbb{I}_p$ is a $d_p \times d_p$ unit matrix.

The structure constants of the Lie algebra are fixed; other constants, including $s(p)$, $c_2(p)$ and d_p, depend on the representation p of the Lie algebra.

The two most important representations of the Lie algebra of $SU(\mathcal{N})$ are the *fundamental* and the *adjoint* representations. The fundamental and adjoint representations are given by the following.

- Fundamental representation.
  The *smallest* size matrices that can provide a faithful representation of $SU(\mathcal{N})$ are $\mathcal{N} \times \mathcal{N}$ Hermitian and traceless matrices. Given their special importance, the generators of the fundamental representation are denoted by T^a. The fundamental representation yields the following

$$T^a T^b = \frac{1}{2}\left[\frac{1}{\mathcal{N}}\mathbb{I}\delta_{ab} + (d^{abc} + iC^{abc})T^c\right] \quad ; \quad Tr(\mathbb{I}) = \mathcal{N} \tag{2.7.5}$$

  where the constants d^{abc} are completely symmetric under the exchange of any two indices. The fundamental representation has the following normalization

$$\mathrm{Tr}(T^a T^b) = \frac{1}{2}\delta_{ab}$$

  Since there are $\mathcal{N}^2 - 1$ generators of the Lie group, taking the trace of the second Casimir operator yields, from Eq. 2.7.4, for the fundamental representation $c_2(F)$ the following

$$\mathrm{Tr}\left(\sum_a T^a T^a\right) = c_2(F)\mathcal{N} \Rightarrow \frac{1}{2}(\mathcal{N}^2-1) = c_2(F)\mathcal{N} \Rightarrow c_2(F) = \frac{1}{2\mathcal{N}}(\mathcal{N}^2-1)$$

- Adjoint representation.
  The Jacobi identity given in Eq. 2.2.3 shows that the structure constants themselves provide a representation of the Lie algebra; from Eq. 2.2.5

$$(F^a)_{ij} = iC_{iaj} \quad ; \quad 1 \le i, a, j \le \mathcal{N}^2 - 1$$

The adjoint representation, $\rho = \exp\{-\omega\} \in SU(\mathcal{N})$, is defined by the action of the group elements on the Lie group generators; for any representation $U \in SU(\mathcal{N})$, from Eq. 2.3.4

$$U^\dagger X^a U = \rho_{a\alpha} X^\alpha$$

For the adjoint representation [Haber (2017)]

$$\mathrm{Tr}(F^a F^b) = \mathcal{N}\delta_{ab}$$

The equation above is also written as

$$C^{abc}C^{abd} = \mathcal{N}\delta_{ab} \quad \Rightarrow \quad F^2 = \sum_{a=1}^{\mathcal{N}^2-1} F^a F^a = \mathcal{N}\mathbb{I} \quad \Rightarrow \quad c_2(A) = \mathcal{N} \quad (2.7.6)$$

In summary, the quadratic Casimir operators are the following:

- Fundamental representation of $SU(\mathcal{N})$: $c_2(F) = \frac{1}{2\mathcal{N}}(\mathcal{N}^2 - 1)$
- Adjoint representation of $SU(\mathcal{N})$: $c_2(A) = \mathcal{N}$

Some identities of the structure constants are the following [Haber (2017)]

$$\mathrm{Tr}(F_a F_b F_c) = \frac{i}{2}\mathcal{N}C^{ade}$$

$$\mathrm{Tr}(F_a F_b F_c F_d) = \delta_{ad}\delta_{bc} + \frac{1}{2}(\delta_{ab}\delta_{cd} + \delta_{ac}\delta_{bd})$$

$$+ \frac{1}{4\mathcal{N}}(C^{ade}C^{bce} + d^{ade}d^{bce}) \quad (2.7.7)$$

2.8   Peter-Weyl Theorem

Let $|U\rangle$ be the coordinate eigenstate for the state space defined on $SU(\mathcal{N})$; let $\langle p, ij|$ be the conjugate eigenstate, which are the tenor product of row vectors of a finite dimension that is determined by the representation p, and with $1 \le i, j \le d_p$. Then

$$\langle p, ij|U\rangle = \sqrt{d_p}D_{ij}^{(p)}(U) \; ; \quad 1 \le i, j \le d_p \quad (2.8.1)$$

Note that

$$\langle U|U'\rangle = \delta(U - U') \quad (2.8.2)$$
$$\langle p, ij|p', i'j'\rangle = \delta^{pp'}\delta_{ii'}\delta_{jj'} \quad (2.8.3)$$

where the δ-function is defined in Eq. 2.5.4 using the metric on group space. The completeness equation on Hilbert space given by

$$\mathbb{I} = \sum_p \sum_{ij} |p, ij\rangle\langle p, ij| = \int dU|U\rangle\langle U| \quad (2.8.4)$$

where the sum on p is over all the irreducible representations.

For $U(1)$ group, the dual basis is $\langle p|$ and

$$\langle p|U\rangle = \sqrt{d_p}D_{ij}^{(p)}(U) = e^{ip\theta} \; ; \quad \langle U|U'\rangle = \delta(\theta - \theta') \; ; \quad \langle p|p'\rangle = \delta_{p-p'}$$

and completeness equation for $U(1)$ is given by

$$\mathbb{I} = \sum_{p=-\infty}^{+\infty} |p\rangle\langle p| = \frac{1}{2\pi} \int_{-\pi}^{+\pi} d\theta |\theta\rangle\langle\theta|$$

Let $f(U)$ be any square-integrable function of $SU(\mathcal{N})$

$$\int_G dU |f(U)|^2 < \infty \tag{2.8.5}$$

The Peter-Weyl theorem for a compact Lie group G states the following [Zelobenko (1973)]

$$\int_G dU D_{ij}^{(\ell)}(U^\dagger) D_{i'j'}^{(m)}(U) = \frac{1}{d_\ell} \delta^{\ell m} \delta^{ii'} \delta^{jj'} \quad : \text{Peter-Weyl theorem}$$

which, for $U(1)$ is given by

$$\frac{1}{2\pi} \int_{-\pi}^{+\pi} d\theta e^{-ip\theta} e^{im\theta} = \delta_{p-m}$$

The character functions for $SU(\mathcal{N})$ is defined by

$$\chi_p(U) = \sum_{i=1}^{d_p} D_{ii}^{(p)}(U) = \text{Tr}(D^{(p)}(U))$$

The Peter-Weyl theorem for the character functions states that

$$\int_G dU \chi_p(U^\dagger) \chi_{p'}(U) = \delta_{pp'} \tag{2.8.6}$$

Let $f(U)$ be an aribtrary function of the group element U. The Fourier theorem states that

$$f(U) = \sum_p d_p \text{Tr}(D^{(p)}(U) a_p) \tag{2.8.7}$$

Using the Peter-Weyl orthogonality theorem gives

$$\text{Tr}(D^{(p)}(U) a_p) = \int dU' f(U') \text{Tr}(D^{(p)}(UU'^\dagger)) \tag{2.8.8}$$

Also

$$\|f\|^2 = \int dU |f(U)|^2 = \sum_p d_p Tr(a_p^\dagger a_p)$$

2.9 Lie Algebra Generators: Differential Operators

The Lie group generators are realized by differential operators acting on functions of the Lie group elements. Consider function $f(U)$ of group element U; any faithful representation can be chosen and the representation is explicitly shown only if needed. To foreground the general case, consider $U(1)$; then, from Taylor expansion

$$f(U) \to f(\Phi U) = f(e^{i\phi} e^{iB}) = e^{i\phi \frac{\partial}{\partial B}} f(U) \; ; \; U = e^{iB} \; ; \; \Phi = e^{i\phi} \tag{2.9.1}$$

The non-Abelian generalization of Eq. 2.9.1, for the action of the group element of $SU(\mathcal{N})$, is given by
$$f(U) \rightarrow f(\Phi U) = e^{i\phi_a E_a} f(U) \; ; \;\; U = e^{iB_a X_a} \; ; \;\; \Phi = e^{i\phi_a X_a}$$
where $B_a, a = 1, 2, .., \mathcal{N}^2 - 1$ are the canonical coordinates of $SU(\mathcal{N})$.

The differential operators E_a obey the Lie algebra commutation equation
$$[E_a, E_b] = iC^{abc} E_c \; ; \;\; a, b, c = 1 \cdots \mathcal{N}^2 - 1 \tag{2.9.2}$$
The differential operators E_a, unlike the generators of the Lie algebra X_a, are *independent* of the representation chosen for X_a, and depend only on the properties of the group manifold. From Eq. 2.3.1
$$e_{ab} X_b = -iU^\dagger \partial_a U$$
$$\Rightarrow \;\; UX_a = -ie_{ab}^{-1} \partial_b U \equiv f_{ab} \frac{\partial}{i\partial B_b} U \tag{2.9.3}$$
where, from Eq. 2.3.10
$$f_{ab}e_{ac} = \delta_{ab} \;\; \Rightarrow \;\; f_{ab} \equiv e_{ab}^{-1}$$
$$\Rightarrow \;\; f = e^{-1} = \frac{\omega}{1 - e^{-\omega}} \; ; \;\; \omega_{ac} = C^{abc} B_b \tag{2.9.4}$$
The Lie algebra generators, from Eq. 2.9.3 have the following differential realization
$$E_a = f_{ab}(B) \frac{\partial}{i\partial B_b} \;\; \Rightarrow \;\; E_a U = UX_a \tag{2.9.5}$$
The commutation given in Eq. 2.9.2 is reproduced since
$$E_a E_b U = E_a U X_b = U X_a X_b$$
$$\Rightarrow [E_a, E_b]U = U[X_a, X_b] = iC^{abc} E_c U \Rightarrow [E_a, E_b] = iC^{abc} E_c \tag{2.9.6}$$
Eq. 2.9.2 also follows directly from the definition given in Eq. 2.9.5 due to the Maurer-Cartan equation given by
$$f_{\alpha i} \partial_i f_{\beta\gamma} - f_{\beta i} \partial_i f_{\alpha\gamma} = -C_{\alpha\beta i} f_{i\gamma} \; : \;\; \partial_i \equiv \partial/\partial B_i \tag{2.9.7}$$
It is shown that the differential generators of the Lie group E_i are Hermitian differential operators. Note that
$$\left(\frac{\partial}{i\partial B_j}\right)^\dagger = -\frac{1}{i}\left(\frac{\partial}{\partial B_j}\right)^\dagger = \frac{\partial}{i\partial B_j} \; : \;\; \text{Hermitian} \tag{2.9.8}$$
hence, one has
$$E_i^\dagger = \frac{\partial}{i\partial B_j} f_{ij} = \frac{\partial f_{ij}}{i\partial B_j(\sigma)} + f_{ij} \frac{\partial}{i\partial B_j(\sigma)} \tag{2.9.9}$$
From Eq. 2.9.4
$$f = \sum_{\ell=0}^{\infty} \omega e^{-\ell\omega} = \sum_{\ell,n=0}^{\infty} \frac{(-1)^n}{n!} \omega(\ell\omega)^n \tag{2.9.10}$$
Note from Eq. 2.2.6
$$\left[\frac{\partial\omega}{i\partial B_j}\right]_{ij} = \sum_{\alpha} C_{i\alpha j} \delta_{j\alpha} = 0 \tag{2.9.11}$$
It follows that, when one evaluates $\sum_j \partial f_{ij}/\partial B_j$, each term in $f(\omega)$ – given in the series expansion in Eq. 2.9.10 – yields zero; hence, from Eq. 2.9.9
$$E_i^\dagger = f_{ij} \frac{\partial}{i\partial B_j} = E_i \; : \;\; \text{Hermitian} \tag{2.9.12}$$
Hence, the generators defined in Eq. 2.9.5 are Hermitian operators.[3]

---

[3] For the Abelian case $E = \partial/i\partial B$ and is Hermitian due to Eq. 2.9.8.

2.10 Vielbein e_{ab} and f_{ab} for $SU(2)$

Consider the adjoint matrix representation of the $SU(2)$ group given by

$$\rho = e^{-\omega} \ ; \ \ \omega_{ab} = \varepsilon_{a\alpha b}\theta_\alpha \ \ : \ \ a, \alpha, b = 1, 2, 3$$

with $\varepsilon_{a\alpha b}$ being the structure constants of $SU(2)$ and is the three-dimensional completely antisymmetric tensor. In matrix notation

$$e = \frac{1 - e^{-\omega}}{\omega} \ ; \ \ f = \frac{\omega}{1 - e^{-\omega}}$$

The identity

$$\varepsilon_{a\alpha b}\varepsilon_{c\alpha d} = \delta_{ac}\delta_{bd} - \delta_{ad}\delta_{bc}$$

yields, in matrix notation

$$\omega^3 = -\theta^2\omega \ ; \ \ \theta^2 = \theta_\alpha\theta_\alpha \tag{2.10.1}$$

From Eq. 2.10.1 only ω, ω^2 are independent matrices since

$$\omega^3 = -\theta^2\omega \ ; \ \ \omega^4 = -\theta^2\omega^2 \ \ \cdots$$

Hence

$$\rho = e^{-\omega} = 1 - \frac{\omega}{\theta}\sin(\theta) + \frac{\omega^2}{\theta^2}(1 - \cos(\theta))$$

Using Eq. 2.10.1, the vielbein is given by

$$e = \frac{1 - e^{-\omega}}{\omega} = \frac{\sin(\theta)}{\theta} - \frac{\omega}{\theta^2}(1 - \cos(\theta))$$

The derivation of f_{ab} is more involved. Consider the equation

$$f - \frac{1}{2}\omega = \frac{\omega}{1 - e^{-\omega}} - \frac{1}{2}\omega = \frac{\omega}{2}\frac{1 + e^{-\omega}}{1 - e^{-\omega}} = \frac{\omega}{2}\coth\left(\frac{\omega}{2}\right) \tag{2.10.2}$$

The $\coth(x)$ function has the expansion

$$\coth(x) = \frac{1}{x} + \frac{x}{3} - \frac{x^3}{45} + \frac{2x^5}{945} - \frac{x^7}{4725} + O(x^9) \tag{2.10.3}$$

From Eqs. 2.10.1 and 2.10.3

$$\frac{\omega}{2}\coth\left(\frac{\omega}{2}\right) = 1 + \left(\frac{\omega}{2}\right)^2\left[\frac{1}{3} + \frac{(\theta/2)^2}{45} + \frac{2(\theta/2)^4}{945} + \frac{(\theta/2)^6}{4725} + \cdots\right]$$

$$= 1 + \left(\frac{\omega}{\theta}\right)^2\left[\frac{(\theta/2)^2}{3} + \frac{(\theta/2)^4}{45} + \frac{2(\theta/2)^6}{945} + \frac{(\theta/2)^8}{4725} + \cdots\right]$$

$$= 1 + \left(\frac{\omega}{\theta}\right)^2\left[1 - (\theta/2)\cot(\theta/2)\right] \tag{2.10.4}$$

where

$$\cot(x) = \frac{1}{x} - \frac{x}{3} - \frac{x^3}{45} - \frac{2x^5}{945} - \frac{x^7}{4725} + O(x^9)$$

Hence, Eqs. 2.10.2 and 2.10.4 yield the final result

$$f = 1 + \frac{1}{2}\omega + \omega^2 A(\theta) \tag{2.10.5}$$

where

$$A(\theta) = \frac{1}{\theta^2}\left(1 - \frac{\theta}{2}\cot\frac{\theta}{2}\right)$$

$$\simeq \frac{1}{12} + \frac{1}{16\cdot 45}\theta^2 + O(\theta^4)$$

For $SU(2)$, the Maurer-Cartan equation ($\partial_i \equiv \partial/\partial\theta_i$) given in Eq. 2.9.7 yields

$$f_{\alpha i}\partial_i f_{\beta\gamma} - f_{\beta i}\partial_i f_{\alpha\gamma} = -\varepsilon_{\alpha\beta i}f_{i\gamma}$$

The Maurer-Cartan equation are satisfied by the explicit form for f given by Eq. 2.10.5 since A satisfies the remarkable differential equation

$$\theta\frac{\partial A}{\partial\theta} + 3A - \theta^2 A^2 = \frac{1}{4} \qquad (2.10.6)$$

Eq. 2.10.6 will be essential in obtaining the Kac-Moody commutation equation for $SU(2)$.

2.11   Left and Right Invariant Generators

The generators for the p irreducible representation, given by $X_a \equiv X^a(p)$, yield the following

$$E_a \exp\{iB^a X_a\} = \exp\{iB^a X_a\}X_a \qquad (2.11.1)$$

From Eq. 2.11.1

$$E_a^n \exp\{iB^a X^a\} = \exp\{iB^a X_a\}X_a^n \qquad (2.11.2)$$

It follows from Eq. 2.11.2 that

$$e^{i\phi_a E_a}e^{iB^a X^a} = e^{iB^a X_a}e^{i\phi_a X_a} \qquad (2.11.3)$$

Let V be a constant $SU(\mathcal{N})$ matrix; then from Eq. 2.11.3

$$e^{i\phi_a E_a}VU = VUe^{i\phi_a X_a} \quad ; \quad U = e^{iB^a X^a} \qquad (2.11.4)$$

Hence, the generator $E_a(U)$ is invariant under a left-multiplication of U by V and hence

$$E_a(U) = E_a(VU) \equiv E_a^L(U) \quad : \quad \text{left-invariant}$$

and E^L, from Eq. 2.9.5, can be written as follows

$$E_a^L \equiv E_a = f_{ab}\frac{\partial}{i\partial B_b} \qquad (2.11.5)$$

A right-invariant generator $E_a^R(U)$ is defined by

$$E_a^R U = X_a U \qquad (2.11.6)$$

From Eq. 2.11.6

$$E_a^R(U) = E_a^R(UV) \quad : \quad \text{right-invariant}$$

Hence

$$[E_a^R, E_b^R]U = [X_b, X_a]U = iC^{bac}E_c^R \quad \Rightarrow \quad [E_a^R, E_b^R] = -iC^{abc}E_c^R$$

From Eq. 2.11.6

$$e^{i\phi_a E_a^R}U = e^{i\phi_a X_a(p)}U \quad ; \quad U = e^{iB^a X^a(p)} \tag{2.11.7}$$

It follows from the definitions that

$$[E_a^L, E_b^R] = 0$$

In summary, the left- and right-invariant generators of the Lie algebra have the following commutation equations

$$[E_a^L, E_b^L] = iC^{abc}E_c^L \quad ; \quad [E_a^R, E_b^R] = -iC^{abc}E_c^R \quad ; \quad [E_a^R, E_b^L] = 0 \tag{2.11.8}$$

Hence, from Eqs. 2.3.4 and 2.11.6

$$E_a^R U = X_a U = UU^\dagger X_a U = \rho_{ab}UX_b = \rho_{ab}E_b^L U$$

$$\Rightarrow E_a^R = \rho_{ab}E_b^L = \rho_{ab}f_{ac}\frac{\partial}{i\partial B_c} = f_{ab}^T\frac{\partial}{i\partial B_b} \tag{2.11.9}$$

since, in matrix notation and from Eq. 2.9.4

$$f = \frac{\omega}{1 - e^{-\omega}} \quad ; \quad \omega^T = -\omega \quad \Rightarrow \quad f^T = e^{-\omega}f = \frac{\omega}{e^{\omega} - 1} \tag{2.11.10}$$

Putting together the left- and right-invariant generators yields, from Eqs. 2.11.4 and 2.11.7, the following

$$e^{i\phi_a E_a^L}e^{i\lambda_a E_a^R}U = e^{i\lambda_a X_a}Ue^{i\phi_a X_a} \quad ; \quad U = e^{iB^a X^a} \tag{2.11.11}$$

The Laplacian on the group space is useful for the Hamiltonian approach to the lattice gauge field, as discussed in Chapter 11. As for any curved space, the Laplacian for functions defined on the group space is given by

$$-\nabla^2 = \frac{1}{g}\frac{\partial}{\partial B^a}\left(\sqrt{g}g^{ab}\frac{\partial}{\partial B^b}\right) \tag{2.11.12}$$

It can be shown that the generators of the Lie algebra E_a yield the following factorization of the Laplacian [Zelobenko (1973)]

$$-\nabla^2 = E_a^L E_a^L = E_a^R E_a^R = E^2 \tag{2.11.13}$$

The Laplacian ∇^2 is a dimensionless second order differential operator defined in Eq. 11.4.5 for the Hamiltonian of the $SU(N)$ lattice gauge field. From Eq. 2.11.2 the Casimir operator is given by[4]

$$E^2 \exp iB^a X^a(p) = \{X^b(p)X^b(p)\} \exp iB^a X^a(p) = c_2(p)\exp iB^a X^a(p)$$

$$\Rightarrow E^2 \exp iB^a X^a(p) = c_2(p)\exp iB^a X^a(p) \quad : \text{eigenfunctions} \tag{2.11.14}$$

Hence, from Eq. 2.11.14

$$-\nabla^2 D_{ij}^{(p)}(U) = E^2 D_{ij}^{(p)}(U) = c_2(p)D_{ij}^{(p)}(U) \tag{2.11.15}$$

Note that

$$X^2(p)|p, ij\rangle = c_2(p)|p, ij\rangle \tag{2.11.16}$$

and for $U(1), c_2(p) = p^2$.

---

[4]Using $\sum_a E_a^R E_a^R = E^2$ yields $E^2 \exp iB^a X^a(p) = \exp iB^a X^a(p) \cdot c_2(p)$ which is equivalent to Eq. 2.11.14.

2.11.1 *Differential realization of SO(3) generators on state space*

The action of group elements define linear transformations on an underlying linear vector space. Let $\mathbf{v}$ be column vector of an d_p-dimensional linear vector space. The action of the group element is defined by matrix multiplication and yields

$$\mathbf{v} \to \mathbf{w} = U\mathbf{v} \qquad (2.11.17)$$

For simplicity, consider the special case of the adjoint representation of $SU(2)$, as this plays a special role in angular momentum theory. The generators and the group element in the adjoint representation of $SU(2)$ are given by

$$(X_i)_{ab} = i\epsilon_{aib} \; ; \quad \rho(\theta) = e^{-\omega} \; ; \quad (\omega)_{ab} = \theta_i\epsilon_{aib} \; : \quad a,i,b = 1,2,3$$

The group element acts on three-dimensional Euclidean space $\mathcal{E}_3$ with (real) three component vectors denoted by $\mathbf{v}$; the group rotates the vector as follows

$$\mathbf{v} \to \rho\mathbf{v}$$

Let $\mathbf{v} = (x_1, x_2, x_3)$ and let $f(\mathbf{x})$ be functions of $\mathcal{E}_3$; the generators on functions of $\mathcal{E}_3$, denoted by L_a, are given by

$$L_a = -i\epsilon_{aij}x_i\frac{\partial}{\partial x_j} \quad \Rightarrow \quad [L_a, L_b] = i\epsilon_{abc}L_c$$

The generators L_a are the angular momentum operators of quantum mechanics. The generators provide the realization of Eq. 2.11.17 for the specific case of the rotation $\mathbf{x}$ into $\mathbf{y}$. Using the identity valid for $\mathcal{O}$ being an operator or a matrix

$$\lim_{N\to\infty} \left(1 + \frac{1}{N}\mathcal{O}\right)^N = e^{\mathcal{O}}$$

yields, in matrix notation

$$\mathbf{x} \to \mathbf{y} = e^{i\theta_a L_a}\mathbf{x} = \lim_{N\to\infty}\left(1 + i\frac{\theta_a}{N}L_a\right)^N\mathbf{x}$$

Note

$$(i\theta_a L_a)x_k = -i^2\theta_a\epsilon_{aik}x_i = -\theta_a\epsilon_{kai}x_i = -(\omega\mathbf{x})_k \quad \Rightarrow \quad (i\theta_a L_a)\mathbf{x} = -\omega\mathbf{x}$$

Hence

$$\mathbf{x} \to \mathbf{y} = \lim_{N\to\infty}\left(1 - \frac{\omega}{N}\right)^N\mathbf{x} = e^{-\omega}\mathbf{x} = \rho(\omega)\mathbf{x}$$

2.12 *SU(N)* Character Function Expansion

For the case when $f(U)$ is a trace function such that $f(U) = f(VUV^\dagger)$, Eq. 2.8.7 yields

$$f(U) = \sum_p d_p a_p \operatorname{Tr}(D^{(p)}(U)) = \sum_p d_p a_p \chi_p(U) \qquad (2.12.1)$$

The character expansion reduces, for the $U(1)$ case, to the discrete Fourier transform. The irreducible representations of $U(1)$ are given by $\exp ip\theta, p \in \mathcal{Z}$ and Eq. 2.12.1 yields

$$d_p = 1 \; ; \; \chi_p(U) = e^{ip\theta} \quad \Rightarrow \quad f(\theta) = \sum_{p=-\infty}^{+\infty} a_p e^{ip\theta}$$

The Fourier representation is useful for performing convolutions. To define block-spins in Migdal's model, discussed in Chapter 10, a four-fold convolution has to be performed. Let e^{A_0} be the action function defined for lattice spacing a; the renormalization transformation $\mathcal{R}$ generates the action A_1 with lattice spacing $2a$ and is given by the following (four-fold convolution)

$$e^{A_0} = \sum_p d_p c_p \chi_p(U) \quad \Rightarrow \quad e^{A_1} = \mathcal{R}(e^{A_0}) = \left\{ \sum_p d_p c_p^4 \chi_p(U) \right\}^{2^{d-2}}$$

The above equation is useful for generating a strong coupling expansion of $\mathcal{R}$, since then the coefficients c_p rapidly go to zero. However, for the weak coupling sector, too many terms have to kept in the series expansion, and working in the group space becomes necessary.

Let X^a be the fundamental representation of the generators of $SU(N)$, and let

$$\mathrm{Tr}\,(X^a X^b) = \frac{1}{2}\delta_{ab} \tag{2.12.2}$$

Define Φ to be the group element in the fundamental representation and let ϕ^a be the canonical coordinates for Φ

$$\Phi = e^{i\phi^a X^a} \tag{2.12.3}$$

It is shown below that[5]

$$\lim_{\epsilon \to 0} \sum_p d_p e^{-\epsilon \frac{g^2}{2} c_2(p)} \chi_p(\Phi) \simeq \exp\left\{ \frac{1}{2g^2\epsilon} \mathrm{Tr}(\Phi + \Phi^\dagger) \right\} \tag{2.12.4}$$

The choice of the fundamental representation for Φ is arbitrary; using any other irreducible representation for Φ would only change the effective coupling constant g. To prove 2.12.4, consider the following Fourier expansion

$$\exp\left\{ \frac{1}{2g^2\epsilon} \mathrm{Tr}(\Phi + \Phi^\dagger) \right\} = \sum_p d_p a_p \chi_p(\Phi) \tag{2.12.5}$$

The orthogonality theorem given in Eq. 2.8.6 yields

$$a_p = \frac{1}{d_p} \int d\Phi \chi_p(\Phi^\dagger) \exp\left\{ \frac{1}{2g^2\epsilon} \mathrm{Tr}(\Phi + \Phi^\dagger) \right\} \tag{2.12.6}$$

---

[5]This identity will be required in the discussion of $SU(N)$ in Chapter 4.

For $\epsilon \to 0$, expanding Φ about unity and using Gaussian integration yields

$$a_p = \frac{1}{d_p} \prod_a \int_{-\infty}^{\infty} d\phi^a \, \text{Tr}(e^{-i\phi^a X_a(p)}) e^{-\frac{1}{2g^2\epsilon} \sum_a \phi^a \phi^a}$$

$$= \frac{1}{d_p} \prod_a \int_{-\infty}^{\infty} d\phi^a \, \text{Tr}\left[\mathbb{I}_p - i\phi^a X_a(p) - \frac{g^2}{2} \sum_{a,b} X^a(p) X^b(p) \phi^a \phi^b \right]$$

$$\times e^{-\frac{1}{2g^2\epsilon} \sum_a \phi^a \phi^a} + O(\epsilon^2)$$

$$= \frac{1}{d_p} \text{Tr}\left[\mathbb{I}_p - \epsilon \frac{g^2}{2} \sum_a X^a(p) X^a(p) \right] + O(\epsilon^2)$$

$$= \frac{1}{d_p} \text{Tr}[\mathbb{I}_p] \times \left(1 - \epsilon \frac{g^2}{2} c_2(p) \right) \simeq e^{-\epsilon \frac{g^2}{2} c_2(p)}$$

where the last line follows from Eq. 2.7.4.

Hence Eq. 2.12.4 is verified by Eq. 2.12.5 due to the result that

$$a_p \simeq e^{-\epsilon \frac{g^2}{2} c_2(p)} + O(\epsilon^2) \tag{2.12.7}$$

2.13   Summary

The emphasis on the review of Lie groups has been on its finite group elements. The exponential mapping of the Lie algebra to the group manifold provides a representation of the finite group elements. The Lie group forms a continuous differentiable space and the properties of this group space were studied in some detail.

The Riemannian metric on the group space was derived and the integration measure was defined. Integration theory over the group space was studied as this forms the basis of the path integral for lattice gauge fields.

The generators of the Lie algebra were realized as differential operators on the group manifold. This realization is required to define the non-Abelian electric field operator for the lattice gauge field. The special case of $SU(2)$ was studied in some detail.

Chapter 3

$SU(\mathcal{N})$ Kac-Moody Algebra

3.1 Introduction

Lie algebras have a natural and nontrivial extension – based on the *synthesis of calculus with the generators of the Lie algebras* – to infinite dimensional Kac-Moody algebras, also called affine Lie algebras. This synthesis is similar in spirit to that of linear algebra with calculus, which leads to the subject of functional analysis.

Compact Lie groups and Lie algebras have a nontrivial generalization to the infinite dimensional Kac-Moody algebras. The Kac-Moody algebra critically hinges on the properties the underlying Lie algebra – as well as the global properties of the underlying Lie group manifold – to form a closed algebra, and this feature will be brought out in the discussion given below.

The properties of the compact Lie group turn out to be crucial in algebraic structure of Kac-Moody algebras, and it is for this reason that Kac-Moody algebras have been included. Although lattice field theories have not yet been defined based on the Kac-Moody algebras, this could nevertheless be a useful future avenue of research.

Consider a loop S^1 parametrized by a *continuous variable* denoted by $\sigma \in [-\pi, +\pi]$; at every point of the loop is a compact Lie group G_σ, which for concreteness is taken to be $SU(\mathcal{N})$; the loop group space is the continuous tensor product of all the Lie groups and is given by

$$LG = \bigotimes_{\sigma=-\pi}^{+\pi} G_\sigma \quad : \text{Loop group}$$

The loop group consists of all maps from S^1 into a compact Lie group G. The continuous 'tensor product' needs to be rendered finite to be well defined. It will be seen later, in Section 3.10, that the loop group is not a simple tensor product. This is because there are *anomalies* that arise due to the continuous tensor product introducing singularities that are rendered finite by appropriately regularizing the loop group generators.

The Kac-Moody generators is realized as functional differential operators in Section 3.4. In Section 3.5, the two-dimensional chiral field is analyzed. The WZW

Lagrangian is obtained in Section 3.6 as an application of the functional differential realization of the Kac-Moody algebra.

The realization of the $SU(2)$ Kac-Moody algebra using functional differentiation has been obtained by Baaquie (1992). The special case of $SU(2)$ Kac-Moody algebra is worked out explicitly to verify the general result. In Section 3.7, the required $SU(2)$ group theory is discussed. Since the KM algebra is realized as the central extension of the loop group, the loop group algebra is analyzed as a preliminary step. In Section 3.8 the $SU(2)$ KM commutator is explicitly evaluated and the $SU(2)$ realization is shown to be exact.

The KM commutation equation is explicitly verified in Section 3.9. The suitably regularized second Casimir of the loop group is constructed and shown to be anomalous. The regularized Virasoro generators are defined essentially as the second Casimir of the KM generators, and in Section 3.10 the Virasoro commutation equations are obtained. All regularizations are done by the technique of point-splitting.

KM algebras have important and diverse application in physics [Goddard and Olive (1988)] and are indispensable in string theories defined on group manifolds [Gepner and Witten (1986)]. Kac-Moody algebras can be used for defining space-time gauge fields, with the dimension coming from the continuous label for the Kac-Moody generators, given by S^1 has an interpretation of extra space dimensions [Baaquie (1991b)]. Supersymmetric gauge fields with the Kac-Moody being the gauge group can be defined [Baaquie (2000)] and the $U(1)$ Kac-Moody gauge field has been shown by Baaquie and Parwani (1996) to be asymptotically free.

3.2 Kac-Moody Algebras

The differential representation E_a of the Lie group generators, discussed in Section 2.9, can be generalized to the loop group generators $E_a \to E_a(\sigma)$ with commutation similar to Eq. 2.9.2 and given by

$$[E_a(\sigma), E_b(\sigma')] = iC^{abc}E_c(\sigma)\delta(\sigma - \sigma') \ ; \ a, b, c = 1 \cdots \mathcal{N}^2 - 1 \ ; \ \sigma \in S^1 \quad (3.2.1)$$

The generators $E_a(\sigma)$ are anomalous due to the δ-function in the commutation equation. Note $E_a(\sigma)$ from Eq. 2.9.12 is a Hermitian differential operator.

The loop group generators $E_a(\sigma)$ have anomalous properties different from those for the underlying compact Lie group. For example, the operator $E_a(\sigma)E_a(\sigma)$ is singular and has to be regularized. Furthermore, unlike the case for a Lie group for which $E_a E_a$ commutes with all the generators, it is shown in Section 3.10 that $E_a(\sigma)E_a(\sigma)$ no longer commutes with all the generators of the loop group [Pressley and Segal (1988)].

One would like to couple the different Lie groups G_σ to go beyond the loop group. This is achieved by a non-trivial fibration of the loop group by the Abelian group $U(1)$ leading to twisted tensor product for the Lie groups G_σ. The key step in enhancing Lie algebras to an infinite dimensional algebra is to combine the finite dimensional Lie algebra of the compact Lie group with the infinite dimensional operator algebra encoded in Heisenberg commutation equation. The extension is made by adding the (infinite dimensional) unit operator $\mathbb{I}$ of the Heisenberg algebra to the generators of the Lie algebra $E_a(\sigma)$.

This gives rise to the infinite dimensional Kac-Moody algebras. The infinitely many generators of the Kac-Moody algebra are given by

$$K_a(\sigma); \; \mathbb{I} \; : \; a = 1 \cdots \mathcal{N}^2 - 1 \; ; \; \sigma \in S^1$$

where $\mathbb{I}$ is the unit operator on the Kac-Moody group. The $SU(\mathcal{N})$ Kac-Moody algebra is given by

$$[K_a(\sigma), K_b(\sigma')] = iC^{abc} K_c(\sigma)\delta(\sigma - \sigma') + ik\delta_{ab}\delta'(\sigma - \sigma')\mathbb{I} \tag{3.2.2}$$

and

$$h'(\sigma) \equiv \frac{\partial h}{\partial \sigma} \; ; \; [K_a(\sigma), \mathbb{I}] = 0$$

To have a well defined path integral for the WZW model, which is based on the Kac-Moody group, one needs to have $k = n/4\pi, n = $ integer.

Let the Lie group coordinates for group G_σ be given by $\theta_a(\sigma)$; the loop group generators are extended by a function to the Kac-Moody generators in the following manner:

$$E_a(\sigma) \to K_a(\sigma) = E_a(\sigma) + kF_{ab}(\sigma)\theta'_b(\sigma)\mathbb{I} \tag{3.2.3}$$

Note for the Abelian case, $C_{abc} = 0$ and the Kac-Moody algebra reduces to the Heisenberg algebra. An element of the infinite dimensional Kac-Moody group is given by

$$\mathcal{U}[\theta] = \exp\left\{ ik\mathbb{I} + i \oint d\sigma \theta_a(\sigma) Q_a(\sigma) \right\}$$

$$[Q_a(\sigma), Q_b(\sigma')] = iC^{abc} Q_c(\sigma)\delta(\sigma - \sigma') + ik\delta_{ab}\delta'(\sigma - \sigma')\mathbb{I} \tag{3.2.4}$$

where $Q_a(\sigma)$ are infinite dimensional constant matrices. The term $k\mathbb{I}$ is usually ignored in $\mathcal{U}[\theta]$ since it gives an Abelian phase that is accounted for only if necessary, as in the discussion of the 2-cocycle in Section 3.4.1.

3.3 Functional Differentiation

The Kac-Moody group has coordinates given by $\theta_a(\sigma)$; partial derivatives by the coordinates of the compact Lie group are replaced by functional differentiation by $\theta_a(\sigma)$. The generators of the Kac-Moody algebra are defined in terms of functional differential operators, which is briefly reviewed.

Consider variables f_n, $n = 0, \pm 1, \pm 2 \ldots \pm N$ that satisfy

$$\frac{\partial f_n}{\partial f_m} = \delta_{n-m}$$

Let $t = n\epsilon$, with $N \to \infty$. The limit $\epsilon \to 0$ yields

$$\frac{\partial f_n}{\partial f_m} \to \frac{\delta f(t)}{\delta f(t')} \equiv \lim_{\epsilon \to 0} \frac{1}{\epsilon} \frac{\partial f_n}{\partial f_m}$$

$$\Rightarrow \frac{\delta f(t)}{\delta f(t')} = \lim_{\epsilon \to 0} \frac{1}{\epsilon} \delta_{n-m} \to \delta(t - t') \tag{3.3.1}$$

In general, the *functional derivative* of $\Omega[f]$ – an arbitrary functional of the path $f(t)$ – is denoted by $\delta/\delta f(t)$ and is defined by

$$\frac{\delta \Omega[f]}{\delta f(t)} = \lim_{\epsilon \to 0} \frac{\Omega[f(t') + \epsilon \delta(t - t')] - \Omega[f]}{\epsilon} \tag{3.3.2}$$

In the notation of state space one has

$$\left\langle f \left| \frac{\delta}{\delta f(t)} \right| \Omega \right\rangle = \frac{\delta}{\delta f(t)} \langle f | \Omega \rangle = \frac{\delta \Omega[f]}{\delta f(t)}$$

Note that ϵ has the dimensions of $[f] \times [t]$.

Examples

- Consider the linear function $\Omega[f] = f(t_0)$; then, from Eq. 3.3.2

$$\frac{\delta \Omega[f]}{\delta f(t)} = \frac{\delta f(t_0)}{\delta f(t)} = \lim_{\epsilon \to 0} \frac{f(t_0) + \epsilon \delta(t - t_0) - f(t_0)}{\epsilon} = \delta(t - t_0)$$

- Let $\Omega[f] = \int d\tau f^n(\tau)$; from above

$$\frac{\delta \Omega[f]}{\delta f(t)} = \int d\tau n f^{n-1}(\tau) \frac{\delta f(\tau)}{\delta f(t)} \tag{3.3.3}$$

$$= \int d\tau n f^{n-1}(\tau) \delta(t - \tau) = n f^{n-1}(t) \tag{3.3.4}$$

3.3.1   *Chain rule*

The chain rule for the calculus of many variables has a generalization to functional calculus. Consider a change of variables from f_n to g_n; the chain rule of calculus yields

$$\frac{\partial}{\partial f_n} = \sum_{m=1}^{N} \frac{\partial g_m}{\partial f_n} \frac{\partial}{\partial g_m}$$

As before, let $t = n\epsilon$, $t' = m\epsilon$; re-write above expression as follows

$$\frac{1}{\epsilon} \frac{\partial}{\partial f_n} = \epsilon \sum_{m=1}^{N} \left[\frac{1}{\epsilon} \frac{\partial g_m}{\partial f_n} \right] \left[\frac{1}{\epsilon} \frac{\partial}{\partial g_m} \right]$$

Taking the limit of $N \to \infty$ and $\epsilon \to 0$ yields

$$\lim_{\epsilon \to 0} \frac{1}{\epsilon} \frac{\partial}{\partial f_n} \to \frac{\delta}{\delta f(t)} = \int dt' \frac{\delta g(t')}{\delta f(t)} \frac{\delta}{\delta g(t')} \quad : \quad \text{Chain rule} \tag{3.3.5}$$

3.4 Kac-Moody Generators

Two well-known realizations of the KM algebras are firstly the algebra of currents bilinear in the free two-dimensional non-Abelian Dirac field, and secondly the field-theoretic construction using the vertex operator of Frenkel and Kac [Francesco et al. (2012)]. The shortcomings of both these realizations are (i) they use fermion-like fields and construct the KM generators as bilinears in these fields, and (ii) normal ordering is essential for obtaining the central extension. Normal-ordering is not required to define the KM algebra; the realizations of KM algebra using fermion bilinears or exponential of free boson introduce spurious singularities and hence need regularization.

The KM generators are considered to be tangent vector fields on the infinite dimensional KM-group manifold. The KM-group manifold is obtained by a non-trivial fibration of the loop group by a $U(1)$ fiber [Pressley and Segal (1988)]. The Kac-Moody generators are explicitly realized as functional differential operators expressed in terms of the coordinates of the (infinite-dimensional) group manifold. The derivation of the Lie group generator E_a given in Eq. 2.9.3 is repeated for the Kac-Moody group. Using Eq. 3.2.4, let

$$Y_{a\sigma} \equiv \mathcal{U}^{\dagger} \frac{\delta \mathcal{U}}{i\delta\theta_a(\sigma)} \tag{3.4.1}$$

Introduce the homotopy parameter τ and define

$$Y_{a\sigma}(\tau) = \mathcal{U}^{\dagger} \frac{\delta \mathcal{U}(\tau)}{i\delta\theta_a(\sigma)} \quad ; \quad \mathcal{U}(\tau) = \exp\left\{i\tau \oint d\sigma \theta_a(\sigma) Q_a(\sigma)\right\} \tag{3.4.2}$$

with boundary conditions

$$Y_{a\sigma}(1) = Y_{a\sigma} \quad ; \quad Y_{a\sigma}(0) = 0$$

Repeating the steps taken to obtain Eq. 2.3.3 yields

$$\frac{dY_{a\sigma}(\tau)}{d\tau} = \mathcal{U}^{\dagger}(\tau) Q_a(\sigma) \mathcal{U}(\tau) \tag{3.4.3}$$

with boundary condition

$$\frac{dY_{a\sigma}(0)}{d\tau} = Q_a(\sigma)$$

Repeating the steps in Eq. 2.3.8 and differentiating Eq. 3.4.3 yields, using Eq. 3.2.4

$$\frac{d}{d\tau}\left(\frac{dY_{a\sigma}(\tau)}{d\tau}\right) = -i\mathcal{U}^{\dagger}(\tau) \oint d\sigma' \theta_\alpha(\sigma') \left[Q_a(\sigma'), Q_a(\sigma)\right] \mathcal{U}(\tau)$$

$$= -i\mathcal{U}^{\dagger}(\tau) \oint d\sigma' \theta_\alpha(\sigma') \left\{iC^{a\alpha b} Q_b(\sigma')\delta'(\sigma' - \sigma) + ik\delta_{\alpha a}\delta'(\sigma' - \sigma)\right\} \mathcal{U}(\tau)$$

$$= -i\mathcal{U}^{\dagger}(\tau)\left\{i\theta_\alpha(\sigma) C^{a\alpha b} Q_b(\sigma) - ik\theta'_\alpha(\sigma)\right\} \mathcal{U}(\tau)$$

$$= -\theta_{ab} \frac{dY_{b\sigma}(\tau)}{d\tau} - k\theta'_a(\sigma) \tag{3.4.4}$$

Writing the non-Abelian indices in Eq. 3.4.4 using matrix and Dirac notation and suppressing the σ index, yields

$$\frac{d}{d\tau}\left|\frac{dY(\tau)}{d\tau}\right\rangle = -\theta\left|\frac{dY(\tau)}{d\tau}\right\rangle - k\left|\theta'\right\rangle \;\; ; \;\; (\theta)_{ab} \equiv \theta_\alpha(\sigma)C^{a\alpha b} \quad (3.4.5)$$

$$\Rightarrow e^{-\tau\theta}\frac{d}{d\tau}\left(e^{\tau\theta}\left|\frac{dY(\tau)}{d\tau}\right\rangle\right) = -k\left|\theta'\right\rangle \quad (3.4.6)$$

Integrating the homotopy parameter from 0 to τ yields, using Eq. 3.4.3

$$\int_0^\tau d\tau'\frac{d}{d\tau}\left(e^{\tau'\theta}\left|\frac{dY(\tau')}{d\tau}\right\rangle\right) = -k\left|\theta'\right\rangle\int_0^\tau d\tau' e^{\tau\theta}$$

$$\Rightarrow e^{\tau\theta}\left|\frac{dY(\tau)}{d\tau}\right\rangle - \left|Q\right\rangle = -k\frac{1}{\theta}\left(e^{\tau\theta} - 1\right)\left|\theta'\right\rangle$$

$$\Rightarrow \left|\frac{dY(\tau)}{d\tau}\right\rangle = e^{-\tau\theta}\left|Q\right\rangle - k\frac{1}{\theta}\left(1 - e^{-\tau\theta}\right)\left|\theta'\right\rangle \quad (3.4.7)$$

Eq. 3.4.7 has an important special case of $\tau = 1$. From Eq. 3.4.3

$$\frac{dY_{a\sigma}(1)}{d\tau} = \mathcal{U}^\dagger(1)Q_a(\sigma)\mathcal{U}(1) = \mathcal{U}^\dagger Q_a(\sigma)\mathcal{U}$$

and hence, from Eq. 3.4.7 [Baaquie (1988)]

$$\mathcal{U}^\dagger Q_a(\sigma)\mathcal{U} = \rho_{ab}Q_b(\sigma) - ke_{ab}\theta'_b(\sigma) \quad (3.4.8)$$

with

$$e = \frac{1}{\theta}\left(1 - e^{-\theta}\right) \;\; ; \;\; \rho = e^{-\theta}$$

Integrating Eq. 3.4.7 with τ from 0 to 1, and using $dY(0)/d\tau = 0$ from Eq. 3.4.3, yields the result

$$\int_0^1 d\tau\left|\frac{dY(\tau)}{d\tau}\right\rangle = \int_0^1 d\tau e^{-\tau\theta}\left|Q\right\rangle - k\left|\theta'\right\rangle\frac{1}{\theta}\int_0^1 d\tau\left(1 - e^{-\tau\theta}\right)$$

$$\Rightarrow \left|Y(1)\right\rangle = \frac{1}{\theta}\left(1 - e^{-\theta}\right)\left|Q\right\rangle - k\frac{1}{\theta}\left(1 - \frac{1}{\theta}\left(1 - e^{-\theta}\right)\right)\left|\theta'\right\rangle$$

$$\Rightarrow \left|Y(1)\right\rangle = e\left|Q\right\rangle - k\frac{1}{\theta}\left(1 - e\right)\left|\theta'\right\rangle \quad (3.4.9)$$

From Eq. 2.9.3

$$Y_{a\sigma}(1) = Y_{a\sigma} = \mathcal{U}^\dagger\frac{\delta\mathcal{U}}{i\delta\theta_a(\sigma)}$$

Writing out Eq. 3.4.9 in components yields from Eq. 2.9.3

$$\mathcal{U}^\dagger\frac{\delta\mathcal{U}}{i\delta\theta_a(\sigma)} = e_{ab}(\sigma)Q_b(\sigma) - k\left\{\frac{1}{\theta}\left(1 - e\right)\right\}_{ab}(\sigma)\theta'_b(\sigma) \quad (3.4.10)$$

The definition of the generator of the loop group is taken from the Lie group, and from Eq. 2.9.5 is given by

$$E_a(\sigma) = f_{ab}\frac{\delta}{i\delta\theta_b(\sigma)} \;\; ; \;\; f = e^{-1} = \frac{\theta}{\left(1 - e^{-\theta}\right)} \quad (3.4.11)$$

From Eqs. 3.4.10 and 3.4.11

$$\frac{\delta \mathcal{U}}{i\delta\theta_a(\sigma)} = \mathcal{U}\left[e_{ab}(\sigma)Q_b(\sigma) - k\left\{\frac{1}{\theta}(1-e)\right\}_{ab}(\sigma)\theta_b'(\sigma)\right]$$

$$\Rightarrow \left[E_a(\sigma) + kF_a\right]\mathcal{U} = \mathcal{U}Q_a(\sigma) = K_a(\sigma)\mathcal{U} \tag{3.4.12}$$

Hence, the Kac-Moody generator, from the defining Eq. 3.4.12, are given by

$$K_a(\sigma) = E_a(\sigma) + kF_a(\sigma) \ ; \ \ F_a(\sigma) = \left\{\frac{1}{\theta}(f-1)\right\}_{ab}(\sigma)\theta_b'(\sigma) \tag{3.4.13}$$

Eq. 3.4.13 is a result quoted, without a proof, in Baaquie (1986, 1988, 1991b).

In components, Eq. 3.4.13 yields

$$K_a(\sigma) = f_{a\beta}(\sigma)\frac{\delta}{i\delta\theta_\beta(\sigma)} + k\theta_{a\beta}^{-1}(\sigma)[f_{\beta\gamma}(\sigma) - \delta_{\beta\gamma}]\frac{\partial\theta_\gamma(\sigma)}{\partial\sigma}$$

The function $F_a(\sigma)$ is well behaved for all values of θ; in particular, for $\theta \simeq 0$

$$\frac{1}{\theta}(f-1) \simeq \frac{1}{2} + \frac{1}{12}\theta + O(\theta^2) \tag{3.4.14}$$

For the Abelian case, $C_{abc} = 0$ and yields

$$F(\sigma) = \frac{k}{2}\theta'(\sigma)$$

Hence the Abelian Kac-Moody algebra, which is isomorphic to the Heisenberg algebra, is given by

$$K(\sigma) = \frac{\delta}{i\delta\theta(\sigma)} + \frac{k}{2}\theta'(\sigma) \tag{3.4.15}$$

Noteworthy 3.1: Kac-Moody Central Extension

The derivation of the central extension is based on solving Eq. 3.4.5; writing the equation in vector notation yields($\theta' = \theta_a'$)

$$\frac{d^2y}{d\tau^2} + \theta\frac{dy}{d\tau} + k\theta' = 0$$

that yields the solution given in Eq. 3.4.9

$$y(1) = y(0) + \int_0^1 d\tau e^{-\tau\theta}\frac{dy(0)}{d\tau} - k\int_0^1 d\tau e^{-\tau\theta}\int_0^\tau d\tau' e^{\tau'\theta}\theta'$$

$$= \int_0^1 d\tau e^{-\tau\theta}Q - k\int_0^1 d\tau e^{-\tau\theta}\int_0^\tau d\tau' e^{\tau'\theta}\theta' \tag{3.4.16}$$

The integrations, in matrix notation, yield the following

$$\int_0^1 d\tau e^{-\tau\theta} = \frac{1}{\theta}(1 - e^{-\tau\theta})$$

and

$$\int_0^1 d\tau e^{-\tau\theta}\int_0^\tau d\tau' e^{\tau'\theta} = \int_0^1 d\tau\frac{1}{\theta}\left(1 - e^{-\tau\theta}\right) = \frac{1}{\theta}\left(1 - \frac{1}{\theta}(1 - e^{-\theta})\right)$$

and we have obtained the result given in Eq. 3.4.10

$$\mathcal{U}^\dagger \frac{\delta \mathcal{U}}{i\delta\theta_a(\sigma)} = e_{ab}(\sigma)Q_b(\sigma) - k\left\{\frac{1}{\theta}(1-e)\right\}_{ab}(\sigma)\theta_b'(\sigma) \ ; \ e_{ab}(\sigma) = \left[\frac{1}{\theta}(1-e^{-\tau\theta})\right]_{ab}$$

Furthermore, the central extension is given by

$$\frac{f}{\theta}\left(1 - \frac{1}{\theta}(1-e^{-\theta})\right) = \frac{1}{\theta}(f-1)$$

There is another set of KM generators $\tilde{K}_\alpha(\sigma)$, similar to the case of Lie groups that have left- and right-invariant generators, that is defined, similar to Eq. 3.4.12, by

$$\tilde{K}_\alpha(\sigma)\mathcal{U} = Q_\alpha(\sigma)\mathcal{U}$$

and hence

$$[\tilde{K}_\alpha(\sigma), \tilde{K}_\beta(\sigma')] = [Q_\beta(\sigma'), Q_\alpha(\sigma)]\mathcal{U} \tag{3.4.17}$$

Note that on the right hand side of Eq. 3.4.17 the order of the commutation equation has been reversed; hence, from Eq. 3.2.4, one obtains the commutation equations (note the minus signs)

$$[\tilde{K}_\alpha(\sigma), \tilde{K}_\beta(\sigma')] = iC^{\beta\alpha\gamma}\tilde{K}_\gamma(\sigma')\delta(\sigma'-\sigma) + ik\delta_{\beta\alpha}\delta'(\sigma'-\sigma)$$
$$= -iC^{\alpha\beta\gamma}\tilde{K}_\gamma(\sigma)\delta(\sigma-\sigma') - ik\delta_{\alpha\beta}\delta'(\sigma-\sigma')$$

Similar to the derivation for $K_\alpha(\sigma)$, the associated generators $\tilde{K}_\alpha(\sigma)$ have a realization given by

$$\tilde{K}_\alpha(\sigma) = f_{\alpha\beta}^T(\sigma)\frac{\delta}{i\delta\theta_\beta(\sigma)} + k(f_{\alpha\beta}^T - \delta_{\alpha\beta})\xi_{\beta\gamma}\theta_\gamma'(\sigma)$$

Furthermore, $\tilde{K}_\alpha(\sigma)$ commutes with $K_\beta(\sigma')$

$$[\tilde{K}_\alpha(\sigma), K_\beta(\sigma')] = 0$$

The commutation equation Eq. 3.2.2 is valid everywhere on the KM-group manifold. Although explicit coordinates $\theta_\alpha(\sigma)$ were used in Eq. 3.4.12 to define K_α, the expression for K_α is coordinate independent in the sense that the form of K_α is the same in all coordinate patches. Similarly, $\tilde{K}_\alpha$ is also coordinate independent.

3.4.1   *Kac-Moody generator and the 2-cocycle*

An exact realization of the Kac-Moody generators was derived in Baaquie (1986). The derivation is based on the theory of projective representation, and the 2-cocycle determines the central extension of the loop group generators [Mickelsson (1985)].

In general, to the infinite set of coordinates of the KM group $\theta_\alpha(\sigma)$ a phase φ is also required to express the KM group element

$$U[\varphi; \theta] = \exp\left(ik\varphi + i\int d\sigma\theta_\alpha(\sigma)Q_\alpha(\sigma)\right)$$

Note the coordinate φ does not appear in K_α since a basis has been chosen for the KM-algebra in which k is proportional to the identity operator.

The projective representation of the KM group element is given by [Mickelsson (1985)]

$$U[\tilde{k};\tilde{\theta}]U[k;\theta] = \exp\{i(\tilde{k}+k+\omega_2[\tilde{\theta},\theta]\}U[\tilde{\theta}\circ\theta] \qquad (3.4.18)$$

where $\tilde{\theta}\circ\theta$, for each σ, is composed according to the composition rules of G.

Consider multiplying group elements with coordinates $\theta_1,\theta_2,\theta_3$. KM group multiplication is associative in that the order of multiplication does not matter. Hence

$$\left(U[\theta_1]U[\theta_2]\right)U[\theta_3] = U[\theta_1]\left(U[\theta_2]U[\theta_3]\right)$$

Associativity implies that the phase factor ω_2 satisfies the following 2-cocycle condition [Mickelsson (1985)]

$$\omega_2(\theta_1,\theta_2) + \omega_2(\theta_1\cdot\theta_2,\theta_3) = \omega_2(\theta_1,\theta_2\cdot\theta_3) + \omega_2(\theta_2,\theta_3) \qquad (3.4.19)$$

The 2-cocycle can be directly obtained using the Campbell-Hausdorff-Baker formula. It can also be obtained from the torsion tensor of the Lie group given in Eq. 3.6.10 using the descent equation of G-cohomology [Mickelsson (1985); Baaquie (1991a)].

Consider the functional derivative

$$-i\mathcal{U}^\dagger[k;\theta]\frac{\delta}{\delta\theta_a(\sigma)}\mathcal{U}[k;\theta] = -\frac{i}{\epsilon}\left\{\mathcal{U}^\dagger[k;\theta]\mathcal{U}[k;\theta+\epsilon\delta_{a\alpha}\delta(\sigma-\sigma')] - 1\right\}$$

Using the group multiplication with 2-cocycle given in Eq. 3.4.18 and the loop group composition equation given in Eq. 3.4.20 yields from above

$$-i\mathcal{U}^\dagger[k;\theta]\frac{\delta}{\delta\theta_a(\sigma)}\mathcal{U}[k;\theta]$$

$$= -\frac{i}{\epsilon}\left[\exp\left\{i\omega_2[\theta,\theta+\epsilon\delta_{a\alpha}\delta(\sigma-\sigma')] + i\epsilon e_{ab}(\theta(\sigma))Q_b(\sigma)\right\} - 1\right]$$

$$= \frac{1}{\epsilon}\omega_2[\theta,\theta+\epsilon\delta_{a\alpha}\delta(\sigma-\sigma')] + e_{ab}(\theta(\sigma))Q_b(\sigma) \qquad (3.4.20)$$

Since

$$\omega_2[\theta,\theta] = 0$$

with notation anticipating later results

$$\frac{1}{\epsilon}\omega[\theta,\theta+\epsilon\delta_{a\alpha}\delta(\sigma-\sigma')] = \left.\frac{\delta\omega_2[\theta,\tilde{\theta}]}{\delta\tilde{\theta}_a(\sigma)}\right|_{\tilde{\theta}=\theta} \equiv ke_{ab}F_\alpha[\theta(\sigma)]$$

Multiplying both sides of Eq. 3.4.20 with $f_{\alpha\beta}$ yields

$$\left[E_\alpha(\sigma) + kF_\alpha(\sigma)\right]\mathcal{U} = \mathcal{U}Q_\alpha(\sigma) = K_\alpha(\sigma)\mathcal{U}$$
$$\Rightarrow K_\alpha(\sigma) = E_\alpha(\sigma) + kF_\alpha(\sigma)$$

Hence, from above, the central extension is given by

$$kF_\alpha(\sigma) = f_{\alpha\beta} \left. \frac{\delta\omega_2[\theta,\tilde{\theta}]}{\delta\tilde{\theta}_\beta(\sigma)} \right|_{\tilde{\theta}=\theta}$$

For the Abelian $U(1)$ group, the Kac-Moody algebra is given by the Heisenberg algebra

$$[Q(\sigma), Q(\sigma')] = ik\delta'(\sigma - \sigma')$$

Using the Campbell-Hausdorff-Baker formula for the Abelian case

$$e^x e^y = e^{X+Y+\frac{1}{2}[X,Y]}$$

yields

$$\exp\left\{i \oint d\sigma\theta(\sigma)Q(\sigma)\right\} \exp\left\{i \oint d\sigma\phi(\sigma)Q(\sigma)\right\}$$

$$= \exp\left\{i \oint d\sigma(\theta+\phi)(\sigma)Q(\sigma) + \frac{i^2}{2} \oint d\sigma d\sigma'\theta(\sigma)\phi(\sigma')[Q(\sigma), Q(\sigma')]\right\}$$

$$= \exp\left\{i\frac{k}{2} \oint d\sigma\theta'(\sigma)\phi(\sigma)\right\} \exp\{i \oint d\sigma(\theta+\phi)(\sigma)Q(\sigma)\} \qquad (3.4.21)$$

Hence, from Eq. 3.4.21, the Abelian 2-cocycle is given by

$$\omega_2[\theta,\phi] = \frac{k}{2} \oint d\sigma\theta'(\sigma)\phi(\sigma)$$

It can be directly verified that the Abelian 2-cocycle satisfies the cocycle condition given in Eq. 3.4.19. Note that, as expected

$$\omega[\theta,\theta] = 0$$

The Abelian central extesion is given by

$$\left. \frac{\delta\omega_2[\theta,\phi]}{\delta\phi(\sigma)} \right|_{\phi=\theta} = \frac{k}{2}\theta'(\sigma)$$

Hence the Abelian Kac-Moody generator is given by

$$K(\sigma) = \frac{\delta}{i\delta\theta(\sigma)} + \frac{k}{2}\theta'(\sigma) \qquad (3.4.22)$$

and reproduces the result obtained earlier in Eq. 3.4.15.

3.5   Chiral Field

The chiral field describes a two-dimensional quantum field theory with the degree of freedom being the nonlinear Lie Group; it has many applications in mathematical physics and is a first step in discussing the WZW Lagrangian. Its essential properties are reviewed.

The chiral field's degree of freedom is the group element given by

$$g(\sigma) = \exp\{X_a\theta_a(\sigma)\} \quad ; \quad [X_a, X_b] = iC^{abc}X_c$$

The Sugawara Hamiltonian for the chiral field is given by [Sugawara (1968)]

$$H_C = \frac{1}{2} \oint d\sigma \sum_a \{\lambda^2 E_a^2(\sigma) + (1/\lambda^2) A_a^2(\sigma)\} \tag{3.5.1}$$

The kinetic term, from Eqs. 3.2.3 and 3.4.11, is given by

$$[E_a(\sigma), E_b(\sigma')] = iC^{abc} E_c(\sigma)\delta(\sigma - \sigma') \tag{3.5.2}$$

$$E_a(\sigma) = f_{ab} \frac{\delta}{i\delta\theta_b(\sigma)} \quad ; \quad f = e^{-1} = \frac{\theta}{(1 - e^{-\theta})} \tag{3.5.3}$$

The potential term is given by

$$A_\alpha(\sigma) = e_{\beta\alpha}(\sigma) \frac{\partial\theta_\beta(\sigma)}{\partial\sigma} \equiv e_{\alpha\beta}^T(\sigma)\theta_\beta'(\sigma)$$

The commutation equation of $E_a(\sigma)$ with $A_\alpha(\sigma)$ is derived.

From Eq. 2.3.1, consider the following

$$g^\dagger(\sigma) \frac{\partial g(\sigma)}{\partial\sigma} = \frac{\partial\theta_\gamma}{\partial\sigma} g^\dagger(\sigma) \frac{\partial g(\sigma)}{\partial\theta_\gamma} = i\theta_\gamma' e_{\gamma\alpha}(\sigma)X_\alpha = iA_\alpha(\sigma)X_\alpha \tag{3.5.4}$$

Using the identity

$$[A, BC] = [A, B]C + B[A, C]$$

yields

$$\left[E_a(\sigma'), g^\dagger(\sigma) \frac{\partial g(\sigma)}{\partial\sigma}\right] = [E_a(\sigma'), g^\dagger(\sigma)] \frac{\partial g(\sigma)}{\partial\sigma} + g^\dagger(\sigma) \frac{\partial}{\partial\sigma}\left([E_a(\sigma'), g(\sigma)]\right) \tag{3.5.5}$$

Note that

$$[E_a(\sigma'), g(\sigma)] = \left(E_a(\sigma')g(\sigma)\right) = g(\sigma)X_a\delta(\sigma - \sigma')$$

Using

$$E_a(\sigma')\left(g^\dagger(\sigma)g(\sigma)\right) = 0 \quad \Rightarrow \quad \left(E_a(\sigma')g^\dagger(\sigma)\right) = -X_a g^\dagger(\sigma)\delta(\sigma - \sigma')$$

yields

$$[E_a(\sigma'), g^\dagger(\sigma)] = \left(E_a(\sigma')g^\dagger(\sigma)\right) = -X_a g^\dagger(\sigma)\delta(\sigma - \sigma')$$

Hence, from Eq. 3.5.5

$$\left[E_a(\sigma'), g^\dagger(\sigma) \frac{\partial g(\sigma)}{\partial\sigma}\right] = -X_a g^\dagger(\sigma) \frac{\partial g(\sigma)}{\partial\sigma}\delta(\sigma - \sigma') + g^\dagger(\sigma) \frac{\partial}{\partial\sigma}\left(g(\sigma)X_a\delta(\sigma - \sigma')\right)$$

$$= -\left[X_a, g^\dagger(\sigma) \frac{\partial g(\sigma)}{\partial\sigma}\right]\delta(\sigma - \sigma') + X_a \frac{\partial}{\partial\sigma}\delta(\sigma - \sigma') \tag{3.5.6}$$

From Eq. 3.5.4

$$\left[X_a, g^\dagger(\sigma) \frac{\partial g(\sigma)}{\partial\sigma}\right] = i\theta_\gamma' e_{\gamma\alpha}(\sigma)[X_a, X_\alpha] = i^2 C^{a\alpha\beta} X_\beta \theta_\gamma' e_{\gamma\alpha}(\sigma) = -C^{a\alpha\beta} X_\beta A_\alpha(\sigma)$$

Hence, from Eqs. 3.5.4 and 3.5.6, one obtains the commutator

$$[E_a(\sigma'), iA_\alpha(\sigma)X_\alpha] = C^{a\alpha\beta} X_\beta A_\alpha(\sigma) + X_a \frac{\partial}{\partial\sigma}\delta(\sigma - \sigma')$$

$$\Rightarrow [E_a(\sigma), A_b(\sigma')] = iC^{ab\beta} A_\beta(\sigma)\delta(\sigma - \sigma') + i\delta_{ab}\delta'(\sigma - \sigma') \tag{3.5.7}$$

In the Sugawara approach, one *postulates* the above commutation equation given in Eq. 3.5.7 and then derives the expression for the kinetic operator $E_a(\sigma)$ given in Eq. 3.5.3.

3.6   The WZW Lagrangian

As an application of the realization of the Kac-Moody algebra, the two-dimensional Wess-Zumino-Witten (WZW) chiral theory [Witten (1983)] is given a Hamiltonian formulation using the KM algebra. The WZW model has wide applications in physics, including the study of superstring theories as well as black holes. The WZW action is derived by exploiting the exact form of the KM realization [Baaquie (1986)].

In the WZW case, unlike the usual sigma-model, the kinetic operator given in Eq. 3.5.2 is replaced by $\pi_\alpha(\sigma)$

$$E_\alpha(\sigma) \to \pi_\alpha(\sigma)$$

and, instead of Eq. 3.5.2, one postulates *anomalous* commutation equations, which for $d = 2$ is given by

$$[\pi_\alpha(\sigma)\,,\,\pi_\beta(\sigma')] = iC_{\alpha\beta\gamma}[\pi_\gamma(\sigma) - kA_\gamma(\sigma)]\delta(\sigma - \sigma') \qquad (3.6.1)$$

Note for $k = 0$, one recovers the chiral field's commutators. Furthermore, as in Eq. 3.5.7

$$[\pi_\alpha(\sigma)\,,\,A_\beta(\sigma')] = iC_{\alpha\beta\gamma}A_\gamma(\sigma)\delta(\sigma - \sigma') + i\delta_{\alpha\beta}\partial_\sigma\delta(\sigma - \sigma')$$

where

$$A_\alpha(\sigma) = e^T_{\alpha\beta}(\sigma)\theta'_\beta(\sigma) = e_{\beta\alpha}(\sigma)\partial_\sigma\theta_\beta(\sigma)\,,\ e_{\alpha\beta} = f^{-1}_{\alpha\beta}$$

and

$$[A_\alpha(\sigma)\,,\,A_\beta(\sigma')] = 0$$

The WZW Hamiltonian has the Sugawara form of the non-linear chiral model given in Eq. 3.5.1; for dimensionless coupling constant λ^2, we have [Sugawara (1968)]

$$H_{WZW} = \frac{1}{2}\int d\sigma[\lambda^2\pi_\alpha(\sigma)\pi_\alpha(\sigma) + (1/\lambda^2)A_\alpha(\sigma)(\sigma)A_\alpha(\sigma)]$$

H_{WZW} together with the anomalous commutation equation for π_α yields, for $k = n/4\pi (n = \text{integer})$ the WZW field equations.

A concrete realization for $\pi_\alpha(\sigma)$ is obtained by choosing $2k$ for the central extension in the KM algebra. Then it can be shown that, for Kac-Moody generators given by $K_\alpha(\sigma)$

$$\pi_\alpha(\sigma) = K_\alpha(\sigma) - kA_\alpha(\sigma) = E_\alpha(\sigma) + 2kF_\alpha(\sigma) - kA_\alpha(\sigma) \qquad (3.6.2)$$

Eq. 3.6.2 yields, using Eq. 3.5.7, the following commutation equation

$$[\pi_\alpha(\sigma)\,,\,\pi_\beta(\sigma')] = [K_\alpha(\sigma)\,,\,K_\beta(\sigma')] - 2k[E_\alpha(\sigma)\,,\,A_\beta(\sigma')]$$

$$= iC_{\alpha\beta\gamma}K_\gamma(\sigma) + 2ki\delta_{\alpha\beta}\partial_\sigma\delta(\sigma - \sigma') - 2k\Big(iC_{\alpha\beta\gamma}A_\beta(\gamma)\delta(\sigma - \sigma') + i\delta_{ab}\delta'(\sigma - \sigma')\Big)$$

$$= iC_{\alpha\beta\gamma}[\pi_\gamma(\sigma) - kA_\gamma(\sigma)]\delta(\sigma - \sigma')$$

and we have recovered the result given in Eq. 3.6.1. Note the Kac-Moody algebra's central extension is necessary for canceling the central extension in the commutation of E_a with A_b, as given in Eq. 3.5.7.

Eq. 3.6.1 yields for the WZW Hamiltonian

$$H_{WZW} = \frac{1}{2} \int d\sigma [\lambda^2 (E_\alpha + 2kF_\alpha - kA_\alpha)^2 + (1/\lambda^2) A_\alpha^2] \qquad (3.6.3)$$

In effect, the WZW central extension term in the action (whose coefficient is k) acts as a *field dependent* background field for the kinetic operator E_α.

The WZW Lagrangian is given by the Dirac-Feynman formula ($N(\epsilon)$ is a normalization constant)

$$\lim_{\epsilon \to 0} N(\epsilon) \exp\{i\epsilon \mathcal{L}_{WZW}[\theta; \widetilde{\theta}]\} = \langle \theta | \exp(-i\epsilon H_{WZW}) | \widetilde{\theta} \rangle \qquad (3.6.4)$$

where $|\theta\rangle = \otimes |\theta(\sigma)\rangle$ is the coordinate eigenstate. For infinitesimal time ϵ, plane wave states can be used to evaluate the Lagrangian using Eq. 3.6.4. The fact that θ_α are curved variables does not enter to leading order.

Let plane wave states be given by

$$\langle \theta | p \rangle = \exp\left\{ i \int d\sigma p_\alpha(\sigma) \theta_\alpha(\sigma) \right\} \equiv e^{ip\theta} \quad ; \quad \int Dp |p\rangle \langle p| = \mathbb{I}$$

The Hamiltonian operator acts on the bra vector; this yields – from Eq. 3.6.3 and suppressing the $\int d\sigma$ integration in the Hamiltonian – the following

$$\langle \theta | \exp(-i\epsilon H_{WZW}) | \widetilde{\theta} \rangle = \int Dp \langle \theta | \exp(-i\epsilon H_{WZW}) | p \rangle \langle p | \widetilde{\theta} \rangle$$

$$= e^{-i\epsilon(1/2\lambda^2) A_\alpha^2 (\theta)} \exp\left\{ -i\epsilon \frac{\lambda^2}{2} \left(f_{\alpha\beta} \frac{\partial}{i\partial\theta_\beta} + 2kF_\alpha(\theta) - kA_\alpha(\theta) \right)^2 \right\} e^{ip(\theta - \widetilde{\theta})}$$

$$\simeq e^{-i\epsilon(1/2\lambda^2) A_\alpha^2 (\theta)} \exp\left\{ -i\epsilon \frac{\lambda^2}{2} (f_{\alpha\beta} p_\beta + 2kF_\alpha(\theta) - kA_\alpha(\theta))^2 \right\} e^{ip(\theta - \widetilde{\theta})} \quad (3.6.5)$$

Make the change of variables

$$p_\beta \to f_{\beta a}^{-1} p_a = e_{\beta a} p_a \quad ; \quad Dp \to \det(e) Dp$$

Ignoring the $\det(e)$ as it does not affect the result to $O(\epsilon)$ yields, from Eq. 3.6.4

$$\langle \theta | e^{-i\epsilon H_{WZW}} | \widetilde{\theta} \rangle = \int Dp \exp\{ -i\epsilon(1/2\lambda^2) A_\alpha^2 (\theta) \}$$

$$\times \exp\{ -i\epsilon(\lambda^2/2)(p_\alpha + 2kF_\alpha(\theta) - kA_\alpha(\theta))^2 e^{i(\theta - \widetilde{\theta}) ep} \}$$

$$= \int Dp \exp\{ -i\epsilon(1/2\lambda^2) A_\alpha^2 (\theta) - i\epsilon(\lambda^2/2)(p_\alpha)^2 + i(\theta - \widetilde{\theta}) e \cdot (p - 2kF + kA) \}$$

$$\qquad (3.6.6)$$

Define

$$\theta - \widetilde{\theta} = \epsilon \dot{\theta} \quad ; \quad \dot{\theta} = \frac{\partial \theta}{\partial t} = \partial_t \theta$$

Performing the Gaussian integration in Eq. 3.6.6 over $\int Dp$, and from Eq. 3.6.4

$$\lim_{\epsilon \to 0} N(\epsilon) \exp\{i\epsilon \mathcal{L}_{WZW}[\theta; \widetilde{\theta}]\} = \langle \theta | \exp\{-i\epsilon H_{WZW}\} | \widetilde{\theta} \rangle$$

$$= \exp \left\{ -i\epsilon (1/2\lambda^2) A_\alpha^2(\theta) + i\epsilon (1/2\lambda^2)(e_{\alpha\beta}\dot{\theta}_\beta)^2 - 2ik\epsilon\dot{\theta}e\left(F - \frac{1}{2}A\right) \right\} \quad (3.6.7)$$

Restoring the $\int d\sigma$ integration and ignoring the normalization, it follows that

$$\mathcal{L}_{WZW} = \int d\sigma (\mathcal{L}_0 + \mathcal{L}_1)$$

where

$$\mathcal{L}_0 = \frac{1}{2\lambda^2} \sum_\alpha [(e_{\alpha\beta}\dot{\theta}_\beta)^2 - (e_{\beta\alpha}\theta'_\beta)^2] = -\frac{1}{2\lambda^2} \sum_\alpha [(e_{\beta\alpha}\partial_t\theta_\beta)^2 - (e_{\beta\alpha}\partial_\sigma\theta_\beta)^2]$$

and

$$\mathcal{L}_1 = -2k\dot{\theta}_\alpha e_{\alpha\beta}\left(F_\beta - \frac{1}{2}A_\beta\right) = -2k\partial_t\theta_\alpha t_{\alpha\beta}\partial_\sigma\theta_\beta \quad (3.6.8)$$

The antisymmetric matrix t is given by

$$t = e\left\{\theta^{-1}(f-1) - \frac{1}{2}e^T\right\} = \theta^{-2}(\theta - \sinh(\theta)) \quad ; \quad e = \frac{1 - e^{-\theta}}{\theta} \quad (3.6.9)$$

The metric on the Lie group manifold and the Minkowski metric is given by

$$g_{\alpha\beta} = e_{\alpha\gamma}e_{\beta\gamma} \quad ; \quad \eta^{\mu\nu} = \mathrm{diag}(1, -1) \quad \text{where} \quad \xi_1 = t \quad ; \quad \xi_2 = \sigma$$

Hence

$$\mathcal{L}_{WZW} = (1/2\lambda^2)\eta^{\mu\nu}\partial_\mu\theta_\alpha\partial_\nu\theta_\beta g_{\alpha\beta}(\theta) + k\epsilon^{\mu\nu}\partial_\mu\theta_\alpha\partial_\nu\theta_\beta t_{\alpha\beta}(\theta)$$

The first term is the σ-model Lagrangian; the second term is the local expression for the WZW term. Compactifying $d = 2$ spacetime into a two-sphere S^2, and using Stokes' theorem gives the WZW action as [Witten (1983)]

$$S_{WZW} = \frac{1}{2\lambda^2} \int d^2\xi \mathcal{L} + \frac{k}{3} \int d^3\xi \epsilon^{\mu\nu\rho}\partial_\mu\theta_\alpha\partial_\nu\theta_\beta\partial_\rho\theta_\gamma T_{\alpha\beta\gamma}(\theta)$$

where the last term is integrated over a 3-ball bounded by S^2; the completely antisymmetric torsion tensor

$$T_{\alpha\beta\gamma} = \frac{1}{2}C_{abc}e_{\alpha a}e_{\beta b}e_{\gamma c} \quad (3.6.10)$$

is given by [Cronstrom and Mickelsson (1983)] $(\partial_\alpha \equiv \partial/\partial\theta_\alpha)$

$$T_{\alpha\beta\gamma} = \partial_\alpha t_{\beta\gamma} + \partial_\beta t_{\gamma\alpha} + \partial_\gamma t_{\alpha\beta}$$

3.7 $SU(2)$ Loop Group

To gain further insight into the Kac-Moody generators $K_i(\sigma)$, the loop generators $E_i(\sigma)$ are analyzed and then their central extension to the Kac-Moody case is considered. The loop group generators can be constructed directly from the underlying compact $SU(2)$ group and all the equations can be studied in full detail.

Furthermore, given the special and important role of the $SU(2)$ KM algebra in conformal field theory, the specific case of $G=SU(2)$ is worked out from first principles and in a completely explicit form [Francesco et al. (2012)].

The $SU(2)$ loop group generators $E_i(\sigma)$ have the commutation equation

$$[E_i(\sigma), E_j(\sigma')] = \varepsilon_{ijk} E_k(\sigma)\delta(\sigma - \sigma') \quad : \quad i, j, k = 1, 2, 3 \tag{3.7.1}$$

Let $\theta_i(\sigma)$, $i=1,2,3$ be all the possible maps from S^1 into $SU(2)$; this defines the coordinates of the KM group manifold [Baaquie (1988)]. Define the adjoint matrix ω by

$$\omega_{ij}(\sigma) = \varepsilon_{i\alpha j}\theta_\alpha(\sigma) \tag{3.7.2}$$

Then, from Eq. 2.9.5

$$E_i(\sigma) = \sum_j f_{ij}(\sigma)\frac{\delta}{i\delta\theta_j(\sigma)} \tag{3.7.3}$$

where, from Eq. 2.9.4, in matrix notation

$$f = \frac{\omega}{1 - e^{-\omega}}$$

Let $\theta = \sqrt{\theta_i\theta_i}$; for $SU(2)$, from Eq. 2.10.5

$$\omega^3 = -\theta^2\omega \quad \Rightarrow \quad f = 1 + \frac{1}{2}\omega + A(\theta)\omega^2 \tag{3.7.4}$$

where, from Eq. 2.10.6

$$A(\sigma) \equiv A[\theta(\sigma)] = \frac{1}{\theta^2}\left(1 - \frac{\theta}{2}\cot\frac{\theta}{2}\right)$$

$$\simeq \frac{1}{12} + \frac{1}{16 \cdot 45}\theta^2 + O(\theta^4)$$

The loop group commutator Eq. 3.7.1 implies the Maurer-Cartan equation ($\partial_i \equiv \partial/\partial\theta_i$) that follows from Eq. 2.9.7

$$f_{\alpha i}\partial_i f_{\beta\gamma} - f_{\beta i}\partial_i f_{\alpha\gamma} = -\varepsilon_{\alpha\beta i}f_{i\gamma} \tag{3.7.5}$$

The explicit form for f given by Eq. 3.7.4 yields Eq. 3.7.5 since A satisfies the remarkable differential equation

$$\theta\frac{\partial A}{\partial\theta} + 3A - \theta^2 A^2 = \frac{1}{4} \tag{3.7.6}$$

Eq. 3.7.6 will be essential in obtaining the KM commutation equation.

3.8 $SU(2)$ Kac-Moody Algebra

The general case of the $SU(\mathcal{N})$ Kac-Moody Algebra is analyzed by first studying the special case of $SU(2)$. The Kac-Moody algebra can be studied explicitly for $SU(2)$, and in full detail, using the exact results for the $SU(2)$ compact Lie group. The $SU(2)$ case provides independent verification for the results obtained [Baaquie (1992)].

The $SU(2)$ KM algebra is given by

$$[K_i(\sigma), K_j(\sigma')] = i\varepsilon_{ijk}K_k(\sigma)\delta(\sigma - \sigma') + ik\delta_{ij}\delta'(\sigma - \sigma')\mathbb{I} \quad : \quad i,j,k = 1,2,3$$

where σ parametrizes a loop and the prime denotes $\partial/\partial\sigma$.

The KM generators are realized as the central extension of the loop group generators $E_i(\sigma)$ given by

$$K_i(\sigma) = E_i(\sigma) + kF_i(\sigma) \tag{3.8.1}$$

where $F_i(\sigma)$ are functions of the $SU(2)$ loop group coordinates.

The Kac-Moody generators are given in the general case by Baaquie (1986)

$$K_i(\sigma) = E_i(\sigma) + kF_i(\sigma) \tag{3.8.2}$$

where from Eq. 3.7.3

$$E_i(\sigma) = f_{ij}(\sigma)\frac{\delta}{i\delta\theta_j(\sigma)}$$

and the central extension is

$$F_i(\sigma) = \omega_{ij}^{-1}(f_{jk} - \delta_{jk})\theta'_k(\sigma) \quad ; \quad \omega_{ij}(\sigma) = \varepsilon_{i\alpha j}\theta_\alpha(\sigma)$$

Hence, for $SU(2)$, from Eqs. 3.7.4 and 3.7.4

$$F_i(\sigma) = \left[\frac{1}{2}\delta_{ij} + A(\sigma)\omega_{ij}(\sigma)\right]\theta'_j(\sigma) \quad ; \quad A[\theta(\sigma)] = \frac{1}{\theta^2}\left(1 - \frac{\theta}{2}\cot\frac{\theta}{2}\right) \tag{3.8.3}$$

The Kac-Moody generators, from Eqs. 3.8.2 and 3.7.1, are given by

$$[K_I(\sigma), K_J(\sigma')] = iC_{IJK}E_k(\sigma)\delta(\sigma - \sigma') + k[E_I(\sigma)F_J(\sigma') - E_J(\sigma')F_I(\sigma)] \tag{3.8.4}$$

Consider the following

$$\Gamma_{IJ} = E_I(\sigma)F_J(\sigma') - E_J(\sigma')F_I(\sigma) \tag{3.8.5}$$

Note

$$iE_I(\sigma)F_J(\sigma') = -\frac{1}{2}f_{IJ}(\sigma)\delta'(\sigma - \sigma') - f_{I\alpha}(\sigma)\omega_{J\alpha}(\omega')A(\sigma')\delta'(\sigma - \sigma')$$
$$+\delta(\sigma - \sigma')f_{I\alpha}(\varepsilon_{J\alpha\beta}A + \omega_{J\beta}\partial_\alpha A)\theta'_\beta(\omega) \tag{3.8.6}$$

Eqs. 3.8.5 and 3.8.6 yield

$$i\Gamma_{IJ} = \Gamma_1 + \Gamma_2 + \Gamma_3 \tag{3.8.7}$$

where

$$\Gamma_1 = \delta_{IJ}\delta'(\sigma - \sigma') - \frac{1}{2}[\omega_{IJ}^2(\sigma)A(\sigma) + \omega_{JI}^2(\sigma')A(\sigma')]\delta'(\sigma - \sigma')$$

$$+\frac{1}{4}\omega'_{IJ}\delta(\sigma - \sigma')$$

$$\Gamma_2 = [f_{Ii}(\sigma)\omega_{iJ}(\sigma')A(\sigma') + f_{Ji}(\sigma')\omega_{iI}(\sigma)A(\sigma)]\delta'(\sigma - \sigma')$$

$$\Gamma_3 = \Gamma_\alpha(\sigma)\theta'_\alpha\delta(\sigma - \sigma')$$

Note that

$$\Gamma_1 + \Gamma_2 = -\delta_{IJ}\delta'(\sigma - \sigma') + \tilde{\Gamma}_\alpha(\sigma)\theta'_\alpha\delta(\sigma - \sigma')$$

where

$$\tilde{\Gamma}_\alpha = \frac{1}{4}\varepsilon_{I\alpha J} + \varepsilon_{I\alpha J}A + \omega_{IJ}\partial_\alpha A + \frac{1}{2}\omega_{Ii}\varepsilon_{i\alpha J}A - \frac{1}{2}\omega_{iJ}\varepsilon_{I\alpha i}A - \omega_{Ii}\omega_{jJ}\varepsilon_{i\alpha j}A^2$$

and which yields

$$\Gamma_\alpha + \tilde{\Gamma}_\alpha = -\frac{1}{4}\varepsilon_{IJ\alpha} - 3\varepsilon_{IJ\alpha}A + \omega_{IJ}\partial_\alpha A + \omega_{J\alpha}\partial_I A - \omega_{I\alpha}\partial_J A$$

$$+(\omega_{Ii}\varepsilon_{i\alpha J} - \omega_{Ji}\varepsilon_{i\alpha I})A + (\omega_{I\alpha}\theta_J - \omega_{J\alpha}\theta_I - \omega_{IJ}\theta_\alpha)A^2 \quad (3.8.8)$$

Using the identity

$$\omega_{I\alpha}\theta_J\theta'_\alpha - \omega_{J\alpha}\theta_I\theta'_\alpha - \omega_{IJ}\theta_\alpha\theta'_\alpha = \varepsilon_{IJ\alpha}\theta^2\theta'_\alpha$$

and

$$\partial_\alpha A = (\theta_\alpha/\theta)\partial A/\partial\theta$$

yields the following

$$\Gamma_4 = (\Gamma_\alpha + \tilde{\Gamma}_\alpha)\theta'_\alpha$$

$$= -\frac{1}{4}\varepsilon_{IJ\alpha}\theta'_\alpha - \varepsilon_{IJ\alpha}\omega_{\alpha\beta}A\theta'_\beta - \varepsilon_{IJ\alpha}\theta'_\alpha\left(3A + \theta\frac{\partial A}{\partial\theta} - \theta^2 A^2\right) \quad (3.8.9)$$

The term in the bracket above is equal to $1/4$ due to Eq. 3.7.6, which is a result that follows from the underlying $SU(2)$ compact Lie group; this fact illustrates the essential role that the underlying compact Lie groups play in the structure and consistency of the Kac-Moody groups and algebras.

Hence, from Eqs. 3.8.3 and 3.8.9

$$\Gamma_4 = -\varepsilon_{IJ\alpha}\left(\frac{1}{2}\delta_{\alpha\beta} + A\omega_{\alpha\beta}\right)\theta'_\beta = -\varepsilon_{IJ\alpha}F_\alpha(\sigma) \quad (3.8.10)$$

and from Eq. 3.8.7, 3.8.8 and 3.8.10

$$i\Gamma_{IJ} = -\delta_{IJ}\delta'(\sigma - \sigma') - \varepsilon_{IJ\alpha}F_\alpha(\alpha)\delta(\sigma - \sigma')$$

which yields from Eqs. 3.8.4 and 3.8.5 the KM commutator

$$[K_I(\sigma), K_J(\sigma')] = i\varepsilon_{IJK}K_k(\sigma)\delta(\sigma - \sigma') + ik\delta_{IJ}\delta'(\sigma - \sigma')$$

There is another set of $SU(2)$ KM generators $\tilde{K}_i(\sigma)$, which commute with $K_i(\sigma)$ and have the algebra (note the negative signs)

$$[\tilde{K}_i(\sigma), \tilde{K}_j(\sigma')] = -i\varepsilon_{ijk}\tilde{K}_k(\sigma)\delta(\sigma - \sigma') - ik\delta_{ij}\delta'(\sigma - \sigma')$$

with the realization [Baaquie (1986)]

$$\tilde{K}_i(\sigma) = f_{ij}^T(\sigma)\frac{\delta}{i\delta\theta_j(\sigma)} + \left[-\frac{1}{2}\delta_{ij} + A(\sigma)\omega_{ij}(\omega)\right]\theta_j'(\sigma)$$

A number of identities specific to $SU(2)$ had to be used for constructing the KM commutator, and they shed new light on the structure of the KM algebra and showed its essential connection with the algebra of the underlying $SU(2)$ compact Lie group.

3.9 Kac-Moody Commutation Equations

The general result derived in Section 3.4 for the generators of the Kac-Moody algebra is verified by an explicit and long calculation. Given the importance of the Kac-Moody algebra and its novelty, an explicit calculation seems to be in order. The calculation shows, as expected, that the central extension – which is given by the constant k – crucially depends on the underlying compact Lie algebra; in particular, the derivation relies on the complete antisymmetry of the structure constants, which holds for compact Lie groups. The results of this Section are based on Baaquie (1988).

The notation being employed is reviewed; $\theta_\alpha(\sigma)$ is the infinite set of real coordinates of the KM-group. C^α are the generators of the adjoint representation of G, and are antisymmetric matrices $C_{\beta\alpha\gamma}$; define the antisymmetric matrix

$$\theta(\sigma) = \theta_\alpha(\sigma)C^\alpha \equiv \theta \ ; \ (C^\alpha)_{\beta\gamma} \equiv C_{\beta\alpha\gamma} \tag{3.9.1}$$

and its inverse

$$\xi(\sigma) = \theta^{-1}(\sigma) \tag{3.9.2}$$

and the matrix

$$f(\sigma) = \theta/[1 - \exp(-\theta)] \tag{3.9.3}$$

To write $K_\alpha(\sigma)$ compactly, define the operator

$$E_\alpha(\sigma) = f_{\alpha\beta}(\sigma)\frac{\delta}{i\delta\theta_\beta(\sigma)} \tag{3.9.4}$$

where $\delta/\delta\theta_\alpha(\sigma)$ is the functional derivative, and the function

$$F_\alpha(\sigma) = [f_{\alpha\beta}(\sigma) - \delta_{\alpha\beta}]\xi_{\beta\gamma}(\sigma)\theta_\gamma'(\sigma) \tag{3.9.5}$$

Then, from Eq. 3.4.13

$$K_\alpha(\sigma) = E_\alpha(\sigma) + kF_\alpha(\sigma) \tag{3.9.6}$$

and, from Eq. 3.2.2

$$[K_a(\sigma), K_b(\sigma')] = iC^{abc}K_c(\sigma)\delta(\sigma - \sigma') + ik\delta_{ab}\delta'(\sigma - \sigma')\mathbb{I} \tag{3.9.7}$$

$K_\alpha(\sigma)$ is a Hermitian functional differential operator acting on functionals of the KM group elements. The KM generators have the commutator

$$[K_I(\sigma), K_J(\sigma')] = [E_I(\sigma)\,,\, E_J(\sigma')] + k\{(E_I(\sigma)F_J(\sigma')) - (E_J(\sigma')F_I(\sigma))\} \tag{3.9.8}$$

The matrix $f_{\alpha\beta}$ given in Eq. 3.9.3 satisfies the Maurer-Cartan equation $\partial_\alpha = \partial/\partial\theta_\alpha$

$$f_{I\beta}\partial_\beta f_{JK} - f_{J\beta}\partial_\beta f_{IK} = -C_{IJ\alpha}f_{\alpha K}\,, \tag{3.9.9}$$

which yields from Eq. 3.9.4

$$[E_I(\sigma)\,,\, E_J(\sigma')] = iC_{IJ\alpha}E_\alpha(\sigma)\delta(\sigma - \sigma') \tag{3.9.10}$$

Define the function Γ_{IJ} by

$$\Gamma_{IJ}(\sigma - \sigma') = (E_i(\sigma)F_J(\sigma')) - (E_J(\sigma')F_I(\sigma)) \tag{3.9.11}$$

Eqs. 3.9.4 and 3.9.5 yield

$$i\Gamma_{IJ}(\sigma - \sigma') = \Gamma_1 + \Gamma_2 + \Gamma_3 + \Gamma_4 + \Gamma_5\,, \tag{3.9.12}$$

where

$$\Gamma_1 = (f_{I\beta}\partial_\beta f_{J\gamma} - f_{J\beta}\partial_\beta f_{I\gamma})\xi_{\gamma\alpha}\theta'_\alpha\delta(\sigma - \sigma') = -C_{IJ\alpha}f_{\alpha\gamma}\xi_{\gamma\alpha}\theta'_\alpha\delta(\sigma - \sigma')$$

and

$$\Gamma_2 = f_{I\beta}f_{J\gamma}(\partial_\beta\xi_{\gamma\alpha} - \partial_\gamma\xi_{\beta\gamma})\theta'_\alpha\delta(\sigma - \sigma')\,,$$

and

$$\Gamma_3 = -(f_{I\beta}\partial_\beta\xi_{J\alpha} - f_{I\beta}\partial_\beta\xi_{I\alpha})\theta'_\alpha\delta(\sigma - \sigma')\,,$$

and

$$\Gamma_4 = f_{I\beta}(\sigma)f_{J\gamma}(\sigma')\{\xi_{\gamma\beta}(\sigma) - \xi_{\gamma\beta}(\sigma')\}\delta'(\sigma - \sigma')\,,$$

and

$$\Gamma_5 = f_{I\beta}(\sigma)\xi_{J\beta}(\sigma')\delta'(\sigma - \sigma') - f_{J\beta}(\sigma')\xi_{I\beta}(\sigma)\delta'(\sigma - \sigma') \tag{3.9.13}$$

Consider the identity

$$\left(x(\sigma') - x(\sigma)\right)\delta'(\sigma - \sigma') = x'(\sigma)\delta(\sigma - \sigma')$$
$$\Rightarrow\; x(\sigma')\delta'(\sigma - \sigma') = x(\sigma)\delta'(\sigma - \sigma') + x'(\sigma)\delta(\sigma - \sigma') \tag{3.9.14}$$

This leads to a simplification of Γ_4 yields

$$\Gamma_2 + \Gamma_4 = 0 \tag{3.9.15}$$

since it can be shown that

$$\partial_\alpha\xi_{\beta\gamma} + \partial_\gamma\xi_{\alpha\beta} + \partial_\beta\xi_{\gamma\alpha} = 0 \tag{3.9.16}$$

From Eqs. 3.9.13 and 3.9.14

$$\Gamma_5 = (f_{I\beta}\xi'_{J\beta} + f'_{J\beta}\xi_{I\beta})\delta(\sigma - \sigma') - \{f\xi + (f\xi)^T\}_{IJ}(\sigma)\delta'(\sigma - \sigma') \qquad (3.9.17)$$

From Eqs. 3.9.2 and 3.9.3 yield

$$f\xi + (f\xi)^T = \frac{1}{1 - e^{-\theta}} + \frac{1}{1 - e^{\theta}} = 1 \qquad (3.9.18)$$

This remarkable identity – which comes from the property of the underlying compact Lie group $SU(\mathcal{N})$ – is the reason why, in spite of the apparently field dependent expression for Γ_{IJ}, one ends up with a *field independent* central extension k for the KM commutation equations.

From Eqs. 3.9.17 and 3.9.18

$$\Gamma_5 = \delta'(\sigma - \sigma') + \Gamma_6 \qquad (3.9.19)$$

After some simplifications,

$$\Gamma_3 + \Gamma_6 = \{f_{I\beta}\partial_J\xi_{\alpha\beta} - f_{J\beta}\partial_I\xi_{\alpha\beta} - \partial_\alpha(f\xi)_{JI}\}\theta'_\alpha\delta(\sigma - \sigma') \qquad (3.9.20)$$

Using the crucial identity given in (A.1)

$$\partial_\alpha(f\xi)_{JI} = f_{\beta I}\partial_J\xi_{\alpha\beta} - f_{J\beta}\partial_I\xi_{\alpha\beta} , \qquad (3.9.21)$$

yields

$$\Gamma_3 + \Gamma_6 = (f - f^T)_{I\beta}\partial_J\xi_{\alpha\beta}\theta'_\alpha\delta(\sigma - \sigma') \qquad (3.9.22)$$

From Eq. 3.9.3

$$f - f^T = \theta \qquad (3.9.23)$$

and hence

$$\Gamma_3 + \Gamma_6 = C_{IJ\beta}\xi_{\beta\alpha}\theta'_\alpha\delta(\sigma - \sigma') \qquad (3.9.24)$$

Collecting Eqs. 3.9.13, 3.9.15, 3.9.19 and 3.9.24 yields

$$i\Gamma_{IJ}(\sigma - \sigma') = -C_{IJ\alpha}(f_{\alpha\gamma} - \delta_{\alpha\gamma})\xi_{\gamma\alpha}\theta'_\alpha\delta(\sigma - \sigma') - \delta_{IJ}\delta'(\sigma - \sigma') , \qquad (3.9.25)$$

or more succinctly, using Eq. 3.9.11

$$(E_I(\sigma)F_J(\sigma')) - (E_J(\sigma')F_I(\sigma)) = iC_{IJ\alpha}F_\alpha(\sigma)\delta(\sigma - \sigma') + i\delta_{IJ}\delta'(\sigma - \sigma') \qquad (3.9.26)$$

The expected commutation equation of the KM-generators, from Eqs. 3.9.8, 3.9.10 and 3.9.26 is given

$$[K_I(\sigma) , K_J(\sigma')] = iC_{IJ\alpha}K_\alpha(\sigma)\delta(\sigma - \sigma') + ik\delta_{IJ}\delta'(\sigma - \sigma') \qquad (3.9.27)$$

3.10 Virasoro Generator: Point-Split Regularization

The Virasoro generator is realized as a second order Hermitian functional differential operator $L(\sigma)$ and is essentially the KM algebra's second Casimir operator $K^2(\sigma)$. $K^2(\sigma)$ is singular due to the singularity arising from $\delta^2/\delta\theta^2(\sigma)$ and hence needs to be regularized. The Virasoro generator is regularized using the point-splitting method [Baaquie (1988)].

Noteworthy 3.2: Representation of the δ function

Consider the following representation of the Dirac-delta function

$$\delta(x) = \lim_{\eta \to 0} \frac{1}{2\pi i}\left[\frac{1}{x - i\eta} - \frac{1}{x + i\eta}\right] = \frac{\eta}{\pi}\frac{1}{x^2 + \eta^2} = \begin{cases} 0 : x \neq 0 \\ \infty : x = 0 \end{cases} \qquad (3.10.1)$$

A proof for the representation, using a Fourier transform, is given by

$$\frac{\eta}{\pi}\int \frac{dx}{x^2 + \eta^2}f(x) = \frac{\eta}{\pi}\int \frac{dp}{2\pi}\int \frac{dx}{x^2 + \eta^2}e^{ipx}f_p$$

$$= \frac{\eta}{\pi}\int dp\frac{1}{2\eta}e^{-\eta|p|}f_p \to \int \frac{dp}{2\pi}f_p = f(0) = \int dx\delta(x)f(x)$$

which is the expected result. Furthermore (note the signs on the $i\eta$'s)

$$\delta'(x) = \frac{1}{2\pi i}\left[\frac{1}{(x + i\eta)^2} - \frac{1}{(x - i\eta)^2}\right] \qquad (3.10.2)$$

The full content of the commutation equation given in Eq. 3.9.10 is realized by the short distance operator product expansion, which for $\sigma \approx \sigma'$ is given by

$$E_\alpha(\sigma)E_\beta(\sigma') \cong \frac{1}{2\pi}\frac{C_{\alpha\beta\gamma}E_\gamma(\sigma)}{\sigma - \sigma' - i\eta} + \text{finite terms} \qquad (3.10.3)$$

To see that Eq. 3.10.3 yields the expected commutation equation, consider

$$[E_\alpha(\sigma), E_\beta(\sigma')] = E_\alpha(\sigma)E_\beta(\sigma') - E_\beta(\sigma')E_\alpha(\sigma)$$

$$= \frac{1}{2\pi}\left\{\frac{C_{\alpha\beta\gamma}E_\gamma(\sigma)}{\sigma - \sigma' - i\eta} - \frac{C_{\beta\alpha\gamma}E_\gamma(\sigma')}{\sigma' - \sigma - i\eta}\right\}$$

$$= \frac{1}{2\pi}C_{\alpha\beta\gamma}E_\gamma(\sigma)\left\{\frac{1}{\sigma - \sigma' - i\eta} - \frac{1}{\sigma - \sigma' + i\eta}\right\} + O(\eta)$$

$$= iC_{\alpha\beta\gamma}E_\gamma(\sigma)\delta(\sigma - \sigma')$$

Similarly, the other commutation equations can be written in terms of the short distance expansion as follows (note the signs of the $i\eta$ terms below)

$$(E_\alpha(\sigma)F_\beta(\sigma')) \cong \frac{1}{2\pi}\frac{\delta_{\alpha\beta}}{(\sigma - \sigma' + i\eta)^2} + \frac{1}{2\pi}\frac{C_{\alpha\beta\gamma}F_\gamma(\sigma)}{\sigma - \sigma' - i\eta} + \text{finite terms} \qquad (3.10.4)$$

Eqs. 3.10.3 and 3.10.4 yield[1]

$$K_\alpha(\sigma)K_\beta(\sigma') \cong \frac{1}{2\pi}\frac{k\delta_{\alpha\beta}}{(\sigma - \sigma' + i\eta)^2} + \frac{C_{\alpha\beta\gamma}K_\gamma(\sigma)}{\sigma - \sigma' - i\eta} + \text{finite terms} \qquad (3.10.5)$$

---

[1]Note that Eq. 3.10.5 also follows directly form Eq. 3.9.27.

The $i\eta$'s in the short distance expansion can be ignored if an explicit point-splitting method is used for regularizing singular operators – for example, by introducing an infinitesimal ε. This is because the limit of $\eta \to 0$ is always taken before the limit of $\varepsilon \to 0$ is taken.

The leading term for expansions given in Eqs. 3.10.4 and 3.10.5 have a positive sign since, for $\sigma = \sigma' + \varepsilon$, the singularity for the regularized positive valued operator should have a positive value – as given in Eq. 3.10.9.

Consider the regularized operator using the *point-splitting* method

$$E_{reg}^2(\sigma) = \lim_{\varepsilon \to 0} \frac{1}{2}\{E_a(\sigma)E_a(\sigma + \varepsilon) + E_a(\sigma + \varepsilon)E_a(\sigma)\} + \text{const} \qquad (3.10.6)$$

Using the identity

$$[AB, C] = A[B, C] + [A, C]C$$

yields

$$[E_a(\sigma)E_a(\sigma + \varepsilon), E_\alpha(\sigma')]$$
$$= E_a(\sigma)[E_a(\sigma + \varepsilon), E_\alpha(\sigma')] + [E_a(\sigma), E_\alpha(\sigma')]E_a(\sigma + \varepsilon)$$
$$= iC^{a\alpha b}\{E_a(\sigma)E_b(\sigma + \varepsilon)\delta(\sigma + \varepsilon - \sigma') - E_a(\sigma)E_b(\sigma + \varepsilon)\delta(\sigma - \sigma')\}$$

The short distance expansion given in Eq. 3.10.3 yields

$$E_a(\sigma)E_b(\sigma + \varepsilon) = -\frac{1}{2\pi\varepsilon}C^{abc}E_c(\sigma)$$

The identity[2]

$$C^{a\alpha b}C^{abc} = -c_2(A)\delta_{\alpha c}$$

where $c_2(A)$ is the second Casimir of the adjoint representation, yields the result

$$[E_{reg}^2(\sigma), E_\alpha(\sigma')] = i\frac{c_2(A)}{2\pi}\frac{E_\alpha(\sigma)}{\varepsilon}\{\delta(\sigma + \varepsilon - \sigma') - \delta(\sigma - \sigma')\}$$
$$= i\frac{c_2(A)}{2\pi}\delta'(\sigma - \sigma')E_\alpha(\sigma) \qquad (3.10.7)$$

Note Eq. 3.10.7 shows that the loop group is not simply a tensor product of the underlying compact Lie group since the second Casimir does not commute with all the generators. It can be shown that for σ parametrizing a loop, $E_\alpha(\sigma)$ are the generators of the loop group [Pressley and Segal (1988)]. The naive commutation of $E^2(\sigma)$ with $E_\alpha(\sigma')$ yields zero, and the anomaly in Eq. 3.10.7 arises from the short distance singularity in Eq. 3.10.3. Eq. 3.10.7 is the *anomaly* of $E^2(\sigma)$ of the loop group [Pressley and Segal (1988)].

Define the regularized Hermitian operator using the point-splitting method

$$K_{reg}^2(\sigma) = \lim_{\varepsilon \to 0} \frac{1}{2}\{K(\sigma)K(\sigma + \epsilon) + K(\sigma + \epsilon)K(\sigma)\} + \frac{a}{\epsilon^2} + b \qquad (3.10.8)$$

---

[2]For $SU(\mathcal{N})$, from Eq. 2.7.6 $c_2(A) = \mathcal{N}$.

To fix the constant a, note from operator product expansion given in Eq. 3.10.5

$$K(0)K(\epsilon) \cong \frac{k\mathrm{dim}G}{2\pi} \cdot \frac{1}{\epsilon^2}$$

and hence

$$K_{reg}^2(0) = \frac{k\mathrm{dim}G}{2\pi} \cdot \frac{1}{\epsilon^2} + \frac{a}{\epsilon^2} + b \qquad (3.10.9)$$

To render $K_{reg}^2(0)$ finite choose

$$a = -\frac{k\mathrm{dim}G}{2\pi} \qquad (3.10.10)$$

All matrix elements of $K_{reg}^2(\sigma)$ are finite; the constant b fixes the value of $K_{reg}^2(0)$, which is left undetermined due to the regularization. In complete analogy to calculation for obtaining Eq. 3.10.7, one obtains

$$[K_{reg}^2(\sigma), K_\alpha(\sigma')] = i\beta\delta'(\sigma - \sigma')K_\alpha(\sigma) \qquad (3.10.11)$$

where

$$\beta = 2k + \frac{c_2(A)}{2\pi} \qquad (3.10.12)$$

In the approach of Knizhnik and Zamolodchikov (1984) the constant β is indirectly evaluated using properties of the null vector and vacuum state of the semi-direct product of the KM and Virasoro algebras. Eq. 3.10.11, however, shows that β results directly from the anomaly in $E^2(\sigma)$ and is independent of the properties of the null and vacuum states.

For the Virasoro commutator, using Eqs. 3.10.8 and 3.10.11 yields

$$[K_{reg}^2(\sigma), K_{reg}^2(\sigma')] = \frac{1}{2}i\beta(K_\sigma K_{\sigma'+\epsilon} + K_{\sigma'+\epsilon}K_\sigma)\delta'(\sigma - \sigma')$$

$$+ \frac{1}{2}i\beta(K_\sigma K_{\sigma'} + K_{\sigma'}K_\sigma)\delta'(\sigma - \sigma' - \epsilon) \qquad (3.10.13)$$

All the operators at σ and σ' have to be isolated, with no cross-terms. To do this Eq. 3.9.14 is used; in particular, it yields the following

$$K_{\sigma'+\epsilon}\delta'(\sigma - \sigma') = K_{\sigma+\epsilon}\delta'(\sigma - \sigma') + K'_{\sigma+\epsilon}\delta(\sigma - \sigma')$$

From the short distance expansion given in Eq. 3.10.5, the leading singularity of the following operator product is given by

$$K_\sigma K'_{\sigma+\epsilon} = \frac{\partial}{\partial\lambda}\left(K_\sigma K_\lambda\right)\Big|_{\lambda=\sigma+\epsilon} = \frac{k\mathrm{dim}G}{2\pi}\frac{2}{(\sigma - \lambda)^3}\Big|_{\lambda=\sigma+\epsilon} = \frac{2a}{\epsilon^3}$$

Hence

$$K_\sigma K_{\sigma'+\epsilon}\delta'(\sigma - \sigma') \cong K_\sigma K_{\sigma+\epsilon}\delta'(\sigma - \sigma') + \frac{2a}{\epsilon^3}\delta(\sigma - \sigma') + O(\epsilon) \qquad (3.10.14)$$

From Eqs. 3.10.8, 3.10.13 and 3.10.14

$$\beta^{-1}[K_{reg}^2(\sigma), K_{reg}^2(\sigma')] = i\left\{K_{reg}^2(\sigma) + K_{reg}^2(\sigma')\right\}\delta'(\sigma - \sigma') - 2ib\delta'(\sigma - \sigma')$$

$$- \frac{ia}{\epsilon^2}\{\delta'(\sigma - \sigma') + \delta'(\sigma - \sigma' - \epsilon)\} + \frac{2ia}{\epsilon^3}\{\delta(\sigma - \sigma') - \delta(\sigma - \sigma' - \epsilon)\} \qquad (3.10.15)$$

$$= i\{K_{reg}^2(\sigma) + K_{reg}^2(\sigma')\}\delta'(\sigma - \sigma') - 2ib\delta'(\sigma - \sigma') - \frac{1}{6}ia\delta'''(\sigma - \sigma') \qquad (3.10.16)$$

The singular terms of $O(\epsilon^{-2})$ and $O(\epsilon^{-3})$ in Eq. 3.10.15 arising from Eqs. 3.10.8 and 3.10.14 *cancel* – as is necessary if point-split regularization is to make sense.

Define the Virasoro generator by

$$L(\sigma) = \frac{1}{\beta} K_{reg}^2(\sigma) \tag{3.10.17}$$

Then, from Eq. 3.10.11

$$[L(\sigma), K_\alpha(\sigma')] = i\delta'(\sigma - \sigma') K_\alpha(\sigma) \tag{3.10.18}$$

The requirement that $L(\sigma)$ generates reparametrizations fixes the normalization in Eq. 3.10.17. The Virasoro algebra is given, from Eq. 3.10.16, by the following

$$[L(\sigma), L(\sigma')] = i(L(\sigma)+L(\sigma'))\delta'(\sigma-\sigma') - i\left(\frac{2a}{\beta}\right)\frac{1}{12}\left\{\frac{12b}{a}\delta'(\sigma-\sigma') + \delta'''(\sigma-\sigma')\right\} \tag{3.10.19}$$

For the Virasoro algebra to contain the subalgebra $SU(1,1)$ of Mobius transformations, b is fixed to be

$$b = \frac{1}{12}a = -\frac{k\dim G}{24\pi} \tag{3.10.20}$$

and hence

$$[L(\sigma), L(\sigma')] = i(L(\sigma)+L(\sigma'))\delta'(\sigma-\sigma') + \frac{i}{2\pi}\frac{c}{12}\{\delta'(\sigma-\sigma') + \delta'''(\sigma-\sigma')\} \tag{3.10.21}$$

The *central charge* c is given by

$$c = -2\pi\left(2\frac{a}{\beta}\right) = \frac{2\tilde{k}\dim G}{2\tilde{k} + c_2(A)} \;\; ; \;\; \tilde{k} = 2\pi k$$

Eq. 3.10.21 agrees with the known result [Goddard and Olive (1988)].

Note that in the realization of K_α the anomaly in $E^2(\sigma)$ is the source of $c_2(A)$ in the central charge, and without this factor c would have been a trivial extension of $U(1)$ factors.

So far, only *local* properties of $K_\alpha(\sigma)$ have been used to obtain the Virasoro algebra; we now further assume that σ parametrizes a (periodic) loop, with $\in [-\pi, \pi]$; then, due to periodicity

$$K_\alpha(\sigma) = \sum_{n=-\infty}^{+\infty} e^{in\sigma} K_{-n}^\alpha$$

and from Eq. 3.9.6

$$[K_n^\alpha, K_m^\beta] = iC_{\alpha\beta\gamma}K_{n+m}^\gamma + i\tilde{k}n\delta_{n+m,0} \tag{3.10.22}$$

For the Virasoro operator, let

$$L(\sigma) = \sum_n e^{in\sigma} L_n$$

Taking the $\varepsilon \to 0$ limit yields from Eqs. 3.10.17 and 3.10.22

$$L_n = \frac{1}{\beta} \sum_{m=-\infty}^{+\infty} K_{n-m} K_m, \;\; n \neq 0,$$

and

$$L_0 = \frac{1}{\beta}\left\{K_0^2 + 2\sum_{m=1}^{\infty} K_{-m}K_m\right\} - \frac{1}{24}c$$

Note that L_n and L_0 are the usual 'normal ordered' expressions for the Virasoro generators except for the constant in L_0. In fact, the constant $c/24$ is the trace anomaly [Goddard and Olive (1988)]. This expression for L_0 is the correct one for computing the character functions of the KM-group and for proving modular invariance of the KM-partition function [Francesco et al. (2012)].

Eq. 3.10.21 yields

$$[L_n, L_m] = (n - m)L_{n+m} + \frac{1}{12}cn(n^2 - 1)\delta_{n+m,0} \qquad (3.10.23)$$

The $SU(1,1)$ closed subalgebra is generated by L_{-1}, L_0, L_1, and fixing b in Eq. 3.10.20 is equivalent to redefining L_0 so as to obtain $L_{\pm 1}$, L_0 as the generators obeying the $SU(1,1)$ algera[3]

$$[L_1, L_{-1}] = 2L_0$$

3.11   Summary

Kac-Moody algebra is the synthesis of the concept algebra with that of calculus, and is achieved by combining the finite dimensional Lie algebra with the infinite dimensional Heisenberg algebra. In the realization considered here, the infinite dimensional KM-group manifold is the starting point of the analysis. The KM commutation equations are a result of the specific properties of the underlying compact group G and of the local properties of the KM group; the KM generators clearly reveal the inner structure of the KM group, including its relation with the loop group.

Kac-Moody generators were realized, for any value of the central extension, as functional differential operators expressed in terms of the coordinates of the KM group manifold. The KM generators do not need any regularization, and the central extension is due to underlying properties of the KM-group manifold.

The derivation of WZW action from the Hamiltonian, using the exact form of the KM realization, showed the connection of the 0-cocycle topological WZW terms with the 2-cocycle, which is algebraic in character since it results from the KM commutation equations [Mickelsson (1985)].

Kac-Moody generators can be defined on the complex plane: the analysis of Sections 3.9 and 3.10 can be carried out on the complex plane by replacing σ by a complex variable z and $\partial/\partial\sigma$ by $\partial/\partial z$. Periodicity in σ of $K_\alpha(\sigma)$ is replaced by assuming $K_\alpha(z)$ is analytic in z.

The realization of Kac-Moody generators was used in defining the Virasoro algebra in parameter space using point splitting and does not require the use of

---

[3]By redefining L_0 to L_0 + const, we can always adjust the central extension such that $L_{\pm 1}$, L_0 forms a closed SU(1,1) sub algebra (see, for example Francesco et al. (2012)).

normal ordering. The Virasoro operator $L(\sigma)$ has a short distance singularity. The derivation given in Section 3.10 shows that point-split regularization of the Virasoro operator is adequate for obtaining the correct result for the central extension [Baaquie (1988)].

3.12   Appendix

A proof is given of

$$\partial_\alpha (f\xi)_{JI} = f_{\beta I} \delta_J \xi_{\alpha\beta} - f_{J\beta} \delta_I \xi_{\alpha\beta} \qquad (A.1)$$

Recall

$$(C^\alpha)_{ij} = C_{i\alpha j} \; ; \;\; \theta = C^\alpha \theta_\alpha \; ; \;\; \xi = \theta^{-1}$$

Hence

$$f = \theta/[1 - \exp(-\theta)] \; ; \;\; e = f^{-1}$$

Let

$$g = f\xi = 1/[1 - \exp(-\theta)] \;\; \Rightarrow \;\; h = g^{-1} = 1 - \exp(-\theta)$$

The definitions above yield

$$\partial_\alpha (f\xi) = \partial_\alpha (g) = -g\partial_\alpha (h) g = -g e^{-\theta} C^\beta e_{\alpha\beta} = -g(1-h) C^\beta e_{\alpha\beta}$$

Using the identity

$$[C^\alpha, g] = (1 - h) C^\beta h_{\alpha\beta}$$

yields, from the two equations above

$$[C^\alpha, g] = -g[C^\alpha, h]g = -g(1-h) C^\beta e_{\alpha\beta} = (hf)_{\alpha\beta} \partial_\beta g = \theta_{\alpha\beta} \partial_\beta g$$

The final result is given by

$$\partial_\alpha (f\xi)_{JI} = \xi_{\alpha\beta}[C^\alpha, g]_{JI} = (\xi C^J \xi)_{\alpha\beta} f_{\beta I} = f_{\beta I} \partial_J \xi_{\alpha\beta} - f_{J\beta} \partial_I \xi_{\alpha\beta}$$

Chapter 4

$SU(\mathcal{N})$ Path Integrals

4.1 Introduction

The study of $SU(\mathcal{N})$ non-Abelian degrees of freedom is central to the study of Yang-Mills gauge fields. Furthermore, lattice gauge fields are defined directly in terms of the finite group elements of $SU(\mathcal{N})$. To gain some insight into the behavior of the non-Abelian degree of freedom and of the lattice gauge field, the Feynman path integral for the $SU(\mathcal{N})$ degrees of freedom is studied.

This Chapter is based on the formalism of *quantum mechanics*, since a system constituted by one nonlinear (curved) degree of freedom is analyzed. The degree of freedom is the unitary matrix, and this Chapter is a precursor to the study of lattice gauge fields, in which infinitely many unitary matrices (degrees of freedom) are coupled.

In quantum mechanics, the degree of freedom is often considered to be the position of the particle, which is a vector of flat Euclidean space. This Chapter studies the quantum mechanics of the $SU(\mathcal{N})$ degree of freedom that takes values in the $SU(\mathcal{N})$ group space. The $SU(\mathcal{N})$ group space is a simply connected compact Riemannian curved metric space – having constant positive curvature. Classical paths consist of trajectories in $SU(\mathcal{N})$ space.

Let H be the Hamilton operator defined by the Laplace-Beltrami operator on $SU(\mathcal{N})$ space. The evolution kernel is the matrix elements of the finite evolution operator $\exp(-TH)$, where T is Euclidean time. The evolution kernel is computed using two different methods, first by using the eigenfunctions of H and then by using the finite time path integral. Comparing these two methods will allow us to calculate the eigenfunctions of H using the finite time path integral and give new insights into the relationship between the eigenfunctions of H and the classical paths given by the Lagrangian L.

Many of the results obtained in this Chapter have direct application to non-Abelian lattice gauge theory. The action of this Chapter is the limit of zero-dimensional space for the lattice gauge field, which is discussed in Chapter 7. The Hamiltonian H discussed in this Chapter is the pure *kinetic term* of the lattice gauge field Hamiltonian studied in Chapter 11.

The Laplace-Beltrami operator on $SU(\mathcal{N})$ has been studied by Wadia (1980) and Menotti and Onofri (1981) using differential techniques. Some of the results derived here coincide with their results and provide a useful check to the path integral calculations.

An exact continuum path integral expression of $\exp(-TH)$ is obtained for the $SU(\mathcal{N})$ using lattice gauge theory methods. The path integrals for $U(1), SU(2)$ and $SU(3)$ are evaluated exactly and it is seen that the $SU(\mathcal{N})$ case is completely reduced to solving the $SU(2)$ path integral. Hence the main focus in doing path integration is on the $SU(2)$ path integral.

By comparing the result of the $SU(3)$ path integral for $\exp(-TH)$ with its eigenfunction expansion, an explicit expression is obtained for the character functions of $SU(3)$. It has been hitherto an intractable problem to exactly evaluate all the character functions of $SU(3)$ using properties of $SU(3)$ Lie algebra. The path integral approach yields this new result.

4.2   Hamiltonian Operator

Define the Hamiltonian operator to be purely kinetic and is given for curved space by the Laplace-Beltrami operator as

$$\hat{H} = -\frac{g^2}{2}\nabla_U^2 \qquad (4.2.1)$$

where g^2 is the coupling constant and carries the dimension of mass. From the results of Section 2.11, the Hamiltonian can be written in terms of the generators of the Lie algebra E_a and yields

$$\hat{H} = \frac{g^2}{2}\sum_a E_a^2 \qquad (4.2.2)$$

The Laplace-Beltrami operator is given in Eqs. 11.4.5 and 2.11.13. Schrödinger's equation for Euclidean time t is given by

$$\hat{H}\psi(U) = -\frac{\partial\psi}{\partial t} \qquad (4.2.3)$$

The Euclidean time Schrödinger's equation is the heat diffusion equation on the $SU(\mathcal{N})$ manifold. The wave functions ψ are elements of the Hilbert space defined on the $SU(\mathcal{N})$ group manifold.

It can be seen from Eq. 2.11.15 that the energy eigenfunctions are $D_{ij}^{(p)}(U)$ with eigenenergy equal to $(1/2)g^2 c_2(p)$, where $c_2(p)$ is the second Casimir eigenvalue and having d_p^2 degeneracy, with d_p being the dimension of the irrep $D_{ij}^{(p)}(U)$.

The probability amplitude or the evolution kernel for finite Euclidean time T, from Eq. 4.2.1, is defined by

$$K(T) = \langle V|e^{-T\hat{H}}|W\rangle = \langle V|\exp\left\{\frac{g^2 T}{2}\nabla_V^2\right\}|W\rangle \qquad (4.2.4)$$

For the case of H being the Laplacian, $K(T)$ is also called the heat kernel. $K(T)$ is the probability amplitude that degree of freedom starting from the point W in

$SU(N)$ will be at the point V in $SU(N)$ after time T, and is the matrix element of the time evolution operator $e^{-T\hat{H}}$.

The completeness equation given in Eq. 2.8.4 yields

$$K(T) = \sum_p \sum_{ij} d_p \langle V|e^{-T\hat{H}}|p,ij\rangle\langle p,ij|W\rangle \tag{4.2.5}$$

$$= \sum_p \sum_{ij} d_p D_{ji}^{(p)}(V^\dagger)e^{-\frac{g^2T}{2}c_2(p)}D_{ij}^{(p)}(W) \tag{4.2.6}$$

$$= \sum_p e^{-\frac{g^2T}{2}c_2(p)} d_p \,\mathrm{Tr}\, D^{(p)}(V^\dagger W) \tag{4.2.7}$$

since, from Eq. 4.2.2, $\langle V|E^2|p,ij\rangle = c_2(p)D_{ji}^{(p)}(V^\dagger)$. The character function of $SU(N)$ is defined by

$$\chi_p(U) = \mathrm{Tr}\, D^{(p)}(U) \tag{4.2.8}$$

and yields

$$K(T) = \langle V|\exp\left\{\frac{g^2T}{2}\nabla_V^2\right\}|W\rangle = \sum_p d_p e^{-\frac{g^2T}{2}c_2(p)}\chi_p(U) \ ; \ U = V^\dagger W \tag{4.2.9}$$

Equation above can be used for the identifying the character functions. If one does an expansion of $K(T)$ in powers of $e^{-g^2T/2}$, the coefficients are the character functions modulo the dimension of the representation.

4.2.1 *U(1) and SU(2) evolution kernel*

For the U(1) Abelian group, $\chi_p(U) = U^p = e^{ipB}$ and hence

$$K(T) = \langle 1|e^{-T\hat{H}}|e^{iB}\rangle = \sum_{p=-\infty}^{\infty} e^{-\frac{g^2T}{2}p^2}e^{ipB} \ ; \ -\pi < B < \pi \tag{4.2.10}$$

For $SU(2)$, we use polar coordinates given in Section 2.5.2. Let $\hat{n}$ be the unit norm vector in three dimensions and $\sigma_i(p)/2$ the $SU(2)$ generators in the pth representation; this yields

$$D^{(p)}(U) = e^{iB\hat{n}\cdot\tilde{\sigma}(p)} \ ; \ 0 < B < \pi \tag{4.2.11}$$

with

$$c_2(p) = p(p+1) \ ; \ d_p = 2p+1 \ ; \ p = 0,1/2,1,3/2\dots$$

For the fundamental representation, with $p=1/2$, the generators are the Pauli matrices. The $SU(2)$ characters are

$$\chi_p(B) = \mathrm{Tr}\, D^{(p)}(U) = \sum_{m=-p}^{p} e^{imB} = \frac{1}{\sin B}\sin[(2p+1)B] \tag{4.2.12}$$

and hence for $SU(2)$

$$K_2 = \sum_{p=0,1/2,1\dots} (2p+1)e^{-\frac{g^2T}{2}p(p+1)}\frac{\sin[(2p+1)B]}{\sin B} \tag{4.2.13}$$

Let

$$d_p = 2p + 1 = m \quad : \quad m = 1, 2, 3, \ldots \quad \Rightarrow \quad c_2(p) = p(p+1) = \frac{1}{4}m^2 - 1$$

Up to an overall constant, Eq. 4.2.13 yields

$$K_2 = \sum_{m=1}^{\infty} m e^{-\frac{g^2 T}{8}m^2} \frac{\sin(mB)}{\sin B} \qquad (4.2.14)$$

$$= \frac{-i}{\sin B} \sum_{m=-\infty}^{\infty} m e^{-\frac{g^2 T}{8}m^2} e^{imB} \qquad (4.2.15)$$

The character functions for $SU(\mathcal{N})$ are formally given by a formula by Weyl. But the explicit evaluation of the character functions for $SU(3)$ and higher order groups, unlike the $SU(2)$ case, has not been tractable using algebraic methods.

4.3   Lattice Action for $SU(\mathcal{N})$

A lattice path integral for the probability amplitude $K(T)$ is derived. Let N be a large integer. Then, using the completeness equation $N - 1$ times, yields the following

$$K(T) = \langle V | \exp(-TH) | W \rangle = \langle V | \exp\left(-N \cdot \left(\frac{T}{N}\right) \hat{H}\right) | W \rangle \qquad (4.3.1)$$

$$= \prod_{n=1}^{N-1} \int dU_n \left\{ \prod_{n=0}^{N-1} \langle U_{n+1} | e^{-\frac{T}{N}\hat{H}} | U_n \rangle \right\} \qquad (4.3.2)$$

with the boundary conditions

$$U_0 = W \quad ; \quad U_N = V \qquad (4.3.3)$$

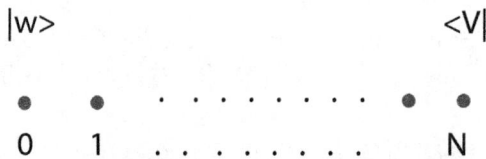

Fig. 4.1   The finite lattice, with boundary values, for the path integral.

The path integral is in effect defined on a one-dimensional finite lattice of $N + 1$ lattice points, as in Figure 4.1. Let $\epsilon = T/N$; from Eq. 4.2.9

$$K(\epsilon) = \langle U_{n+1} | e^{-\epsilon \hat{H}} | U_n \rangle = \sum_p d_p e^{-\frac{\epsilon g^2}{2}c_2(p)} \chi_p(U_n U_{n+1}^\dagger) \qquad (4.3.4)$$

From Eqs. 2.12.4 and 4.2.4, defining $\epsilon = \frac{T}{N}$, the sum over the character functions, for $N \to \infty$, yields

$$K(\epsilon) = \langle U_{n+1} | \exp\left\{\frac{g^2 T}{2N}\nabla_U^2\right\} | U_n \rangle \simeq \exp\left\{\frac{N}{2g^2 T} \text{Tr}(U_n U_{n+1}^\dagger + U_{n+1} U_n^\dagger)\right\} \qquad (4.3.5)$$

where U_n stands for the fundamental representation. Hence the path integral is given by

$$K(T) = \lim_{N \to \infty} \prod_{n=1}^{N-1} \int dU_n \left[\prod_{n=0}^{N-1} \exp\left\{ \frac{N}{2g^2T} \operatorname{Tr}(U_n U_{n+1}^\dagger + U_{n+1} U_n^\dagger) \right\} \right] \qquad (4.3.6)$$

$$\equiv \int DU \exp(A)$$

Note that the product over the integrations and product in Eq. 4.3.6 have *different* limits reflecting the presence of the boundary conditions, which are given in Eq. 4.3.3. Doing a change of variable $U_n \to U_n W^\dagger$, yields the equivalent boundary condition

$$U_0 = I, U_N = V^\dagger W = U \qquad (4.3.7)$$

The finite time lattice action is given by

$$A = \frac{N}{2g^2T} \sum_{n=0}^{N-1} \operatorname{Tr}(U_n U_{n+1}^\dagger + U_{n+1} U_n^\dagger) \qquad (4.3.8)$$

The action in Eq. 4.3.8 is the rudimentary version – for discrete time and zero dimension space lattice – of the lattice gauge field action given in Eq. 7.1.5; the action is also that of the nonlinear sigma model for a time lattice [Zinn-Justin (1993)].

4.4   Classical Paths and Winding Number

One can try to naively take the continuum limit of the path integral given in 4.3.6. As shown in Appendix 4.10, one immediately runs into the following two difficulties: a) the continuum action is nonlinear and intractable and b) there are apparent divergences that yield an ill-defined path integral.

The continuum limit, instead, is taken by first finding all the classical trajectories about which the lattice action A is stationary, expanding A about these stationary trajectories and then taking the $N \to \infty$ limit about these classical stationary paths; this is discussed in Appendix 4.10. The expansion about the classical solutions yields a well defined path integral that can be solved exactly using Gaussian path integration.

To derive the classical field equations on the lattice, note that the path integral given in Eq. 4.3.6 is unchanged under the following infinitesimal change of variables

$$U_n \to U_n(1 + i\epsilon_n^a X_a) + O(\epsilon^2) \ ; \ \ \epsilon_0^a = \epsilon_N^a = 0 \qquad (4.4.1)$$

This gives the change in the action A as given by

$$A \to A + \delta A + O(\epsilon^2) \qquad (4.4.2)$$

where

$$\delta A = i \sum_n \sum_a \epsilon_n^a \delta A_n^a \qquad (4.4.3)$$

For A to be stationary under arbitrary variations given by ϵ_n^a requires

$$\delta A_n^a = 0 = \operatorname{Tr} X_a (U_{n+1}^\dagger U_n - U_n^\dagger U_{n+1} + U_{n+1}^\dagger U_n - U_n^\dagger U_{n+1}) \tag{4.4.4}$$

which is the classical Euler-Lagrange field equation. The boundary conditions are given by Eq. 4.3.7. Solving Eq. 4.4.4 with boundary conditions given by Eq. 4.3.7 and denoting the classical solution by e_n, yields for the lattice

$$e_n = U^{n/N} \; ; \; 0 \le n \le N \tag{4.4.5}$$

Define the time derivative by

$$\frac{\partial U(t)}{\partial t} = \lim_{N \to \infty} \frac{N}{T} (U_{n+1} - U_n) \tag{4.4.6}$$

The continuum version of Eq. 4.4.4 is given by

$$\frac{\partial}{\partial t} \left\{ \operatorname{Tr} \left(X_a U^\dagger \frac{\partial U}{\partial t} \right) \right\} = 0 \tag{4.4.7}$$

with the boundary conditions for Eq. 4.3.7 becoming

$$U(0) = I, U(T) = U \tag{4.4.8}$$

The continuum solution is

$$e_t = U^{t/T} \; ; \; 0 \le t \le T \tag{4.4.9}$$

If the variable U was non-compact, the solution for Eq. 4.4.5 would yield only one classical solution. However, since U is a compact variable, the field equation 4.4.4 has many solutions; the different classical solution however gives different values for the corresponding classical actions.

The classical solutions involve the center of the group. The center of the Lie Algebra is defined as the subset of generators which can be simultaneously diagonalized, and for $SU(\mathcal{N})$ the center contains $\mathcal{N} - 1$ diagonal generators out of the total $\mathcal{N}^2 - 1$ generators. Let $X_I, I = 1, 2, \cdots \mathcal{N} - 1$ be the diagonal generators of the center. For example, $X_I = \sigma_3/2$ for $SU(2)$ and $X_I = \lambda_3/2, \lambda_8/2$ for $SU(3)$, where λ_α are the Gell-Mann matrices.

For $U \in SU(\mathcal{N})$ in the fundamental representation

$$U = S^\dagger \exp(ib^I X_I) S \tag{4.4.10}$$

$$= S^\dagger \begin{pmatrix} e^{iA_1} & & & 0 \\ & e^{iA_2} & & \\ & & \ddots & \\ 0 & & & e^{-i \sum_I A_I} \end{pmatrix} S \tag{4.4.11}$$

where the A_I are linearly related to the b^I. The matrix S is independent of A_I and is an element of $SU(\mathcal{N})$. The subgroup spanned by $\exp(ib^I X_I)$ is the center and forms the maximal Abelian toral subgroup of $SU(\mathcal{N})$ [2].

To simplify matters, consider a further change of variables $U_n \to SU_n S^\dagger$ in the path integral given by Eq. 4.3.6, and obtain the diagonal boundary conditions

$$U_0 = I, U_N = \exp(ib^I X_I) \equiv U \tag{4.4.12}$$

The multiple classical solutions of equation Eq. 4.4.4 for $SU(\mathcal{N})$, from Eq. 4.4.5, are

$$e_n(\ell_1, \ell_2, \cdots \ell_{\mathcal{N}-1}) = \begin{pmatrix} e^{i(A_1 + 2\pi\ell_1)\frac{n}{N}} & & 0 \\ & \ddots & \\ 0 & & e^{-i\sum_I(A_I + 2\pi\ell_1)\frac{n}{N}} \end{pmatrix} \tag{4.4.13}$$

$$= \exp\left(i\frac{n}{N}(b^I + 2\pi\ell^I)X_I\right) \tag{4.4.14}$$

with distinct paths given by

$$\ell_1, \ell_2, \cdots \ell_{\mathcal{N}-1} = 0, \pm 1, \pm 2 \cdots \pm(N-1) \tag{4.4.15}$$

All the classical solutions obey boundary conditions

$$e_0(\ell) = I, e_N(\ell) = \exp(ib^I X_I) \tag{4.4.16}$$

Note the number of distinct classical solutions for the finite lattice is $(2N+1)^{\mathcal{N}-1}$. In the continuum limit, the distinct classical solutions are given by

$$e_t(\ell) = \exp\left(i\frac{t}{T}(b^I + 2\pi m^I)X_I\right) \tag{4.4.17}$$

with[1]

$$m_1, m_2, \cdots m_{\mathcal{N}-1} = 0, \pm 1, \pm 2 \cdots \pm \infty \tag{4.4.18}$$

The classical solutions, both for the lattice and the continuum, are trajectories in the $SU(\mathcal{N})$ group space. The countably infinite classical solutions given by $e_t(\ell)$ will be used to expand the continuum path integral.

To clarify the meaning of the integers m_I consider first the case of $U(1)$; the continuum solutions are, from Eq. 4.4.17

$$e_t(m) = e^{i\frac{t}{T}(A + 2\pi\ell)}, m = 0, \pm 1, \pm 2 \cdots \pm \infty \tag{4.4.19}$$

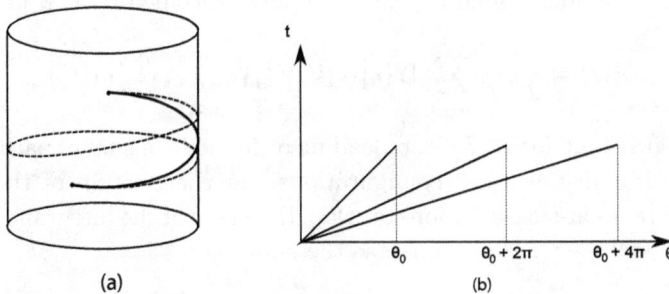

Fig. 4.2   Multiple classical solutions for a periodic variable.

Representing elements of $U(1)$ by points on the unit circle, the classical solutions for the continuum case are then paths on cylinder of unit radius given by the space

---
[1]The integers m^I are linearly related to the ℓ_I's

$R \times U(1)$ where R is the real line. The case of $m = 0, e_t(0)$ represents a straight path from 1 to e^{iA}, for $m = 1, e_t(1)$ represents the path from 1 to e^{iA} with one winding around the cylinder, and general $e_t(m)$ represents the path between 1 to e^{iA} with m winding. See Figure 4.2(a). For the winding solutions, the effective range of the angle increases, as shown in Figure 4.2(b).

For $SU(\mathcal{N})$, the circle of $U(1)$ is replaced by the $\mathcal{N} - 1$ dimensional torus given by

$$T = \underbrace{U(1) \times U(1) \times \cdots \times U(1)}_{(\mathcal{N}-1)}$$

The classical solutions are paths in the space $R \times T$ and are characterized by the winding numbers $m_1, m_2, \cdots m_{n-1}$. The winding numbers are directly related to the index of irreducible representations given by the composite label p, and this will be seen in the derivation of the $SU(3)$ character functions.

The paths with different windings are all topologically in-equivalent and are characterized by the winding numbers ℓ. The set of all paths starting and ending at same point, say the identify element, form the homotopy group of $SU(\mathcal{N})$.

The countably infinite classical solutions are configurations of local maxima of the action A. The time continuum limit of the path integral is taken by first expanding of the path integral about these classical paths. Define the change of integration variables around each of the classical paths $e_n(\ell)$ by

$$U_n = e_n(\ell)V_n \;\; ; \;\; dU_n = dV_n \tag{4.4.20}$$

with boundary conditions

$$V_0 = I \;\; ; \;\; V_N = I \tag{4.4.21}$$

Note that the new variables V_n are independent of boundary values U. The boundary values now appear in the path integral via the classical solutions as external parameters. The action around the classical path, specified by ℓ, is given by

$$A(\ell) = \frac{N}{2g^2T} \sum_{n=0}^{N-1} \text{Tr}\{e_1^\dagger(\ell)V_n V_{n+1}^\dagger + V_{n+1}V_n^\dagger e_1(\ell)\} \tag{4.4.22}$$

The path integral 4.3.6 for $K(T)$ is, to leading order in N, a sum of path integrations around the stationary classical configurations. In the domain of the integration where the different classical solutions overlap, the value of the integrand is of $O(e^{-N})$ and vanishes as $N \to \infty$. Hence one has the expansion

$$K(T) = \sum_\ell \int DV e^{A(\ell)}|_{V_0=V_N=1} + O(e^{-N}) \tag{4.4.23}$$

where

$$\sum_\ell = \prod_{I=1}^{\mathcal{N}-1} \sum_{\ell_I=-N}^{+N} , \int DV = \prod_{n=1}^{N-1} \int dV_n$$

4.5 $U(1)$ Path Integral

From Eq. 4.3.8, for $U_n = \exp\{iA_n\}$, the finite time lattice action for the Abelian group $U(1)$ is given by

$$A = \frac{N}{2g^2T} \sum_{n=0}^{N-1} \mathrm{Tr}(U_n U_{n+1}^\dagger + U_{n+1} U_n^\dagger) = \frac{N}{g^2T} \sum_{n=0}^{N-1} \cos(A_{n+1} - A_n). \qquad (4.5.1)$$

From Eq. 4.3.6 $K_1(T)$ yields the following path integral

$$K_1(T) = \lim_{N\to\infty} \prod_{n=1}^{N-1} \int_{-\pi}^{+\pi} dA_n \exp\left\{ \frac{N}{g^2T} \sum_{n=0}^{N-1} \cos(A_{n+1} - A_n) \right\} \Big|_{A_0=0, A_N=A} \qquad (4.5.2)$$

$$= \int DA e^A$$

To solve Eq. 4.5.2 exactly, consider the expansion

$$e^{\alpha \cos\theta} = \sum_{p=-\infty}^{+\infty} I_p(\alpha) e^{ip\theta} \qquad (4.5.3)$$

where I_p is the associated Bessel function. To prove Eq. 4.5.3, note that

$$\int_{-\pi}^{+\pi} \frac{d\theta}{2\pi} e^{\alpha \cos A} e^{-ip\theta} = \int_0^{+\pi} \frac{d\theta}{\pi} e^{\alpha \cos\theta} \cos(p\theta) = I_p(\alpha)$$

From Eqs. 4.5.1 and 4.5.3, using $A_0 = 0, A_N = A$, the action functional has the expansion

$$e^A = \prod_{n=0}^{N-1} e^{\frac{N}{g^2T}\cos(A_{n+1}-A_n)} = \prod_{n=0}^{N-1} \left[\sum_{p_n=-\infty}^{+\infty} I_{p_n}(\alpha) e^{ip_n(A_{n+1}-A_n)} \right]$$

$$= \prod_{n=0}^{N-1} \sum_{p_n=-\infty}^{+\infty} I_{p_0}(\alpha) e^{ip_0 A_1} I_{p_1}(\alpha) e^{ip_1(A_2-A_1)} \cdots I_{p_{N-1}}(\alpha) e^{ip_{N-1}(A-A_{N-1})} \qquad (4.5.4)$$

Starting the integration from A_1 and going on to A_{N-1}, each integration yields a delta function $\delta(p_n - p_{n-1})$ and the path integral is given by

$$K_1(T) = \int DA e^A = \mathcal{N}(1) \sum_{p=-\infty}^{+\infty} \lim_{N\to\infty} I_p^N\left(\frac{N}{g^2T}\right) e^{ipA} \qquad (4.5.5)$$

where $\mathcal{N}(1)$ is a normalization constant.

The expansion for the associated Bessel function is given by Gradshteyn and Ryzhik (1980)

$$\lim_{z\to\infty} I_p(z) = \frac{e^z}{\sqrt{2\pi z}}\left[1 - \frac{4p^2-1}{8z} + O(1/z^2)\right]$$

Hence, up to constants

$$\lim_{N\to\infty} I_p^N\left(\frac{N}{g^2T}\right) = \mathrm{const}\left(1 - \frac{g^2T}{2N}\left(p^2 - \frac{1}{2}\right)\right)^N = \mathrm{const} \cdot e^{-\frac{g^2T}{2}p^2} + O(N^{-1})$$

$\mathcal{N}(1)$ cancels the overall constants and yields the exact answer

$$K_1(T) = \sum_{p=-\infty}^{+\infty} e^{-\frac{g^2T}{2}p^2} e^{ipA} \qquad (4.5.6)$$

which agrees with Eq. 4.2.10.

4.6 $U(1)$ Path Integral: Classical Paths

The expansion about the multiple classical solutions, given in general by Eq. 4.4.23, is studied for the special case of $U(1)$ so as to better understand the $SU(\mathcal{N})$ path integral. Recall from Eq. 4.5.2 $K_1(T)$ is given by the following

$$K_1(T) = \lim_{N \to \infty} \prod_{n=1}^{N-1} \int_{-\pi}^{+\pi} dA_n \exp\left\{ \frac{N}{g^2 T} \sum_{n=0}^{N-1} \cos(A_{n+1} - A_n) \right\}\Big|_{A_0=0, A_N=A}$$

Consider the $N \to \infty$ limit for the Abelian path integral using the expansion about the classical solutions. For U(1), Eq. 4.4.20 yields

$$U_n = e^{iA_n} = e^{i\frac{n}{N}(A+2\pi\ell)} e^{i\xi_n}, \quad -\pi < \xi_n < \pi \tag{4.6.1}$$

with boundary conditions

$$U_0 = 1 \; ; \;\; U_N = e^{iA} \quad \Rightarrow \quad \xi_0 = \xi_N = 0 \tag{4.6.2}$$

Using Eqs. 4.4.17, 4.4.22 and 4.4.23 yields

$$K_1(T) = \lim_{N \to \infty} \sum_{\ell=-N}^{+N} \prod_n \int_{-\pi}^{+\pi} d\xi_n \exp\left\{ \frac{N}{g^2 T} \sum_n \cos\left(\xi_{n+1} - \xi_n + \frac{1}{N}(A+2\pi\ell) \right) \right\} \tag{4.6.3}$$

Expanding the action A about the ℓth local maximum, Eq. 4.6.3 yields, to leading order in N

$$K_1(T) = \sum_{\ell=-N}^{+N} \int D\xi \exp\left\{ -\frac{N}{2g^2 T} \sum_{n=0}^{N-1} (\xi_{n+1} - \xi_n + \frac{1}{N}(A+2\pi\ell))^2 \right\} \tag{4.6.4}$$

In obtaining Eq. 4.6.4 the action in $O(1)$ has been used; using $\delta\xi_n = \xi_{n+1} - \xi_n$, yields the following

$$O(\delta\xi_n) = \frac{1}{\sqrt{N}} \tag{4.6.5}$$

Using the boundary condition 4.6.2 yields

$$\sum_{n=0}^{N-1} \delta\xi_n = 0 \tag{4.6.6}$$

Hence, from Eqs. 4.6.4 and 4.6.6

$$A(\ell) = -\frac{N}{2g^2 T} \sum_{n=0}^{N-1} \left(\delta\xi_n + \frac{1}{N}(A+2\pi\ell) \right)^2$$

$$= -\frac{1}{2g^2 T}(A+2\pi\ell)^2 - \frac{N}{2g^2 T} \sum_n (\delta\xi_n)^2$$

The ξ_n integrations decouple from the boundary condition and, upto an overall constant, the evolution kernel is given by

$$K_1(T) = \lim_{N \to \infty} \sum_{\ell=-N}^{+N} e^{-\frac{1}{2g^2 T}(A+2\pi\ell)^2} = \sum_{\ell=-\infty}^{+\infty} e^{-\frac{1}{2g^2 T}(A+2\pi\ell)^2} \tag{4.6.7}$$

The sum consists of the exponential of the classical actions for classical paths of different winding number ℓ.

Note that, for $U(1)$, the path integral over the variables ξ_n completely decouples from the boundary variable e^{iA}. The crucial difference between the Abelian $U(1)$ and the non-Abelian $SU(N)$ case is that for the latter case the path integration over the ξ_n variables directly couple to boundary condition given by U.

Using the Poisson summation formula

$$\sum_{n=-\infty}^{+\infty} \delta(x-n) = \sum_{n=-\infty}^{+\infty} e^{2\pi inx}$$

yields

$$\sum_{\ell=\infty}^{+\infty} f_\ell = \sum_{p=-\infty}^{+\infty} \int_{-\infty}^{+\infty} dx e^{2\pi ipx} f_x \tag{4.6.8}$$

Hence, from Eq. 4.6.7

$$K_1(T) = \sum_{p=-\infty}^{+\infty} e^{-\frac{g^2 T}{2} p^2} e^{ipA} \tag{4.6.9}$$

which agrees with Eq. 4.2.10. In fact, the two expression for $K_1(T)$ can be taken to be a proof of the Poisson formula given in Eq. 4.6.8.

Note that the form of $K_1(T)$ given in Eq. 4.6.9 is a typical expression yielded directly by the Hamiltonian (as in Eq. 4.2.10) and is related to the action form of solution given in Eq. 4.6.7 via the non-trivial and non-perturbative transformation given by Poisson's summation formula.

The expansion of the path integral about the classical paths exists, in general, for path integrals for compact degrees of freedom [Baaquie (2014)].

4.7 $SU(N)$ Continuum Action

Consider the expansion of the path integral about the classical solution given, from Eqs. 4.4.22 and 4.4.23, by

$$K(T) = \sum_\ell \int DV \exp\{A(\ell)\} \tag{4.7.1}$$

where

$$A(\ell) = \frac{N}{g^2 T} \sum_n \mathrm{Tr}\{e_1^\dagger(\ell) V_n V_{n+1}^\dagger + V_{n+1} V_n^\dagger e_1(\ell)\} \tag{4.7.2}$$

The classical solution $e_1(\ell)$ labeled by ℓ is given by

$$e_1(\ell) = \exp\left\{\frac{i}{N}(b^I + 2\pi\ell^I)X_I\right\} \equiv \exp\left\{\frac{i}{N}b^I(\ell)X_I\right\} \tag{4.7.3}$$

Define canonical coordinates by

$$V_n = e^{i\xi_n^a X_a} \quad ; \quad \xi_0^a = 0 = \xi_N^a \tag{4.7.4}$$

The continuum limit is discussed in the Appendix 4.10 and Eq. 4.10.18. Using subscripts to denote the non-Abelian index yields the following ($\partial \xi_a \equiv \partial \xi_a / \partial t$)

$$A(\ell) = -\frac{1}{2g^2} \int_0^T dt \left(\partial \xi_a \partial \xi_a + \frac{C_{Iab}}{T}(b_I + 2\pi \ell_I) \xi_a \partial \xi_b \right)$$
$$-\frac{1}{2g^2 T} \sum_{I=1}^{\mathcal{N}-1} (b_I + 2\pi \ell_I)^2 \tag{4.7.5}$$

with boundary conditions: $\xi^a(0) = \xi^a(T) = 0$.

The classical action about the ℓth path is given by

$$A_c(b(\ell)) = -\frac{1}{2g^2 T} \sum_{I=1}^{\mathcal{N}-1} (b_I(\ell))^2 \quad ; \quad b^I(\ell) \equiv b_I + 2\pi \ell_I \tag{4.7.6}$$

The finite time continuum Lagrangian for the ℓth classical path, from Eq. 4.10.18, is given by

$$L(\ell) = -\frac{1}{2g^2} \left[(\partial \xi^a)^2 + C_{Iab} \frac{b_I(\ell)}{T} \xi^a \partial \xi^b \right] \tag{4.7.7}$$

The Lagrangian is quadratic in ξ and is negative definite since it is an expansion about the ℓth maxima. The second term in $L(\ell)$ is non-zero solely due to the antisymmetry of $C_{Iab} = -C_{Iba}$; for a symmetric coefficient this term is a zero-valued surface term. The variables $\xi^a(t)$ are fixed to be zero at $t = 0$ and $t = T$, and the finite time T also appears explicitly in the Lagrangian. The boundary values b_I appear explicitly in the Lagrangian. The continuum Feynman path integral, from Eq. 4.10.18, is given by

$$K(T) = \sum_\ell e^{A_c(\ell)} \int D\xi e^{\int_0^T dt L(\ell)} \tag{4.7.8}$$

where

$$\sum_\ell \equiv \prod_{I=1}^{\mathcal{N}-1} \sum_{\ell_I=-\infty}^{+\infty} \quad ; \quad \int D\xi = \prod_{t=0}^{T} \prod_a \int_{-\infty}^{+\infty} d\xi_t^a \Big|_{\xi_0^a = \xi_T^a = 0} \tag{4.7.9}$$

Since the Lagrangian is quadratic, $K(T)$ can be explicitly evaluated Gaussian integration. The $U(1)$ case is trivial, since the structure constants are zero and the ξ path integral decouples from the boundary conditions, with the complete result for the $U(1)$ case given by the classical action A_c as in Eq. 4.6.7.

4.8 $SU(2)$ Path Integration

The $SU(2)$ case is the most important since it will be shown later that the path integral for $SU(\mathcal{N})$ reduces to a product of the $SU(2)$ case – when expressed in the appropriate coordinates.

For $SU(2)$, the structure constants are

$$C_{ijk} = \epsilon_{ijk} \quad ; \quad \epsilon_{123} = 1 \tag{4.8.1}$$

where ϵ_{ijk} is the completely antisymmetric tensor. The group generators are

$$X_i = \frac{1}{2}\sigma_i \ ; \ \ X_I = X_3 = \frac{1}{2}\sigma_3 \tag{4.8.2}$$

The boundary value of Eq. 4.4.12 is given by

$$U = \begin{pmatrix} e^{iA} & 0 \\ 0 & e^{-iA} \end{pmatrix} = \exp(i2AX_3) \tag{4.8.3}$$

yielding for $SU(2)$ the classical solutions

$$e_n(\ell) = \exp\left\{i\frac{n}{N}2(A+2\pi\ell)X_3\right\} \tag{4.8.4}$$

and hence from Eq. 4.7.3

$$b^I(\ell) = 2(A+2\pi\ell) \tag{4.8.5}$$

The Lagrangian for $SU(2)$ is from Eqs. 4.7.7 and 4.8.5

$$
\begin{aligned}
L(\ell) &= -\frac{1}{2g^2T}\left[\sum_{a=1}^{3}(\partial\xi_a)^2 + 2\epsilon_{3ab}\frac{(A+2\pi\ell)}{T}\xi_a\partial\xi_b\right] \\
&= -\frac{1}{2g^2T}\left[\sum_{a=1}^{3}(\partial\xi_a)^2 + 2\alpha(\xi_1\partial\xi_2 - \xi_2\partial\xi_1)\right]
\end{aligned} \tag{4.8.6}
$$

where

$$\alpha = \frac{(A+2\pi\ell)}{T} \tag{4.8.7}$$

Using the fixed boundary conditions of ξ_a given by Eq. 4.10.16 and an integration by parts for the action given in Eq. 4.8.6 yields

$$L(\ell) = -\frac{1}{2g^2T}[(\partial\xi_1)^2 + (\partial\xi_2)^2 + (\partial\xi_3)^2 + 4\alpha\xi_1\partial\xi_2] \tag{4.8.8}$$

Using

$$(\partial\xi_2)^2 + 4\alpha\xi_1\partial\xi_2 = (\partial\xi_2 + 2\alpha\xi_1)^2 - (2\alpha\xi_1)^2 \tag{4.8.9}$$

the Lagrangian is given by

$$\mathcal{L}(\ell) = -\frac{1}{2g^2T}(\partial\xi_3)^2 - \frac{1}{2g^2}(\partial\xi_2 + 2\alpha\xi_1)^2 - \frac{1}{2g^2}[(\partial\xi_1)^2 - (2\alpha\xi_1)^2] \tag{4.8.10}$$

Note that ξ_3, which belongs to the center of $SU(2)$, decouples from the other variables and yields

$$K_2(T) = \sum_{\ell} e^{A_c}\tilde{K}_2 \tag{4.8.11}$$

where, from Eqs. 4.7.6 and 4.8.5

$$A_c = -\frac{2}{g^2T}(A+2\pi\ell)^2 \tag{4.8.12}$$

From Eq. 4.8.10

$$A_1 = -\frac{1}{2g^2 T} \int_0^T dt[(\partial \xi_1)^2 - (2\alpha \xi_1)^2] \;\; ; \;\; A_2 = -\frac{1}{2g^2 T} \int_0^T dt(\partial \xi_2 + 2\alpha \xi_1)^2$$

and

$$\tilde{K}_2 = \int D\xi_1 e^{A_1(\xi_1)} \left\{ \int D\xi_2 e^{A_2(\xi_1,\xi_2)} \right\} = \int D\xi_1 e^{A_1(\xi_1)} K_2'(\xi_1) \quad (4.8.13)$$

A normalization function arising from the ξ_3-path integration has been dropped. The ξ_2 path integration is first performed. Make the change of variables

$$W_t = \partial \xi_{2t} + 2\alpha \xi_{1t} \quad (4.8.14)$$

giving

$$\xi_{2t} = \int_0^t dt' (W_t' - 2\alpha \xi_{1t'}) \quad (4.8.15)$$

Eq. 4.8.15 yields the identity

$$\xi_{20} = 0 \quad (4.8.16)$$

The boundary conditions on ξ_2 yield

$$\xi_{20} = 0 = \xi_{2T} \quad (4.8.17)$$

From Eqs. 4.8.17 and 4.8.15, one constraint equation is

$$0 = \xi_{2T} = \int_0^T dt(W_t - 2\alpha \xi_{1t}) \quad (4.8.18)$$

There are no constraints on W_0 or W_T. The integration measure is given by

$$D\xi_2 = \prod_{t=0}^T dW_t \delta(\xi_{2T}) = DW \int_{-\infty}^{\infty} dx \exp\left\{ ix \int_0^T dt(W_t - 2\alpha \xi_{1t}) \right\}$$

yielding the result

$$K_2'(\xi_1) = \int D\xi_2 e^{A_2(\xi_1,\xi_2)}$$

$$= \int DW\,dx \exp\left[ix \int_0^T dt(W_t - 2\alpha \xi_{1t}) - \frac{1}{2g^2 T} \int_0^T dt W_t^2 \right]$$

$$= \int_{-\infty}^{\infty} dx \exp\left[-\frac{g^2 T}{2} x^2 - 2ix\alpha \int_0^T \xi_{1t} dt \right]$$

Dropping the subscript on ξ_1, the probability amplitude (ignoring normalization terms) for $\tilde{K}$, from Eq. 4.8.13, is given by the path integral

$$K_2' = \int_{-\infty}^{\infty} dx \int D\xi e^{A'(\xi,x)}\big|_{\xi_0=\xi_T=0} \quad (4.8.19)$$

where, from Eq. 4.8.13

$$A' = -\frac{g^2 T}{2} x^2 - 2ix\alpha \int_0^T dt\xi_t - \frac{1}{2g^2} \int_0^T dt[(\partial \xi_t)^2 - (2\alpha \xi_t)^2] \quad (4.8.20)$$

To perform the ξ-integration, since ξ is zero on the boundaries consider the Fourier sine expansion

$$\xi_t = \sum_{k=1}^{\infty} \sin\left(\frac{\pi k t}{T}\right)\xi_k \qquad (4.8.21)$$

Using the orthogonality relation

$$\int_0^T dt \sin\left(\frac{\pi k t}{T}\right)\sin\left(\frac{\pi k' t}{T}\right) = \frac{T}{2}\delta_{kk'} \qquad (4.8.22)$$

yields

$$A' = -\frac{1}{2g^2}\frac{T}{2}\sum_{k=1}^{\infty}\left[\left(\frac{\pi k}{T}\right)^2 - (2\alpha)^2\right]\xi_k^2 - 2ix\alpha\sum_{k=1}^{\infty}\frac{2T}{2k-1}\xi_k - \frac{g^2 T}{2}x^2 \qquad (4.8.23)$$

Hence Eq. 4.8.19 yields

$$\tilde{K}_2 = \int_{-\infty}^{\infty} dx\, e^{-\frac{g^2 T}{2}x^2}\prod_{k=1}^{\infty}\int_{-\infty}^{\infty} d\xi_k e^{A'} \qquad (4.8.24)$$

$$= \int_{-\infty}^{\infty} dx\, e^{f(x)}\prod_{k=1}^{\infty}\frac{1}{\sqrt{(\frac{\pi k}{T})^2 - (2\alpha)^2}} \qquad (4.8.25)$$

The ξ_k-Gaussian integrations yield, from Eq. 4.8.23

$$f(x) = -\frac{g^2}{2\alpha}\tan(\alpha T)x^2 \qquad (4.8.26)$$

Also from Gradshteyn and Ryzhik (1980)

$$\prod_{k=1}^{\infty}\left(1 - \left(\frac{2\alpha T}{\pi k}\right)^2\right) = \frac{1}{2\alpha T}\sin(2\alpha T) \qquad (4.8.27)$$

Hence, from Eqs. 4.8.24, 4.8.26 and 4.8.27, doing the x-integration yields

$$\tilde{K}_2 = \sqrt{\frac{2\alpha}{\sin(2\alpha T)}}\sqrt{\frac{\alpha}{\tan(\alpha T)}} = \frac{1}{T}\frac{|\alpha T|}{|\sin(\alpha T)|} = \frac{1}{T}\frac{A + 2\pi\ell}{\sin(A + 2\pi\ell)}$$

where, from Eq. 4.8.7

$$\alpha = \frac{(A + 2\pi\ell)}{T}$$

and finally

$$\tilde{K}_2(A) = N_2(T)\frac{A + 2\pi\ell}{\sin(A)} \qquad (4.8.28)$$

where $N_2(T)$ is the overall normalization function. Using Eqs. 4.7.6 and 4.8.5 yields, from Eq. 4.8.12 for $SU(2)$

$$A_c = -\frac{2}{g^2 T}(A + 2\pi\ell)^2 \qquad (4.8.29)$$

Using Eqs. 4.7.8, 4.8.28 and 4.8.29 yields the following result

$$K_2(T) = \sum_{\ell_1} e^{A_c} \tilde{K}_2 = \frac{N_2(T)}{\sin(A)} \sum_{\ell=-\infty}^{\infty} (A + 2\pi\ell) e^{-\frac{2}{g^2 T}(A+2\pi\ell)^2} \qquad (4.8.30)$$

To compare the result given in Eq. 4.8.30 with exact result given by Eq. 4.2.14 the Poisson's summation 4.6.8 is used. For this the function of integer ℓ has to be continued to the continuous variable x. Make the following continuation

$$\frac{A + 2\pi\ell}{\sin(A)} e^{-\frac{2}{g^2 T}(A+2\pi\ell)^2} \rightarrow \frac{A + 2\pi x}{\sin(A)} e^{-\frac{2}{g^2 T}(A+2\pi x)^2} \qquad (4.8.31)$$

Hence from Eqs. 4.6.8, 4.8.29, 4.8.30 and 4.8.31

$$K_2(T) = \frac{N_2(T)}{\sin(A)} \sum_{p=-\infty}^{\infty} \int_{-\infty}^{\infty} dx e^{ipx} (A + x) e^{-\frac{2}{g^2 T}(A+x)^2} \qquad (4.8.32)$$

$$= \frac{-i}{\sin(A)} \sum_{p=-\infty}^{\infty} p e^{-\frac{g^2 T}{8} p^2} e^{ipA} \qquad (4.8.33)$$

and which agrees with Eq. 4.2.14 as expected.

Note the path integral expression for $SU(2)$ given in Eq. 4.8.30 could be directly obtained from Eq. 4.2.14 since all the character functions of $SU(2)$ are known. The path integral was nevertheless evaluated directly as this computation illustrates the role played by the classical paths and provides independent confirmation that the expansion of the lattice path integral about the classical paths for taking the continuum limit is valid.

The exact character functions of $SU(3)$ are not known and hence one does not have an exact expression for the evolution kernel; instead, the $SU(3)$ path integration will be directly evaluated in Section 4.8.1 and then 'inverted' in Chapter 5 using the Poisson summation formula to yield the character functions for $SU(3)$.

4.8.1 *SU(3) path integration*

To evaluate the $SU(3)$ path integral, the $SU(3)$ structure constants f_{ijk} and generators of the Lie Algebra are given by

$$c_{ijk} = f_{ijk} \; ; \quad X_i = \frac{1}{2}\lambda_i \; : \quad i, j, k = 1, 2, ..., 8$$

where λ_i are the Gell-Mann matrices. For the commuting center of the group, denoted by index 3 and 8, the diagonal generators in the fundamental representation are given by

$$X_3 = \frac{1}{2}\lambda_3 = \frac{1}{2} \begin{pmatrix} 1 & 0 & 0 \\ 0 & -1 & 0 \\ 0 & 0 & 0 \end{pmatrix} \; ; \; X_8 = \frac{1}{2}\lambda_8 = \frac{1}{2\sqrt{3}} \begin{pmatrix} 1 & 0 & 0 \\ 0 & 1 & 0 \\ 0 & 0 & -2 \end{pmatrix} \qquad (4.8.34)$$

The non-zero structure constants for the center are given by

$$f_{312} = 1 \; ; \; f_{345} = \frac{1}{2} \; ; \; f_{367} = -\frac{1}{2} \; ; \; f_{845} = \frac{\sqrt{3}}{2} \; ; \; f_{867} = \frac{\sqrt{3}}{2} \qquad (4.8.35)$$

The boundary value $SU(3)$ group element is given by

$$U = \begin{pmatrix} e^{iA} & 0 & 0 \\ 0 & e^{iB} & 0 \\ 0 & 0 & e^{-i(A+B)} \end{pmatrix} = \exp\{i(b_3 X_3 + b_8 X_8)\}$$

Eqs. 4.8.34 and 4.8.36 yield

$$b_3 = A - B \; ; \; b_8 = \sqrt{3}(A + B) \tag{4.8.36}$$

The classical solutions are labeled by two winding numbers ℓ, m and are given by

$$e_t(\ell, m) = \begin{pmatrix} e^{it(A+2\pi\ell)/T} & 0 & 0 \\ 0 & e^{it(B+2\pi m)/T} & 0 \\ 0 & 0 & e^{-it(A+B+2\pi\ell+2\pi m)/T} \end{pmatrix} ; \ell, m \in \mathcal{Z}$$

Let

$$A(\ell) = A + 2\pi\ell \; ; \; B(m) = B + 2\pi m$$

For $SU(3)$, the Lagrangian given in Eq. 4.7.7 yields

$$\mathcal{L}(\ell, m) = -\frac{1}{2g^2}\left[\sum_{a=1}^{8}(\partial\xi_a)^2 + \frac{1}{T}f_{Iabb_I}\xi_a\partial\xi_b\right]$$

$$= -\frac{1}{2g^2}\left[(\partial\xi_3)^2 + (\partial\xi_8)^2\right] - \frac{1}{2g^2}\left[(\partial\xi_1)^2 + (\partial\xi_2)^2 + \frac{2}{T}(A(\ell) - B(m))\xi_1\partial\xi_2\right]$$

$$- \frac{1}{2g^2}\left[(\partial\xi_4)^2 + (\partial\xi_5)^2 + \frac{2}{T}(2A(\ell) + B(m))\xi_4\partial\xi_5\right]$$

$$- \frac{1}{2g^2}\left[(\partial\xi_6)^2 + (\partial\xi_7)^2 + \frac{2}{T}(A(\ell) + 2B(m))\xi_6\partial\xi_7\right] \tag{4.8.37}$$

where Eqs. 4.8.35 and 4.8.36 have been used. The Lagrangian in Eq. 4.8.37 shows that the center variables ξ_3, ξ_8 have decoupled from the boundary conditions carried by A, B and from the other variables as well. The remaining six variables are grouped into three $SU(2)$-like calculations given by (ξ_1, ξ_2), (ξ_4, ξ_5) and (ξ_6, ξ_7). The classical action is given by

$$\mathcal{A}_c(\ell, m) = -\frac{1}{2g^2}[(b_3)^2 + (b_8)^2] = -\frac{2}{g^2}[A^2(\ell) + B^2(\ell) + A(\ell)B(\ell)] \tag{4.8.38}$$

Using the $SU(2)$ results given in Eqs. 4.8.27 and 4.8.30, the $SU(3)$ kernel is given by

$$K_3(T) = \sum_{\ell,m=-\infty}^{+\infty} e^{\mathcal{A}_c(\ell,m)}\tilde{K}_2\left[\frac{1}{2}(A(\ell) - B(m))\right]\tilde{K}_2\left[\frac{1}{2}(2A(\ell) + B(m))\right]$$

$$\times \tilde{K}_2\left[\frac{1}{2}(A(\ell) + 2B(m))\right] \tag{4.8.39}$$

$$= \frac{1}{s(A, B)} \sum_{\ell,m=-\infty}^{+\infty} e^{\mathcal{A}_c(\ell,m)}(A(\ell) - B(m))(2A(\ell) + B(m))(A(\ell) + 2B(m))$$

where the denominator $s(A, B)$ is independent of ℓ, m and given by

$$s(A, B) = 8 \sin \frac{1}{2}(A - B) \sin \frac{1}{2}(2A + B) \sin \frac{1}{2}(A + 2B) \qquad (4.8.40)$$

The Poisson summation formula given in Eq. 4.6.8 to invert g^2 to $1/g^2$ in the action $\mathcal{A}_c(\ell, m)$ given in Eq. 4.8.39. To carry out this inversion note the following.

$$A(\ell) \to A + 2\pi x \; ; \quad B(m) \to B + 2\pi y \; ; \quad x \to \frac{x}{2\pi} - A \; ; \quad y \to \frac{y}{2\pi} - B$$

This yields

$$K_3 = \frac{1}{s(A, B)} \sum_{p,q=-\infty}^{+\infty} e^{-ipA} e^{-ipA} K_3'(p, q) \qquad (4.8.41)$$

where, upto a normalization constant

$$K_3'(p, q) = \int_{-\infty}^{+\infty} dx dy\, e^{ipx+iqy}(x - y)(2x + y)(x + y) e^{-\frac{2}{g^2 T}(x^2+y^2+xy)}$$

$$= -ipq(p - q) \exp\left\{ -\frac{g^2 T}{2} \cdot \frac{1}{3}(p^2 + q^2 - pq) \right\} \qquad (4.8.42)$$

Eqs. 4.8.41 and 4.8.42 yield the result

$$K_3(T) = \frac{i}{s(A, B)} \sum_{p,q=-\infty}^{+\infty} e^{ipA-ipB} pq(p + q) e^{-\frac{g^2 T}{2} \cdot \frac{1}{3}(p^2+q^2+pq)} \qquad (4.8.43)$$

In Section 5.2, it will be shown how the exact values for all the $SU(3)$ character functions can be obtained from the expression for the evolution kernel given in Eq. 4.8.43 above.

4.8.2   *SU($\mathcal{N}$) path integration*

The general case of the evolution kernel for $SU(\mathcal{N})$ is solved by generalizing the result for $SU(3)$.

The continuum Feynman path integral, from Eq. 4.7.8, is given by

$$K_{\mathcal{N}}(T) = \sum_{\ell} e^{\mathcal{A}_c(\ell)} \int D\xi e^{\int_0^T dt L(\ell)}$$

From Eq. 4.7.7 the continuum finite time Lagrangian around the ℓth classical path is given by

$$L(\ell) = -\frac{1}{2g^2}[(\partial \xi^a)^2 + C_{Iab} \frac{b^I(\ell)}{T} \xi^a \partial \xi^b \; ; \quad \mathcal{A}_c(\ell) = -\frac{1}{2g^2 T} \sum_I (b_I + 2\pi \ell_I)^2$$

For $SU(\mathcal{N})$ there are $\mathcal{N}^2 - 1$ generators and $\mathcal{N} - 1$ commuting generators that form the center of the group. Hence there are $\mathcal{N}(\mathcal{N} - 1)$ off-diagonal generators. A double index notation is used for the off-diagonal generators; let $K; L = 1, 2, \ldots, \mathcal{N}$; for $K < L$ there are $\mathcal{N}(\mathcal{N} - 1)/2$ distinct labels. The off-diagonal generators of

$SU(N)$ are labeled by $a = \pm(KL)$ and hence all the off-diagonal generators are accounted for. The diagonal generators are labeled by $I = 1, 2, \ldots, N - 1$.

The nonzero values of C_{Iab} are evaluated. The double index notation yields the following [Zelobenko (1973)]

$$C_{Iab} = [\lambda_I(K) - \lambda_I(L)]\delta_{a,-b} \quad ; \quad I = 1, 2, \ldots, N - 1 \qquad (4.8.44)$$

where $\lambda_I(K)$ are the *weight vectors* of $SU(N)$.[2] Using Eqs. 4.7.7 and 4.8.44 yields

$$L(l) = -\frac{1}{2g^2}\bigg\{ \sum_{l=1}^{N-1} \dot{\xi}_l^2 + \sum_{K<L} [(\dot{\xi}_{(KL)})^2 + (\dot{\xi}_{-(KL)})^2]$$

$$+ \frac{2}{T} \sum_{K<L} \Big(A_K + 2\pi\ell_K - A_L - 2\pi\ell_L \Big) \xi_{(KL)}\dot{\xi}_{-(KL)} \bigg\} \qquad (4.8.45)$$

since it can be shown that [Zelobenko (1973)]

$$\frac{1}{2} \sum_{I=1}^{N-1} \lambda_I(K)b^I(\ell) = A_K + 2\pi\ell_K \qquad (4.8.46)$$

Recall

$$e^{ib^I X_I} = \mathrm{diag}(e^{iA_1}, e^{iA_2}, \ldots, e^{iA_N}) \quad ; \quad \sum_{I=1}^{N} A_I = 0$$

Eq. 4.8.45 shows that the center variables of $SU(N)$, given by ξ_I, decouple from the boundary values encoded in A_I, as was the case for $SU(3)$. The remaining $N(N-1)$ integration variables split into $N(N-1)/2$ separate $SU(2)$ path integration, again as was the case for $SU(3)$. Each path integral involves only the variables $\xi_{(KL)}$ and $\xi_{-(KL)}$ and is identical to the $SU(2)$ path integration evaluated in Section 4.8. The result for $SU(N)$, up to an overall constant, is given by

$$K_N(\alpha) = \prod_{I=1}^{N} \sum_{l_I=-\infty}^{+\infty} \left(\prod_{I<J} \frac{A_I - A_J + 2\pi l_I - 2\pi l_J}{\sin\frac{1}{2}(A_I - A_J + 2\pi l_I - 2\pi l_J)} \right) \delta\Big(\sum_I \ell_I\Big)$$

$$\times \exp\left(-\frac{1}{2g^2 T} \sum_{I=1}^{N} (A_I + 2\pi l_I)^2 \right) \quad ; \quad \sum_{I=1}^{N} A_I = 0 \qquad (4.8.47)$$

The result obtained above agrees with the result of Menotti and Onofri (1981) that was derived using the Schrödinger differential equation.

4.9 Summary

The Hamiltonian for $SU(N)$ degree of freedom is given by the Laplace-Beltrami differential operator on the group manifold. In the later study of the Hamiltonian for the lattice gauge field, it will be seen that the Hamiltonian for the field theory at every lattice site is the Hamiltonian for $SU(N)$ degree of freedom. This was

---

[2]The weight vectors of a Lie algebra are the root vectors for the adjoint representation.

the main motivation for studying the quantum mechanics of the $SU(N)$ degree of freedom.

The evolution kernel for the $SU(N)$ Hamiltonian was defined using the path integral. The evolution kernel for $SU(N)$ was evaluated exactly by expanding the action about the classical paths. The general case of $SU(N)$ was seen to be reducible to the evaluation of the $SU(2)$ path integral. The $SU(2)$ group seems to play a key role in the general theory of the $SU(N)$ group, since the irreducible representations of $SU(N)$ also show the key role played by the $SU(2)$ algebra.

4.10   Appendix: Continuum Limit of $SU(N)$ Path Integral

Naively taking the $N \to \infty$ limit for the path integral, given by defining

$$\partial_t U = (N/T)(U_{n+1} - U_n)$$

yields

$$K(T) = \int DU \exp\left(-\frac{1}{2g^2}\int_0^T dt\,\mathrm{Tr}(\partial_t U^\dagger \partial_t U)\right) \qquad (4.10.1)$$
$$U(0) = V, \qquad U(T) = U$$

$$\int DU \equiv \lim_{N\to\infty}\prod_{n=1}^{N-1}\int dU_n \qquad (4.10.2)$$

The expressions 4.10.1 and 4.10.2 are the ones derived by Marinov and Terentyev (1979) and, as they stand, are ill defined. The action in Eq. 4.10.1 is highly nonlinear and there is no apparent reason, as recognized by Marinov and Terentyev (1979), why the semi-classical approximation should be exact for $K(T)$. A more serious problem is the divergences arising from the nonlinear pieces of the action, as well as from the measure dU_n.

Using the canonical coordinates for $U_n = \exp(iB_n^a X_a)$, where X_a are the generators, yields from Eq. 2.5.3

$$\int DU = \lim_{N\to\infty}\prod_{n=1}^{N-1}\int dU_n = \lim_{N\to\infty}\int\prod_{na}dB_n^a\exp\left(-\frac{N}{24}\sum_{na}(B_n^a)^2 + 0(B^4)\right)$$

$$= \lim_{N\to\infty}\prod_{t=0}^{T}\prod_a\int dB_t^a\exp\left(-\frac{N}{24}\frac{N}{T}\int_0^T B^2 dt\right) \qquad (4.10.3)$$

The measure has linear divergence in N for the coefficient of the mass-like B^2 term in Eq. 4.10.1, which arises from the invariant measure. Since there is no mass term in the action given by Eq. 4.10.1, the only way that a finite path-integral expression can exist for the heat kernel $K(T)$ is for the divergent nonlinear terms from the action to exactly cancel the divergence due to the measure term, and in effect give zero-mass renormalization.

The mass-like quadratic divergent term that comes from the measure undergoes a renormalization due to the nonlinear interaction terms contained in the action

$A(\xi, b)$. It is explicitly shown in this Appendix that for a finite lattice of size N, in the one-loop approximation, mass renormalization goes as $1/N$. Using Feynman diagrams, it is shown that for $N = \infty$, when expanded about the classical solutions, the action $A(\xi, b(\ell))$ reduces to its quadratic part, with all the nonlinear terms going to zero as $1/N$.

To take the continuum limit, make the change of variables

$$U_n = e_n(l) \exp(i\xi_n^a X_a) \equiv e_n(l) V_n \tag{4.10.4}$$

The multiple classical solutions for $SU(\mathcal{N})$, from Eq. 4.4.13, are[3]

$$e_n(l) = \begin{pmatrix} e^{i(A_1+2\pi\ell_1)\frac{n}{N}} & & 0 \\ & \ddots & \\ 0 & & e^{-i\sum_I(A_I+2\pi\ell_I)\frac{n}{N}} \end{pmatrix} = e^{i\frac{n}{N}(b_I+2\pi\ell_I)X_I}$$

The boundary conditions are

$$\xi_0^a = \xi_N^a = 0, \quad \text{and} \quad dU_n = dV_n \tag{4.10.5}$$

The action about the ℓth classical path is given by

$$A(\ell) = \frac{N}{2g^2T} \sum_n \text{Tr}\{e_1^\dagger(\ell)V_n V_{n+1}^\dagger + V_{n+1}V_n^\dagger e_1(\ell)\} \tag{4.10.6}$$

Expanding the path integral about the stationary classical paths $e_n(\ell)$ yields

$$K(T) = \sum_l \prod_n \int dV_n \exp[A(\ell)] + 0(e^{-N}) \tag{4.10.7}$$

The measure term yields, from Eq. 2.5.3, the following

$$\prod_n dV_n = \prod_{na} d\xi_n^a \exp\left(-\frac{N}{24}\sum_{na}(\xi_n^a)^2 + 0(\xi^4)\right) \tag{4.10.8}$$

Let $\delta\xi_n^a \equiv \xi_{n+1}^a - \xi_n^a$; using Eq.4.4.13 for the classical paths and the Campbell-Hausdorff formula given in Eq. 2.6.1 yields[4]

$$A(\ell) = -\frac{N}{2g^2T}\sum_n\left\{\sum_a(\delta\xi_n^a)^2 - \frac{1}{12}\sum_a(C_{abc}\xi_n^b\delta\xi_n^c)^2 + 0\left((\xi^2\delta\xi)^2\right)\right.$$
$$\left. + \frac{1}{N}(b_I+2\pi l_I)C_{Iab}\xi_n^a\delta\xi_n^b\right\} - \frac{N}{24}\sum_n(\xi_n^a)^2 + A_c(\ell) + 0(1/N)$$

---

[3] $e_n(l) \equiv e_n(\ell_1, \ell_2, \cdots \ell_N)$ with $\sum_I \ell_I = 0$.
[4] The pure ξ action is approximately given, from Eq. 2.4.5, by the following

$$\text{Tr}(\delta U \delta U^\dagger) = \frac{1}{2}\sum_{ab} g_{ab}(\xi)\delta\xi^a\delta\xi^b$$
$$= \frac{1}{2}\sum_{ab}\left[\delta_{ab} + \frac{1}{12}C_{aac}C_{c\beta b}\xi^\alpha\xi^\beta\right]\delta\xi^a\delta\xi^b = \frac{1}{2}\sum_a\left[(\delta\xi^a)^2 - \frac{1}{12}(C_{abc}\xi_n^b\delta\xi_n^c)^2\right]$$

where

$$A_c(\ell) = -\frac{1}{2g^2T} \sum_I (b_I + 2\pi\ell_I)^2$$

Since the coupling of the ξ_n^a field to the classical solutions is down by a factor of $1/N$, only the pure ξ part of the action, given by A_{eff} below, will contribute to possible divergences in N. Hence

$$A(\ell) = A_{\text{eff}} + \tilde{A} + 0(1/N) \tag{4.10.9}$$

where

$$A_{\text{eff}} = -\frac{N}{2g^2T} \sum_n \left\{ \sum_a (\delta\xi_n^a)^2 - \frac{1}{12} \sum_a (C_{abc}\xi_n^b \xi_{n+1}^c)^2 \right\} - \frac{N}{24} \sum_n (\xi_n^a)^2$$

and

$$\tilde{A} = -\frac{1}{2g^2T} \sum_{nI} (b_I + 2\pi l_I) C_{Iab} \xi_n^a \delta\xi_n^b - \frac{1}{2g^2T} \sum_I (b_I + 2\pi l_I)^2$$

To evaluate mass renormalization δm, the propagator of the ξ field is calculated to one-loop, which is to leading order in g^2T.

The orthogonality of the sine function on a finite lattice is given by

$$\sum_{n=1}^{N-1} \sin\left(\frac{\pi kn}{N}\right) \sin\left(\frac{\pi pn}{N}\right) = \frac{N}{2}\delta_{p-k}$$

and yields the following Fourier expansion for the degrees of freedom ξ_n^a

$$\xi_n^a = \frac{2}{N} \sum_{n=1}^{N-1} \sin\left(\frac{\pi kn}{N}\right) \xi_k^a$$

This yields the following for A_{eff}

$$A_{\text{eff}} = A_0 + A_{\text{int}}$$

with

$$A_0 = -\frac{1}{2g^2T} \sum_k f_k (\xi_k^a)^2 \quad ; \quad f_k = 4\left(1 - \cos\left(\frac{\pi k}{N}\right)\right)$$

and

$$A_{\text{int}} = A_{\text{int}}^{(1)} + A_{\text{int}}^{(2)}$$
$$= \frac{2C_{a\gamma b}C_{c\gamma d}}{3N^3 g^2 T} \sum_n \sum_{k_1,k_2,k_3,k_4} \sin\left(\frac{\pi nk_1}{N}\right) \sin\left(\frac{\pi(n+1)k_2}{N}\right)$$
$$\times \sin\left(\frac{\pi nk_3}{N}\right) \sin\left(\frac{\pi(n+1)k_4}{N}\right) \xi_{k_1}^a \xi_{k_2}^b \xi_{k_3}^c \xi_{k_4}^d - \frac{N}{12N} \sum_k (\xi_k^a)^2 \tag{4.10.10}$$

The propagator, to $O(g^4T^2)$, is given by

$$G_p^{ab} = E\left[\xi_k^a \xi_k^b\right] = \frac{1}{Z} \int D\xi e^{A_0} \left\{1 + A_{\text{int}}^{(1)} + A_{\text{int}}^{(2)}\right\} \xi_p^a \xi_p^b \quad ; \quad Z = \int D\xi e^{A_{\text{eff}}}$$

The free propagator is given by

$$\frac{1}{Z_0}\int D\xi e^{A_0}\xi_p^a\xi_q^b = g^2T\frac{\delta^{ab}}{f_p}\delta_{p-q} \quad ; \quad Z_0 = \int D\xi e^{A_0}$$

Hence, Feynman perturbation yields

$$G_p^{ab} = g^2T\frac{\delta^{ab}}{f_p} + (g^2T)^2\frac{\delta^{ab}}{f_p^2}(\Pi_1(p) + \Pi_2(p)) \simeq g^2T\frac{\delta^{ab}}{f_p - \Pi(p)} \qquad (4.10.11)$$

where (see Figure 43)

$$\Pi(p) = \Pi_1(p) + \Pi_2(p)$$

The effective renormalized mass-like term is given by[5]

$$\delta m = \lim_{p\to 0} N\Pi(p) = \lim_{p\to 0} N[\Pi_1(p) + \Pi_2(p)] = \delta m_1 + \delta m_2 \qquad (4.10.12)$$

The quartic term is symmetrized only under exchange of k_1, k_2 and k_3, k_4; for the calculation of δm this is sufficient. Hence

$$A_{int}^{(1)} = \frac{C_{a\gamma b}C_{c\gamma d}}{6N^3g^2T}\sum_n\sum_{k_1,k_2;k_3,k_4}\Gamma_n(k_1,k_2,k_3,k_4)\xi_{k_1}^a\xi_{k_2}^b\xi_{k_3}^c\xi_{k_4}^d$$

$$\Gamma_n(k_1,k_2;k_3,k_4) = \left\{\sin\left(\frac{\pi n k_1}{N}\right)\sin\left(\frac{\pi(n+1)k_2}{N}\right) - \sin\left(\frac{\pi n k_2}{N}\right)\sin\left(\frac{\pi(n+1)k_1}{N}\right)\right\}$$

$$\times\left\{\sin\left(\frac{\pi n k_3}{N}\right)\sin\left(\frac{\pi(n+1)k_4}{N}\right) - \sin\left(\frac{\pi n k_4}{N}\right)\sin\left(\frac{\pi(n+1)k_3}{N}\right)\right\}$$

Fig. 4.3 A one-loop diagram for determining δm.

The Feynman diagrams for $\Pi_1(p), \Pi_2(p)$ are given in Figure 4.3; the vertex $\Gamma_n(k_1, k_2; k_3, k_4)$ is indicated by the wavy line connecting the two straight lines. Note the identity

$$C_{a\gamma c}C_{b\gamma c} = N\delta^{ab}$$

The combinatorial factor for $\Pi_1(p)$ is $4\times 2 = 8$ and yields, from Eq. 4.10.11, the following

$$\Pi_1(p) = \frac{4N}{3N^3g^2T}\sum_{n,k}\frac{1}{f_k}\Gamma_n(p,k;p,k) \quad ; \quad f_k = 4\left(1-\cos\left(\frac{\pi k}{N}\right)\right)$$

$$\Pi_1(p) = \frac{4N}{3N^3g^2T}\sum_{n,k}\frac{1}{f_k}\left(\sin\left(\frac{\pi n k}{N}\right)\sin\left(\frac{\pi n p}{N}\right)\left\{\cos\left(\frac{\pi k}{N}\right) - \cos\left(\frac{\pi p}{N}\right)\right\}\right.$$

$$\left. - \sin\left(\frac{\pi n k}{N}\right)\cos\left(\frac{\pi n p}{N}\right)\left\{\sin\left(\frac{\pi k}{N}\right) - \sin\left(\frac{\pi p}{N}\right)\right\}\right)^2$$

---

[5]The factor of the total number of lattice sites N appears in the definition of δm since N is part of the integration measure and appears in the limit of $N\to\infty$.

To obtain δm_1 as given in Eq. 4.10.12, let $p = \pi/N \simeq 0$; then

$$\lim_{p \to 0} \Pi_1(p) = \frac{4N}{3N^3 g^2 T} \sum_{n,k} \frac{1}{f_k} \left(\sin\left(\frac{\pi n k}{N}\right) \sin\left(\frac{\pi k}{N}\right) \right)^2 + O\left(\frac{1}{N^2}\right)$$

$$= \frac{2N}{3N^2 g^2 T} \sum_k \frac{1}{f_k} \sin^2\left(\frac{\pi k}{N}\right)$$

Note

$$\sum_k \frac{1}{f_k} \sin^2\left(\frac{\pi k}{N}\right) = \frac{1}{2} \sum_k \cos^2\left(\frac{\pi k}{2N}\right) = \frac{1}{4} N$$

Hence

$$\delta m_1 = N\Pi_1(0) = N \frac{2N}{3N^2 g^2 T} \cdot \frac{1}{4} N = \frac{N}{6g^2 T} \qquad (4.10.13)$$

The Feynman diagram for $\Pi_2(p)$ is given in Figure 4.3 and has a combinatorial factor of 2; hence

$$\delta m_2 = N\Pi_2(0) = -\frac{N}{6g^2 T} \qquad (4.10.14)$$

The renormalized mass (as an expansion in $g^2 T$), from Eqs. 4.10.13 and 4.10.14

$$\delta m = \delta m_1 + \delta m_2 = 0 + 0(1/N)$$

Note the first term in Eq. 4.10.15 comes from the measure given in Eq. 4.10.8 and the second term arises from a single contraction of quartic term in the action 4.10.9. A nontrivial cancellation of the terms independent of N yields

$$\delta m \to 0 \quad \text{as} \quad N \to \infty \qquad (4.10.15)$$

Hence, as $N \to \infty$, the coefficient of the divergent mass-like term arising from the measure is exactly canceled by the term arising from a one-loop diagrams as given in Figure 4.3. This cancellation is generic for massless $S(N)$ degrees of freedom. The same generic cancellation of the quadratic mass-like term arising from the measure by the one-loop terms occurs for lattice gauge theory in Chapter 9.

By power counting, no other nonlinear terms contribute to δm since all of these vanish as $N \to \infty$, leaving an action which is only quadratic in ξ_n^a. Also, the range of ξ_n^a extends from the group manifold to the real line as $N \to \infty$.

To define the continuum limit let

$$\xi_t^a = \xi_n^a \; ; \quad -\infty \le \xi_t^a \le \infty \; ; \quad \xi_{t=0}^a = \xi_{t=T}^a = 0 \qquad (4.10.16)$$

and

$$\int_0^T dt f(t) = \lim_{N \to \infty} \frac{T}{N} \sum_n f_n$$

Define

$$\partial \xi_a \equiv \frac{\partial \xi_t^a}{\partial t} = \lim_{N \to \infty} (N/T) \delta \xi_n^a \qquad (4.10.17)$$

From Eqs. 4.10.9 and 4.10.15

$$A(\ell) = -\frac{N}{2g^2 T} \sum_n \sum_a (\delta \xi_n^a)^2 + \tilde{A}$$

Taking the limit $N \to \infty$ yields the final answer

$$A(\ell) = -\frac{1}{2g^2} \int_0^T dt \left(\partial \xi_a \partial \xi_a + \frac{C_{Iab}}{T} (b_I + 2\pi \ell_I) \xi_a \partial \xi_b \right)$$

$$-\frac{1}{2g^2 T} \sum_{I=1}^{N-1} (b_I + 2\pi \ell_I)^2 \qquad (4.10.18)$$

and yields the evolution kernel

$$K(T) = \prod_{l_I=1}^{N-1} \sum_{l_I=-\infty}^{+\infty} \prod_{t=0}^{T} \prod_{a=1}^{N^2-1} \int_{-\infty}^{\infty} d\xi^a(t) \exp[A(\ell)] \qquad (4.10.19)$$

Boundary condition : $\xi^a(0) = \xi^a(T) = 0$

Chapter 5

$SU(3)$ Character Functions

The character functions of $SU(3)$ are evaluated exactly from the evolution kernel and is shown to reproduce the first Weyl formula. The Weyl chamber for $SU(3)$ is evaluated and its connection to the character functions is shown.

The general result obtained in Eq. 4.8.47 for $SU(\mathcal{N})$ is shown to reduce to the earlier result derived for $SU(3)$. For $SU(3)$, Eq. 4.8.47 has $A_1 = A, A_2 = B$ and $A_3 = -(A+B)$; hence, it yields the following

$$K_3(\alpha) = \frac{1}{s(A,B)} \sum_{l=-\infty}^{+\infty} \sum_{m=-\infty}^{+\infty} [A(l) - B(m)][A(l) + 2B(m)][2A(l) + B(m)]$$

$$\times \exp\left\{ -\frac{2}{g^2 T}[A^2(l) + B^2(m) + A(l)B(m)] \right\}$$

where

$$A(l) = A + 2\pi l \qquad B(m) = B + 2\pi m$$

and

$$s(A,B) = 8 \sin\frac{1}{2}(A-B) \sin\frac{1}{2}(2A+B) \sin\frac{1}{2}(A+2B)$$

The result for $K_3(\alpha)$ and $s(A,B)$ obtained earlier in Eqs. 4.8.39 and 4.8.40 has been recovered from the general result for $SU(\mathcal{N})$, as expected.

5.1 Casimir Operator for $SU(3)$

The evolution kernel on $SU(N)$ comprises the matrix elements of $\exp(+\frac{1}{2}\alpha\nabla^2)$ where ∇^2 is the $SU(\mathcal{N})$ Laplace-Beltrami operator and α is a constant. The matrix elements of the pth irreducible representation are eigenfunctions of ∇^2.

Let U be an element of $SU(\mathcal{N})$. Then

$$-\nabla^2 \mathcal{D}_{ij}^{(p)}(U) = c_2(p)\mathcal{D}_{ij}^{(p)}(U) \tag{5.1.1}$$

where $c_2(p)$ is the value of the quadratic Casimir operator in the pth irreducible representation.

Let $|U\rangle$ be the coordinate eigenstate of the Hilbert space on $SU(\mathcal{N})$; then the conjugate eigenstate $\langle p, ij|U\rangle$ satisfies

$$\langle p, ij|U\rangle = \sqrt{d_p}\mathcal{D}_{ij}^{(p)}(U) \qquad 1 \le i, j \le d_p \tag{5.1.2}$$

where d_p is the dimension of $\mathcal{D}^{(p)}$. Hence, from Eqs. 5.1.1 and 5.1.2, the $SU(\mathcal{N})$ evolution kernel is given by

$$K_{\mathcal{N}}(\alpha) = \langle W \mid \exp\left(+\frac{1}{2}\alpha\nabla^2\right) \mid V\rangle \tag{5.1.3}$$

$$= \sum_p \exp\left[-\frac{1}{2}\alpha c_2(p)\right] d_p \chi_p(W^+ V) \tag{5.1.4}$$

where

$$\chi_p(U) = \mathrm{Tr}\mathcal{D}^{(p)}(U) \tag{5.1.5}$$

is the character function for the pth irreducible representation, and the sum in Eq. 5.1.4 is over all p. Eq. 5.1.4 is the key equation for identifying the character functions.

In the fundamental representation, let

$$U \equiv W^\dagger V = R \ \mathrm{diag}(e^{iA_1}, \ldots, e^{iA_N})R^\dagger \tag{5.1.6}$$

with R being an element of $SU(\mathcal{N})$ and

$$\sum_{i=1}^{\mathcal{N}} A_I = 0 \tag{5.1.7}$$

For $SU(3)$, Eq. 5.1.6 yields

$$U = R \ \mathrm{diag}(e^{iA}, e^{iB}, e^{-i(A+B)})R^\dagger \tag{5.1.8}$$

The irreducible representations of $SU(3)$ are labeled by two positive integers (p, q); choose the convention that $(1, 1)$ is the one-dimensional (trivial) representation, the fundamental three-dimensional representation and its complex conjugage is **3** are $(2, 1)$ and **3*** is $(1,2)$, respectively. For $SU(3)$ [Zelobenko (1973)]

$$d_{p,q} = \frac{1}{2}pq(p+q) \quad : p, q = 1, 2, \ldots, \infty$$

$$c_2(p,q) \equiv c(p,q) = \frac{1}{3}(p^2 + q^2 + pq) - 1 \tag{5.1.9}$$

and

$$\chi_{p,q}^*(A, B) = \chi_{q,p}(A, B) \tag{5.1.10}$$

$$\chi_{p,q}(A, B) = -\chi_{p,q}(B, A) \tag{5.1.11}$$

Hence, Eqs. 5.1.4, 5.1.8 and 5.1.11, yield, up to an overall constant

$$K_3 = \sum_{p,q=1}^{\infty} \frac{1}{2}pq(p+q)\frac{1}{2}\left[\chi_{p,q}(A, B) + \chi_{p,q}^*(A, B)\right] e^{-\frac{1}{6}\alpha(p^2+q^2+pq)} \tag{5.1.12}$$

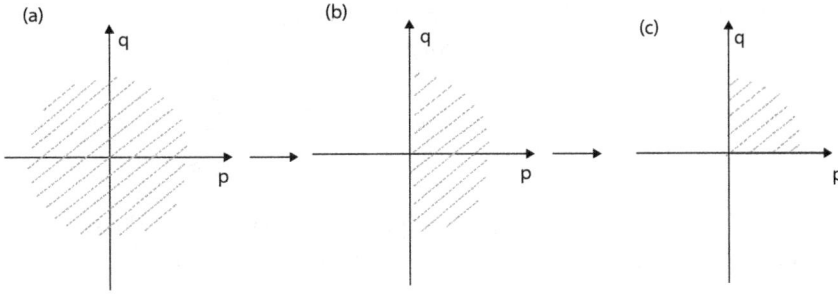

Fig. 5.1   Symmetry of the summation of the character functions of $SU(3)$.

5.2   Evolution Kernel for $SU(3)$

Eq. 4.8.43, for $g^2 T = \alpha$, yields

$$K_3(T) = \frac{i}{s(A,B)} \sum_{p,q=-\infty}^{+\infty} e^{ipA-ipB} pq(p+q) e^{-\frac{\alpha}{6}(p^2+q^2+pq)} \tag{5.2.1}$$

with

$$s(A,B) = 8\sin\frac{1}{2}(A-B)\sin\frac{1}{2}(2A+B)\sin\frac{1}{2}(A+2B)$$

Eq. 5.1.9 yields the following simplification to Eq. 5.2.1

$$K_3(\alpha) = \sum_{p,q=-\infty}^{+\infty} d_{p,q} \exp\left[-\frac{1}{2}\alpha c(p,q)\right]\Omega_{p,q}(A,B) \tag{5.2.2}$$

with

$$\Omega_{p,q}(A,B) = \frac{1}{s(A,B)}[\sin(pA-qB)+\sin(qA-pB)] \tag{5.2.3}$$

Note that the p, q summations in Eq. 5.2.2, as shown in Figure 5.1(a), are over four quadrants in the pq plane, whereas the p, q summations in Eq. 5.1.12, as shown in Figure 5.1(c), are only over the first quadrant. The main obstacle in converting Eq. 5.2.2 into Eq. 5.1.12 is that in Eq. 5.1.12, the summation is only over integers p, q such that $p, q > 0$, whereas in Eq. 5.2.2 the sum in p, q is over **all** integers.

The sum in Eq. 5.2.2 needs to *restricted* to the first quadrant, as shown in Figure 5.1. The reduction of the four-quadrant summation to the first quadrant will be possible due to certain symmetries of $d_{p,q}$ and $c(p,q)$. Note that

$$d_{p,q} = d_{q,p} = -d_{-p,-q} \tag{5.2.4}$$

and

$$c(p,q) = c(q,p) = c(-p,-q) \tag{5.2.5}$$

Using Eqs. 5.2.4 and 5.2.5 reduces Eq. 5.2.2 to a sum over the first and fourth quadrants as shown in Figure 5.1(b). Hence

$$K_3(\alpha) = \sum_{p=1}^{\infty}\left(\sum_{q=1}^{\infty} + \sum_{q=-\infty}^{-1}\right) d_{p,q}\Omega_{p,q}\exp\left[-\frac{1}{2}\alpha c(p,q)\right] \tag{5.2.6}$$

$$\equiv K_3^{(1)} + K_3^{(2)} \tag{5.2.7}$$

The first sum in Eq. 5.2.6 is over the first quadrant in (p,q), but the second sum is over the fourth quadrant and has to be further reduced. Hence

$$K_3^{(2)} = \sum_{p=1}^{\infty}\sum_{q=-\infty}^{-1} d_{p,q}\exp\left[-\frac{1}{2}\alpha c(p,q)\right]\Omega_{p,q}$$

$$= \sum_{p=1}^{\infty}\sum_{q=1}^{\infty} d_{p,-q}\exp\left[-\frac{1}{2}\alpha c(p,-q)\right]\Omega_{p,-q} \tag{5.2.8}$$

Using the fact that

$$c(p,-q) = \begin{cases} c(p-q,q), & p > q \\ c(p,q-p), & p < q \end{cases} \tag{5.2.9}$$

Eq. 5.2.8 yields

$$K_3^{(2)} = \sum_{p=1}^{\infty}\left(\sum_{q=1}^{p-1} + \sum_{q=p+1}^{\infty}\right) d_{p,-q}\exp\left[-\frac{1}{2}\alpha c(p,-q)\right]\Omega_{p,-q}$$

$$= \sum_{q=1}^{\infty}\sum_{p=q+1}^{\infty} d_{p,-q}\exp\left[-\frac{1}{2}\alpha c(p-q,q)\right]\Omega_{p,-q}$$

$$+ \sum_{p=1}^{\infty}\sum_{q=p+1}^{\infty} d_{p,-q}\exp\left[-\frac{1}{2}\alpha c(p,q-p)\right]\Omega_{p,-q} \tag{5.2.10}$$

Rearranging the summations in 5.2.10, as shown in Figure 5.1(c), yields

$$K_3^{(2)} = \sum_{p=1}^{\infty}\sum_{q=1}^{\infty} d_{p,q}\exp\left[-\frac{1}{2}\alpha c(p,q)\right][-\Omega_{p+q,-q} + \Omega_{p,-(p+q)}] \tag{5.2.11}$$

Eqs. 5.2.7, 5.2.8 and 5.2.11 yield

$$K_3(\alpha) = \sum_{p,q=1}^{\infty} d_{p,q}\exp\left[-\frac{1}{2}\alpha c(p,q)\right][\Omega_{p,q} - \Omega_{p+q,-q} + \Omega_{p,-(p+q)}] \tag{5.2.12}$$

5.3   Character Functions of $SU(3)$

Comparing Eqs. 5.1.12 and 5.2.12, the $SU(3)$ character functions are given by

$$\chi_{p,q}(A,B) + \chi_{p,q}^*(A,B) = \Omega_{p,q} - \Omega_{p+q,-q} + \Omega_{p,-(p+q)}$$

$$\Rightarrow \chi_{p,q}(A,B) = -\frac{i}{s(A,B)}\Big[\exp(ipA - iqB) - \exp(-iqA + ipB)$$

$$+ \exp[-ip(A+B)]\big\{\exp(-iqA) - \exp(-iqB)\big\}$$

$$+ \exp[iq(A+B)]\big\{\exp(ipB) - \exp(ipA)\big\}\Big] : \quad p,q = 1,2,\cdots\infty \tag{5.3.1}$$

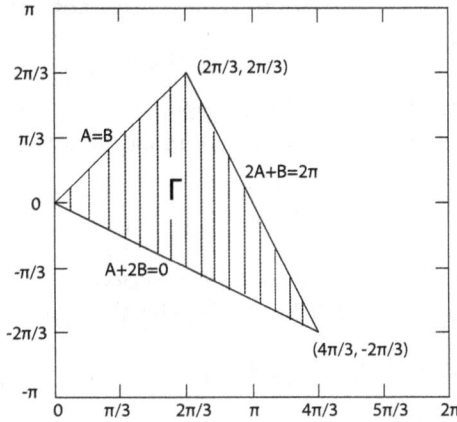

Fig. 5.2 Domain of the Weyl chamber Γ, indicated by the shaded portion, in the torus spanned by (A, B). The three boundaries are given by $A = B$, $A + 2B = 0$ and $2A + B = 2\pi$.

where, from Eq. 4.8.40

$$s(A, B) = 8 \sin \frac{1}{2}(A - B) \sin \frac{1}{2}(A + 2B) \sin \frac{1}{2}(2A + B)$$

Note that $\chi_{p,q}$ satisfies Eqs. 5.1.10 and 5.1.11; the **3** (fundamental) representation, as is expected from Eq. 5.1.8, yields

$$\chi_{2,1}(A, B) = e^{iA} + e^{iB} + e^{-i(A+B)} = \chi_{1,2}^*(A, B)$$

$\chi_{p,q}$ yields the non-trivial identity

$$\chi_{p,q}(0, 0) = \frac{1}{2}pq(p + q) = d_{p,q}$$

The expression for $\chi_{p,q}$ in Eq. 5.3.1 is a ratio of two determinants and is the first Weyl formula for the character functions. Weyl's derivation uses the integral properties of the group. A purely algebraic derivation of $\chi_{p,q}$ can be given using the Lie algebra [Zelobenko (1973)]. The derivation given here is independent of these two derivations.

The invariant measure on $SU(3)$ for trace class functions is given by

$$dU = s^2(A, B)dAdB \qquad -\pi \le A, B \le \pi \qquad (5.3.2)$$

where U is an element of $SU(3)$. This measure yields the expected orthonormality theorem given in Eq. 2.8.6

$$\int_{-\pi}^{+\pi} \frac{dAdB}{(2\pi)^2} s^2(A, B)\chi_{(p,q)}^*(A, B)\chi_{(p',q')}(A, B) = \delta_{pp'}\delta_{qq'} \qquad (5.3.3)$$

Further simplification of Eq. 5.3.3 can be made. Since $\chi_{p,q}$ depends only on the trace of U, it is invariant under the Weyl group W; this discrete and finite subgroup of $SU(\mathcal{N})$ consists of reflections of root vectors in the root space, and for $SU(3)$ consists of six elements. For an element $T \in W$

$$W : U \to T^{-1}UT \qquad (5.3.4)$$

The Weyl chamber in the $SU(3)$ group space consists of points A, B such that

$$s(A, B) \geq 0 \tag{5.3.5}$$

The two-dimensional torus $U(1) \otimes U(1)$ – spanned by A and B – is split into six disjoint chambers, each a reflection of the compact domain Γ defined from Eq. 5.3.5 by (see Figure 5.2)

$$\Gamma : A \geq B \quad 0 \leq A + 2B \leq 2\pi \quad 0 \leq 2A + B \leq 2\pi \tag{5.3.6}$$

The orbit of the Weyl chamber Γ under the action of the Weyl group is the total space $U(1) \otimes U(1)$ [Zelobenko (1973)]. Using the Weyl group yields

$$\frac{1}{A_\Gamma} \int_\Gamma \mathrm{d}A \mathrm{d}B s^2(A, B) \chi^*_{(p,q)}(A, B) \chi_{(p,q)}(A, B) = \delta_{pp'} \delta_{qq'} \tag{5.3.7}$$

$$A_\Gamma = \text{area of } \Gamma = \frac{4}{6}\pi^2 \tag{5.3.8}$$

The form Eq. 5.3.7 with integration over A_Γ, which is one sixth the area of the torus, is more suitable for numerical calculations. Also, to prove Eq. 5.3.7 directly without using the Weyl group is cumbersome and complicated.

5.4   Summary

The $SU(3)$ character functions were derived from the $SU(3)$ evolution kernel, a derivation independent from Weyl's classic derivation. Certain non-trivial symmetries of the dimensionality and Casimir function of $SU(3)$ were central to the derivation.

Both $SU(2)$ and $SU(3)$ require only two invariants, namely $c_2(p)$ and d_p for obtaining the character functions, and hence the evolution kernel can yield the character functions for these two groups. It is not possible to derive the $SU(\mathcal{N})$ character functions from the $SU(\mathcal{N})$ evolution kernel, since one needs $\mathcal{N} - 1$ algebraic invariants for it, whereas the evolution kernel uses only two of Lie algebra invariants.

Chapter 6

Fermion Calculus

A quantitative description of a quantum entity is provided by the degree of freedom, which is an indeterminate quantity – taking a range of possible values [Baaquie (2013)]. The entire edifice of quantum mechanics and quantum field theory is built on the *degree of freedom*.

The degrees of freedom come in many varieties, and can be either a discrete or continuous real (or complex) variable. Let a, b be two degrees of freedom; then they *commute* under multiplication, in the sense that, similar to two numbers, a, b satisfy $ab = ba$. Commuting variables are generically called *bosonic variables*, or bosonic degrees of freedom. Typical of the bosonic case are the degrees of freedom for a collection of quantum mechanical particles.

Unlike bosonic variables, *fermionic variables*, also called fermionic degrees of freedom, *anticommute*; namely, if ζ, η are two fermionic variables, then $\zeta\eta = -\eta\zeta$.

Interactions of fundamental particles are generally mediated by bosonic fields such as the Maxwell electromagnetic field, whereas mass is usually carried by particles that are fermions, the most familiar being the electron.

The following are two key features that distinguish fermionic from bosonic variables [Baaquie (2014)].

- Fermions obey the Pauli exclusion principle, which entails that no two fermions can occupy the same quantum state. This is the reason the concept of intensity does not apply to a fermion. A high intensity electric field is a reflection of the presence of large number of photons, which are bosons, in the same quantum state; for photons, any number of photons can be in the same quantum state. In contrast, a fermionic quantum state can either have only one fermion (electron) or no fermion; in particular, a fermion exists at a point or there is no fermion there.
- The state function of a multi-bosonic system is totally symmetric in that the exchange of any two bosonic degrees of freedom yields the same state function. In contrast, a multi-fermion system is totally anti-symmetric: the exchange of any two fermion degrees of freedom gives the same state – *but* with a *negative sign*.

The Pauli exclusion principle implies a *discrete* nature for fermions since a fermion degree of freedom has only two possibilities, either occupying a state or not occupying a state. In discussing the fermion Hilbert space in Section 6.3, it will be seen that the Hilbert space of a single fermion is identical (isomorphic) to the Hilbert space of a *discrete degree of freedom* that takes only two values.

Both the key features of fermions, namely obeying the Pauli exclusion principle and the state function being anti-symmetric, can be mathematically realized by introducing a new type of variable, namely fermionic variables; similar to the bosonic case fermionic degrees of freedom can be described by either real or complex fermionic variables.

Fermionic variables are defined in Section 6.1 and fermion integration is discussed in Section 6.2. The fermionic Hilbert space as well as its dual space is defined in Section 6.3. The concept the antifermionic Hilbert space is discussed in Section 6.4.

Gaussian integration for real and complex fermions is discussed in Section 6.6. The fermionic path integral is defined for the particle and anti-particle system in Section 6.7, and the transition probability amplitude is obtained in Section 6.12. A simple one-dimensional toy model is constructed in Section 6.13 to show how quark confinement can arise by coupling the fermions to a gauge field.

The discussion on fermion calculus generalizes the one given in [Baaquie (2014)] – by including a metric on the fermionic state space. This metric is required for defining the lattice Dirac field, and is discussed in Chapter 12.

6.1   Fermionic Variables

The defining property of fermions is the Pauli exclusion principle, namely that at most only one fermion can occupy a quantum state. In other words, for fermions, a state either has no fermions, or at most one fermion.

Let $|0\rangle_F$ be the fermion vacuum state, and let $a_F^\dagger$ be the fermion creation operator. Then

$$|0\rangle_F : \text{ground state; no fermions}$$
$$a_F^\dagger|0\rangle_F : \text{one fermion}$$
$$(a^\dagger)_F^2|0\rangle = 0 : \text{null state} \tag{6.1.1}$$

The second defining property of fermions is that two different fermions must give an anti-symmetric state function on being exchanged; hence two distinct fermionic creation operators, represented by say $a_1^\dagger, a_2^\dagger$, must satisfy the following anti-symmetry

$$a_1^\dagger a_2^\dagger|0\rangle = -a_2^\dagger a_1^\dagger|0\rangle$$

and which is realized by imposing the following anti-commutation relation

$$a_1^\dagger a_2^\dagger = -a_2^\dagger a_1^\dagger \quad \Rightarrow \quad \{a_1^\dagger, a_2^\dagger\} = 0$$

where the anti-commutator is defined for any two quantities A, B by

$$\{A, B\} \equiv AB + BA \tag{6.1.2}$$

Instead of working with fermionic creation and annihilation operators acting on the ground state $|0\rangle$, one can instead describe fermions using a calculus distinct from the calculus based on real numbers, which is used for describing bosons.

An independent and self-contained formalism for realizing all the defining properties of fermions is provided by a set of *anti-commuting fermionic variables* $\psi_1, \psi_2, \ldots \psi_N$ and its conjugate $\bar{\psi}_1, \bar{\psi}_2, \ldots \bar{\psi}_N$ defined by the following properties

$$\{\psi_i, \psi_j\} = -\{\psi_i, \psi_j\}$$
$$\{\bar{\psi}_i, \psi_j\} = -\{\psi_j, \bar{\psi}_i\}$$
$$\{\bar{\psi}_i, \bar{\psi}_j\} = -\{\bar{\psi}_j, \bar{\psi}_i\}$$

Hence, it follows that

$$\psi_i^2 = 0 = \bar{\psi}_i^2$$

Fermionic differentiation is defined by

$$\frac{\delta}{\delta \psi_i} \psi_j = \delta_{i-j} \quad ; \quad \frac{\delta}{\delta \psi_i} \bar{\psi}_j = 0$$

and

$$\frac{\delta^2}{\delta \psi_i \delta \psi_j} = -\frac{\delta^2}{\delta \psi_j \delta \psi_i} \quad \Rightarrow \quad \frac{\delta^2}{\delta \psi_i^2} = 0 = \frac{\delta^2}{\delta \bar{\psi}_i^2}$$

Similarly all the fermionic derivative operators $\delta/\delta \psi_i$, $\delta/\delta \bar{\psi}_i$ anti-commute.

6.2   Fermion Integration

Similar to the case of

$$\int_{-\infty}^{+\infty} dx\, f(x)$$

which is invariant under $x \to x + a$, that is

$$\int_{-\infty}^{+\infty} dx\, f(x) = \int_{-\infty}^{+\infty} dx\, f(x + a),$$

define fermion integration by the following

$$\int d\bar{\psi}\, f(\bar{\psi}) = \int d\bar{\psi}\, f(\bar{\psi} + \bar{\eta}) \tag{6.2.1}$$

Since $\bar{\psi}^2 = 0$, Taylors expansion shows that the most general function of the variable $\bar{\psi}$ is given by

$$\bar{\psi}^2 = 0 \quad \Rightarrow \quad f(\bar{\psi}) = a + b\bar{\psi}$$

It follows that rules of fermion integration that yield Eq. 6.2.1 are given by the following

$$\int d\bar{\psi} = 0 = \int d\psi$$

$$\int d\bar{\psi}\bar{\psi} = 1 = \int d\psi\psi \qquad (6.2.2)$$

$$\int d\bar{\psi}d\psi\psi\bar{\psi} = 1 = -\int d\bar{\psi}d\psi\bar{\psi}\psi$$

For N fermionic variables ψ_i, with $i = 1, 2, ...N$, one has the generalization

$$\left[\prod_{n=1}^{N} \int d\psi_n\right] \psi_{i_1}\psi_{i_2}...\psi_{i_n} = \epsilon_{i_1,i_2,...i_n} \qquad (6.2.3)$$

where $\epsilon_{i_1,i_2,...i_n}$ is the completely antisymmetric epsilon tensor.

Consider a change of variables for a single variable, namely

$$\psi = a\chi + \zeta$$

where a is a constant and ζ is a constant fermion. From Eq. 6.2.2, the non-zero fermion integral yields

$$1 = \int d\psi\psi = \int d\psi(a\chi + \zeta) = \int d\psi a\chi \quad \Rightarrow \quad d\psi = \frac{1}{a}d\chi \qquad (6.2.4)$$

Note this is the inverse for the case of real variables, since $x = ay$ yields $dx = ady$.

For the case of N fermions, the anti-symmetric matrix $M_{ij} = -M_{ji}$ yields the following change of variables

$$\psi_i = \sum_{j=1}^{N} M_{ij}\chi_j \quad \Rightarrow \quad \psi = M\chi$$

Similar to Eq. 6.2.4, it follows that

$$\prod_{i=1}^{N} d\psi_i = \frac{1}{\det M}\prod_{j=1}^{N} d\chi_j \quad \Rightarrow \quad D\psi = \frac{1}{\det M}D\chi \qquad (6.2.5)$$

where $D\psi = \prod_{i=1}^{N} d\psi_i$ and so on.

6.3   Fermion Hilbert Space

There are two distinct fermionic variables, namely the variable ψ and its *dual* $\bar{\psi}$. Comparing with the bosonic case, if one takes the fermionic variable $\bar{\psi}$ to be the analog of the coordinate x, then ψ is the analog of the conjugate momentum variable p. The two different state spaces $V_{\bar{\psi}}$ and V_ψ are based on the coordinate variable $\bar{\psi}$ and its conjugate ψ, respectively. There are consequently *two* Hilbert spaces, namely $V_{\bar{\psi}}$ and V_ψ, that are dual to each other, as shown in Figure 6.1.

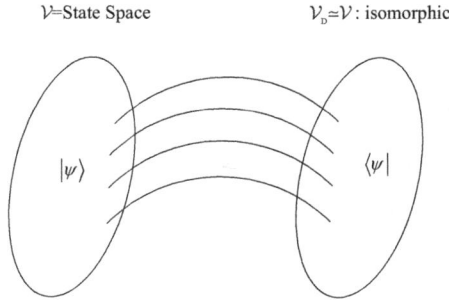

$\mathcal{V}$=State Space $\mathcal{V}_{\mathrm{D}} \simeq \mathcal{V}$: isomorphic

$|\psi\rangle$ $\langle\psi|$

Fig. 6.1   The Hilbert state space V and its dual V_D.

In analogy with bosonic variables x for which $\langle\phi|\phi\rangle = \int dx \phi^*(x)\phi(x) \geq 0$, the norm for the fermionic Hilbert space needs to be defined so as to yield a positive norm fermionic Hilbert space.

Choose $V_{\bar\psi}$ to be the state space of the fermionic degree of freedom $\bar\psi$. A fermion state function is then given by scalar product of the dual coordinate state vector $\langle\bar\psi| \in V_\psi$ with $|f\rangle \in V_{\bar\psi}$ and yields

$$f(\bar\psi) = \langle\bar\psi|f\rangle = a + b\bar\psi \qquad (6.3.1)$$

The dual state of $|f\rangle$, denoted by $\langle f|$, and the metric on fermionic Hilbert space is defined such that

$$\langle f|f\rangle = r|a|^2 + |b|^2 > 0 \ ; \ r > 0 \qquad (6.3.2)$$

The *real number* $r > 0$ is a metric on the fermion state space. Physical (normalizable) state functions have $\langle f|f\rangle = 1$ and yield the following interpretation

$$\langle f|f\rangle = 1 \ \Rightarrow \ r|a|^2 + |b|^2 = 1$$
$$r|a|^2 = \text{probability, there is no fermion}$$
$$|b|^2 = \text{probability, there is one fermion}$$

Since a, b are complex numbers, the space of all physical state functions is equal to a three-dimensional sphere S^3. Note a fermion state function, similar to the boson case, is equivalent to all state functions related to it by a global phase $e^{i\phi}$. Factoring out the phase from the physically distinct state functions yields the following Hilbert space [Baaquie (2013)]

$$V_{\bar\psi} \equiv S^3/S^1 = S^2 : \text{Bloch sphere}$$

Hence, similar to a spin $1/2$ system, the distinct physical states of a single fermion Hilbert space are parametrized by the points of a two-dimensional sphere. Or more formally, each state vector of the single fermion space of states corresponds to one point on the two-dimensional sphere.

The single fermion state space is seen to be isomorphic to the state space of the Ising spin discussed in Baaquie (2014) and shows that, in essence, a fermion is a *discrete* degree of freedom.

Consider the following metric on state space ($r > 0$)

$$\exp\{-r\bar{\psi}\psi\} \tag{6.3.3}$$

Define the dual state by

$$\bar{\psi} \to \psi \quad \Rightarrow \quad f^D(\psi) = \langle f|\psi\rangle = a^* + b^*\psi$$

To achieve the required scalar product, note that

$$
\begin{aligned}
\langle f|f\rangle &= \int d\bar{\psi}d\psi f^D(\psi)e^{-r\bar{\psi}\psi}f(\psi) \\
&= \int d\bar{\psi}d\psi(a^* + b^*\psi)e^{-r\bar{\psi}\psi}(a + b\bar{\psi}) \\
&= \int d\bar{\psi}d\psi\Big(|a|^2 + |b|^2\psi\bar{\psi} + a^*b\bar{\psi} + ab^*\psi\Big)(1 - r\bar{\psi}\psi)
\end{aligned}
$$

Using the rules for fermion integration given in Eq. 6.2.2, the positive definite scalar product is given by

$$\langle f|f\rangle = r|a|^2 + |b|^2$$

6.3.1   *Fermionic completeness equation*

The fermionic variables have a phase space representation with $(\bar{\psi}, \psi)$ being analogous to the coherent state representation of the creation and destruction operators $(a^\dagger, a)$. Let $|\psi\rangle$ be the fermionic eigenstate and $\langle\bar{\psi}|$ be the dual fermionic eigenstate. The completeness equation for the fermion degree of freedom is given by

$$\mathbb{I} = \int d\bar{\psi}d\psi|\psi\rangle e^{-r\bar{\psi}\psi}\langle\bar{\psi}| \quad ; \quad r > 0 \tag{6.3.4}$$

The fermion completeness equation is similar to the completeness equation for coherent states. The fermion basis states are overcomplete – as is the case for the bosonic coherent basis states – and the metric on the fermion state space that accounts for the overcompleteness is given by $\exp\{-r\bar{\psi}\psi\}$.

The inner product of the basis with its dual is given by the self-consistency equation that follows from the completeness equation and yields

$$\langle\bar{\psi}|\psi\rangle = \frac{1}{r}e^{r\bar{\psi}\psi} \tag{6.3.5}$$

To verify that Eq. 6.3.5 is indeed consistent with the completeness equation given in Eq. 6.3.4, let $\bar{\zeta}$, ζ be fermionic variables; then, from Eq. 6.3.5

$$
\langle\bar{\zeta}|\zeta\rangle = \int d\bar{\psi}d\psi\langle\bar{\zeta}|\psi\rangle e^{-r\bar{\psi}\psi}\langle\bar{\psi}|\zeta\rangle = \frac{1}{r^2}\int d\bar{\psi}d\psi\left(1 + r\bar{\zeta}\psi\right)\left(1 - r\bar{\psi}\psi\right)\left(1 + r\bar{\psi}\zeta\right)
$$

$$
= \frac{1}{r^2}\int d\bar{\psi}d\psi\left(r^2\bar{\zeta}\psi\bar{\psi}\zeta - r\bar{\psi}\psi\right) = \frac{1}{r}\int d\bar{\psi}d\psi\left(1 + r\bar{\zeta}\zeta\right)\psi\bar{\psi} = \frac{1}{r}e^{r\bar{\zeta}\zeta}
$$

as expected.

To verify that Eq. 6.3.4, consider the following

$$\langle \bar{\psi} | \mathbb{I} | f \rangle = \int d\bar{\chi} d\chi \langle \bar{\psi} | \chi \rangle e^{-r\bar{\chi}\chi} \langle \bar{\chi} | f \rangle = \frac{1}{r^2} \int d\bar{\chi} d\chi e^{r\bar{\psi}\chi} e^{-r\bar{\chi}\chi} (a + b\bar{\chi})$$

$$= \frac{1}{r} \int d\bar{\chi} d\chi (1 + r\bar{\psi}\chi)(1 - r\bar{\chi}\chi)(a + b\bar{\chi})$$

$$= \frac{1}{r} \int d\bar{\chi} d\chi (-ar\bar{\chi}\chi + br\bar{\psi}\chi\bar{\chi}) = a + b\bar{\psi}$$

$$\Rightarrow \quad \langle \bar{\psi} | \mathbb{I} | f \rangle = \langle \bar{\psi} | f \rangle \tag{6.3.6}$$

Hence, as expected, the component expression given in Eq. 6.3.1 has been recovered.

The inner product of the basis states and completeness are self-consistently determined since one requires the other: to prove completeness one needs the inner product. In particular, another proof of the consistency of the resolution of the identity as given in Eq. 6.3.4 is the following.

$$\mathbb{I}^2 = \int d\bar{\psi} d\psi d\bar{\zeta} d\zeta |\psi\rangle e^{-r\bar{\zeta}\psi} \langle \bar{\zeta} | \zeta \rangle e^{-r\bar{\psi}\zeta} \langle \bar{\psi} |$$

$$= \int d\bar{\psi} d\psi d\bar{\zeta} d\zeta |\psi\rangle e^{-r\bar{\zeta}\psi} \frac{1}{r} e^{r\bar{\zeta}\zeta} e^{-r\bar{\psi}\zeta} \langle \bar{\psi} |$$

$$= \int d\bar{\psi} d\psi |\psi\rangle e^{-r\bar{\psi}\psi} \langle \bar{\psi} | = \mathbb{I}$$

where the result follows from the inner product given in Eq. 6.3.5 and performing the $\bar{\zeta}, \zeta$ integrations.

6.3.2 *Fermionic momentum operator*

The state space depends on $\bar{\psi}$ so we need to determine what is the representation of the dual coordinate ψ on the state space. The fermion coordinate operator $\hat{\psi}$ has the fermionic coordinate eigenstate given by

$$\hat{\psi} | \psi \rangle = \psi | \psi \rangle$$

where the coordinate eigenvalue ψ is fermionic. The scalar product yields

$$\langle \bar{\psi} | \hat{\psi} | \psi \rangle = \psi \langle \bar{\psi} | \psi \rangle = \psi e^{r\bar{\psi}\psi} = \frac{1}{r} \frac{\delta}{\delta\bar{\psi}} (e^{\bar{\psi}\psi})$$

$$\Rightarrow \psi \langle \bar{\psi} | \psi \rangle = \frac{1}{r} \frac{\delta}{\delta\bar{\psi}} e^{\bar{\psi}\psi} \quad \Rightarrow \quad \psi = \frac{1}{r} \frac{\delta}{\delta\bar{\psi}}$$

Hence, on state space $V_{\bar{\psi}}$ the dual coordinate ψ yields

$$\psi f(\bar{\psi}) = \frac{1}{r} \frac{\delta}{\delta\bar{\psi}} f(\bar{\psi}) \tag{6.3.7}$$

Hence, as mentioned in Section 6.3, the variable ψ is the analog of the momentum operator p and is evidenced by its action on state space, as given in Eq. 6.3.7.

In terms of the fermionic variables the creation and annihilation operators have the following realization

$$a^\dagger = r\bar{\psi}, \quad a = \frac{1}{r} \frac{\delta}{\delta\bar{\psi}} \quad \text{with} \quad \{a^\dagger, a\} = 1$$

The identification of the fermion variables with the fermion creation destruction operators is consistent with the identification made earlier, in the discussion on Eq. 6.3.4, of the fermion completeness equation with the completeness equation for the coherent state space.

6.4 Antifermion State Space

Let χ, $\bar{\chi}$ be a set of fermionic variables; the dual coordinate eigenstate is defined by $\langle\chi| \in V_\chi$; hence

$$\langle\chi|f\rangle = a + b\chi$$

This definition for the coordinate of the state space will lead to the conclusion that V_χ is the state space for antifermions. The completeness equation *continues* to be

$$\mathbb{I} = \int d\bar{\chi}d\chi |\bar{\chi}\rangle e^{-r\bar{\chi}\chi}\langle\chi| \tag{6.4.1}$$

Consistency with the completeness equation for the antifermionic variables requires that the inner product *continues* to be the same as the particle case, namely

$$\langle\chi|\bar{\chi}\rangle = \frac{1}{r}\exp\{r\bar{\chi}\chi\} \tag{6.4.2}$$

The following derivation provides a consistency check for the above completeness equation

$$\langle\eta|\mathbb{I}|\bar{\eta}\rangle = \frac{1}{r^2}\int d\bar{\chi}d\chi e^{r\bar{\chi}\eta}e^{-r\bar{\chi}\chi}e^{r\bar{\eta}\chi}$$

$$= \frac{1}{r^2}\int d\bar{\chi}d\chi \left(1 + r\bar{\chi}\eta\right)\left(1 - r\bar{\chi}\chi\right)\left(1 + r\bar{\eta}\chi\right)$$

$$= \frac{1}{r}\int d\bar{\chi}d\chi \left(-\bar{\chi}\chi + r\bar{\chi}\eta\bar{\eta}\chi\right) = \frac{1}{r}\int d\bar{\chi}d\chi \left(-\bar{\chi}\chi\right)\left(1 + \bar{\eta}\eta\right)$$

$$= \frac{1}{r}\exp\{r\bar{\eta}\eta\}$$

In the completeness equation given in Eq. 6.4.1, the variables $\chi, \bar{\chi}$ have been interchanged for the state space vectors – as compared to the completeness equation for variables $\psi, \bar{\psi}$, as in Eq. 6.3.4 – but with the metric being unchanged, namely $\exp\{-r\bar{\chi}\chi\}$ and $\exp\{-r\bar{\psi}\psi\}$.

To compensate for the difference in the basis states for these two cases, an extra minus sign needs to be introduced in the conjugation of the state space vector $f(\chi)$. Conjugation is defined by[1]

$$\chi \to \chi^c = -\bar{\chi} \quad \Rightarrow \quad f^D(\chi) = f^*(\chi^c) = f^*(-\bar{\chi}) \equiv a^* - b^*\bar{\chi}$$

---

[1] This is because the order of integration in the scalar product is *reversed* compared to the fermion case, for which, under conjugation $\psi \to \bar{\psi}$.

To verify that the rule for conjugation yields a positive definite norm for the state space V_χ, consider the following

$$\langle f|f\rangle = \int d\bar{\chi}d\chi f^D(\chi)e^{-r\bar{\chi}\chi}f(\chi)$$

$$= \int d\bar{\chi}d\chi(a^* - b^*\bar{\chi})(a+b\chi)(1-r\bar{\chi}\chi)$$

$$= \int d\bar{\chi}d\chi(r|a|^2 + |b|^2)\chi\bar{\chi} = r|a|^2 + |b|^2 > 0$$

The fermion momentum operator is defined by

$$\langle\chi|\bar{\chi}|\bar{\chi}\rangle = \bar{\chi}e^{r\bar{\chi}\chi} = -\frac{1}{r}\frac{\delta}{\delta\chi}e^{\bar{\chi}\chi}$$

and yields

$$\bar{\chi}f(\chi) = -\frac{1}{r}\frac{\delta}{\delta\chi}f(\chi)$$

The fact that anticommuting variables $\chi, \bar{\chi}$ represent antifermions becomes clear when it is combined with fermions, as is done in the following Section.

6.5 Fermion and Antifermion Hilbert Space

One can choose either of the Hilbert spaces $V_{\bar{\psi}}$ or V_χ to be fermionic state space; once a choice is made for the fermion state space, it automatically allows for the introduction of the concept of antifermions. The representation of antifermions is fixed by the choice that is made for the fermion state space. The normal convention is to choose $V_{\bar{\psi}}$ to be the fermionic state space and $\bar{\psi}$ the fermionic degree of freedom and V_χ to be the antifermionic state space with χ being the anitfermionic degree of freedom.

The state space and path integral for the fermion and antifermion system is discussed. A system containing both particle and anti-particle has a state space given by the tensor product of the fermion and antifermion state spaces, namely $V_{\bar{\psi}} \otimes V_\chi$.

The most general state vector is given by

$$f(\bar{\psi}, \chi) = \langle\bar{\psi}, \chi|f\rangle = a + b\bar{\psi} + c\chi + d\bar{\psi}\chi \tag{6.5.1}$$

Hermitian conjugation for the fermion and antifermion state space is defined by the following operations.

(1) Complex conjugate all the coefficients.
(2) Reverse the order of all the fermion variables.
(3) The fermion variables are conjugated.

$$\begin{pmatrix}\psi\\\chi\end{pmatrix} \to \begin{pmatrix}\psi^c\\\chi^c\end{pmatrix} = \begin{pmatrix}1 & 0\\0 & -1\end{pmatrix}\begin{pmatrix}\bar{\psi}\\\bar{\chi}\end{pmatrix} \tag{6.5.2}$$

The completeness for the fermion-antifermion degreess of freedom is

$$\mathbb{I} = \int d\bar{\psi}d\psi d\bar{\chi}d\chi |\psi, \bar{\chi}\rangle e^{-r\bar{\psi}\psi - r\bar{\chi}\chi} \langle \bar{\psi}, \chi|$$

Consider the state vector $|f\rangle$ given in Eq. 6.5.1; its dual state vector, using the rules for fermion and antifermion conjugation, is given by

$$f^D(\bar{\psi}, \chi) = (\langle \bar{\psi}, \chi|f\rangle)^D$$
$$= f^*(\bar{\psi}^c, \chi^c) = \langle f|\psi, -\bar{\chi}\rangle = a^* + b^*\psi - c^*\bar{\chi} - d^*\bar{\chi}\psi \qquad (6.5.3)$$

The rule for conjugation yields the following positive definite inner product

$$\langle f|f\rangle = \int d\bar{\psi}d\psi d\bar{\chi}d\chi \, \langle f|\psi, \bar{\chi}\rangle \, e^{-r\bar{\psi}\psi - r\bar{\chi}\chi} \, \langle \bar{\psi}, \chi|f\rangle$$
$$= \int d\bar{\psi}d\psi d\bar{\chi}d\chi \, (a^* + b^*\psi - c^*\bar{\chi} - d^*\bar{\chi}\psi) \left(1 - r\bar{\psi}\psi - r\bar{\chi}\chi + r^2\bar{\psi}\psi\bar{\chi}\chi\right)$$
$$\times \left(a + b\bar{\psi} + c\chi + d\bar{\psi}\chi\right)$$
$$= \int d\bar{\psi}d\psi d\bar{\chi}d\chi \left(r^2 |a|^2 + r |b|^2 + r |c|^2 + |d|^2\right) \bar{\psi}\psi\bar{\chi}\chi$$
$$= r^2 |a|^2 + r |b|^2 + r |c|^2 + |d|^2$$

For a probabilistic interpretation of the state vector it needs to be normalized, and yields the following.

- $\langle f|f\rangle = C$
- $r^2|a|^2/C =$ probability of the system having no fermion or antifermion
- $r|b|^2/C =$ probability of the system having one fermion
- $r|c|^2/C =$ probability of the system having one antifermion
- $|d|^2/C =$ probability of the system having one fermion and one antifermion

6.6   Real and Complex Fermions: Gaussian Integration

The rules of fermion integration are used for performing real and complex fermion Gaussian integration.

Consider N real fermion variables χ_n with $n = 1, 2, ..., N$. Define the partition function

$$Z[J] = \prod_{n=1}^{N} \int d\chi_n \exp\left\{ -\frac{1}{2}\chi_n M_{nm}\chi_m + J_n\chi_n \right\} \qquad (6.6.1)$$

The external source J_n is fermionic, and M is a real anti-symmetric matrix.

For N odd, $Z[J]$ is always zero. To see this consider the case of $J = 0$; on expanding the exponential, one always has powers of the even product of fermions; hence, term by term, the partition function is given by terms that are all zero, since all terms in the expansion have an odd number of fermion integrations.

$$\prod_{i=1}^{N} \int d\chi_i \, (\text{even products of } \chi_j) = 0 \; ; \; N : \text{odd}$$

Let N be even. Since all the fermions are real, one can only use a real transformation for simplifying M. Matrix algebra yields the result that every real antisymmetric matrix M can be brought into a block-diagonal Jordan canonical form by an orthogonal transformation O in the following manner

$$M = O^T \begin{pmatrix} 0 & \lambda_1 & & & & \\ -\lambda_1 & 0 & & & & \\ & & 0 & \lambda_2 & & \\ & & -\lambda_2 & 0 & & \\ & & & & \ddots & \\ & & & & & 0 & \lambda_{\frac{N}{2}} \\ & & & & & -\lambda_{\frac{N}{2}} & 0 \end{pmatrix} O = O^T \Lambda O \qquad (6.6.2)$$

where

$$O^T O = 1 : \text{orthogonal}$$

Perform a change of variables; from Eq. 6.2.5

$$O\chi = \xi : \text{real}$$

$$\Rightarrow \prod_{i=1}^{N} d\chi_i = \det(O) \prod_{i=1}^{N} d\xi_i = \prod_{i=1}^{N} d\xi_i$$

The partition function factors into a product of $N/2$ terms and yields

$$Z[0] = \prod_{i=1}^{N} \int d\xi_i \exp\left\{ -\frac{1}{2}\xi_n \Lambda_{nm}\xi_m \right\}$$

$$= \prod_{i=1}^{N/2} \left[\int d\xi_1 d\xi_2 \exp\left\{ -\frac{1}{2} (\xi_1\ \xi_2) \begin{pmatrix} 0 & \lambda_i \\ -\lambda_i & 0 \end{pmatrix} \begin{pmatrix} \xi_1 \\ \xi_2 \end{pmatrix} \right\} \right]$$

$$= \prod_{i=1}^{N/2} \int d\xi_1 d\xi_2 e^{+\lambda_i \xi_1 \xi_2} = \prod_{i=1}^{N/2} \lambda_i \qquad (6.6.3)$$

From Eq. 6.6.2

$$\det M = \prod_{i=1}^{N/2} \det \begin{pmatrix} 0 & \lambda_i \\ -\lambda_i & 0 \end{pmatrix} = \prod_{i=1}^{N/2} \lambda_i^2$$

Hence, from Eq. 6.6.3, the partition function is given by

$$Z[0] = \sqrt{\det M} \qquad (6.6.4)$$

$$(6.6.5)$$

The square root of an antisymmetric matrix is known as a Pfaffian and has many remarkable mathematical properties.

The general case of the partition function in the presence of an external fermionic source J_i can be derived from the results obtained. Rewriting the action given in Eq. 6.6.1 yields

$$Z[J] = \int D\chi \exp\left\{ -\frac{1}{2}\left(\chi + JM^{-1}\right)M\left(\chi + M^{-1}J\right) + \frac{1}{2}J^{-1}MJ \right\} \qquad (6.6.6)$$

Using the fundamental invariance of fermion integration under the shift of the fermion integration variable given in Eq. 6.2.1 allows us to shift the integration variable as follows

$$\chi \to \chi - JM^{-1}$$

and yields, from Eqs. 6.6.6 and 6.6.4

$$Z[J] = \int D\chi e^{-\frac{1}{2}\chi M\chi} e^{\frac{1}{2}J^T M^{-1}J} = Z[0]\, e^{\frac{1}{2}J^T M^{-1}J}$$

$$\Rightarrow Z[J] = \sqrt{\det M}\, e^{\frac{1}{2}J^T M^{-1}J}$$

The propagator can be obtained by fermionic differentiations on J_i as follows

$$\langle \chi_i \chi_j \rangle = \frac{\delta^2}{\delta J_i \delta J_j} \cdot \frac{1}{Z} \int D\chi e^{S+J\chi}\Big|_{J=0} = \frac{\delta^2}{\delta J_i \delta J_j} \exp\left\{ \frac{1}{2}\sum_{mn=1}^{N} J_m^T M_{mn}^{-1} J_n \right\}\Big|_{J=0}$$

$$= \frac{\delta}{\delta J_i}\left[\left(\frac{1}{2}M_{jk}^{-1}J_k - \frac{1}{2}J_k M_{kj}^{-1}\right)e^S \right]\Big|_{J=0} = \frac{1}{2}M_{ji}^{-1} - \frac{1}{2}M_{ij}^{-1}$$

$$\Rightarrow \langle \chi_i \chi_j \rangle = M_{ji}^{-1}$$

6.6.1   *Complex Gaussian fermions*

Consider the N-dimensional Gaussian integral for fermions ψ_n and $\bar{\psi}_n$

$$Z[J] = \prod_{n=1}^{N}\int d\bar{\psi}_n d\psi_n \exp\{-\bar{\psi}_n M_{nm}\psi_m + \bar{J}_n\psi_n + \bar{\psi}_n J_n\}$$

where $M_{nm} = -M_{mn}$ is an antisymmetric matrix. For real fermions $\psi = \psi^*$. For complex fermions $\psi = \psi_1 + i\psi_2$.

An antisymmetric matrix $M = -M^T$ can be diagonalized by a *unitary* transformation

$$M = U^\dagger \begin{pmatrix} \lambda_1 & & \\ & \ddots & \\ & & \lambda_N \end{pmatrix} U, \quad U^\dagger U = 1$$

In matrix notation

$$M = U^\dagger \Lambda U, \qquad UU^\dagger = 1, \qquad \det(UU^\dagger) = 1 \qquad (6.6.7)$$

where $\Lambda = \mathrm{diag}(\lambda_1, ...\lambda_N)$. Since the fermions ψ, $\bar{\psi}$ are complex, define the change of variables using the unitary matrix U, and from Eq. 6.2.5[2]

$$\bar{\psi}U^\dagger = \bar{\eta}, \qquad \eta = U\psi$$

$$D\eta D\bar{\eta} = \frac{1}{\det(U)}D\psi \times \frac{1}{\det(U^\dagger)}D\bar{\psi} = D\psi D\bar{\psi} \qquad (6.6.8)$$

---

[2]Note for real fermions one could not use a unitary transformation for a change of variables as this would lead to the transformed fermions being complex.

Hence, the fermion integrations completely factorize and yield

$$Z[0] = \prod_n \int d\bar{\eta}_n d\eta_n \exp\left\{ -\sum_{n=1}^N \lambda_n \bar{\eta}_n \eta_n \right\}$$

$$= \prod_n \left[\int d\bar{\eta}_n d\eta_n e^{-\lambda_n \bar{\eta}_n \eta_n} \right] = \prod_n \lambda_n = \det M \qquad (6.6.9)$$

The partition function with an external source is given as before by a shift of fermion integration variables. Write the partition function as

$$Z[J] = \int D\bar{\psi} D\psi \exp\{-(\bar{\psi} - \bar{J}M^{-1})M(\psi - M^{-1}J) + \bar{J}M^{-1}J\}$$

Using the fundamental property of fermion integration – that it is invariant under a constant shift of fermion variables as given in Eq. 6.2.1 – yields

$$\bar{\psi} \to \bar{\psi} + \bar{J}M^{-1} \; ; \; \psi \to \psi + M^{-1}J$$

and hence

$$Z[J] = \int D\bar{\psi} D\psi \exp\{-\bar{\psi}M\psi + \bar{J}M^{-1}J\} = (\det M) \exp\{\bar{J}M^{-1}J\} \qquad (6.6.10)$$

Consider the propagator given by

$$G_{mn} = E[\bar{\psi}_m \psi_n] = \frac{1}{Z} \int D\bar{\psi} D\psi \bar{\psi}_m \psi_n \exp\{-\bar{\psi}M\psi\}$$

Using the anticommuting rules of fermion differentiation yields, from Eq. 6.6.10

$$G_{mn} = E[\bar{\psi}_m \psi_n] = \frac{\delta^2}{\delta \bar{J}_n \delta J_m} Z[J]\bigg|_{J=0=\bar{J}} = -M_{nm}^{-1} = M_{mn}^{-1} \qquad (6.6.11)$$

The fermion Gaussian integration obtained in Eq. 6.6.9 can be directly done using the rules of fermion integration. On expanding the exponential term $\exp\{\bar{\psi}M\psi\}$, only *one term* – namely $(\bar{\psi}M\psi)^N/N!$ containing the product of all the fermion variables – is non-zero inside the integrand. Using the notation of summing over repeated indexes, Eq. 6.2.3 yields the following

$$\int D\bar{\psi} D\psi e^{\bar{\psi}M\psi} = \int D\bar{\psi} D\psi \left[\frac{1}{N!} (\bar{\psi}M\psi)^N \right]$$

$$= \frac{1}{N!} M_{i_1 j_1} M_{i_2 j_2} \cdots M_{i_N j_N} \int D\bar{\psi} D\psi \bar{\psi}_{i_1} \psi_{j_1} \bar{\psi}_{i_2} \psi_{j_2} \cdots \bar{\psi}_{i_N} \psi_{j_N}$$

$$= \frac{1}{N!} M_{i_1 j_1} M_{i_2 j_2} \cdots M_{i_N j_N} \epsilon_{i_1 i_2 \cdots i_N} \epsilon_{j_1 j_2 \cdots j_N}$$

$$= \det M$$

6.7 Fermionic Path Integral and Hamiltonian

One way of understanding the difference between a fermion and an antifermion is to examine the evolution of fermions in (Euclidean) time. For clarity, consider a time lattice with spacing ϵ and let lattice time be denoted by $n\epsilon$. The fermion degrees of freedom are defined on the lattice and denoted by $\bar{\psi}_n$, ψ_n. Consider a typical action for the lattice fermions given by

$$S_P = -r \sum_n \bar{\psi}_n \psi_n + 2K \sum_n \bar{\psi}_{n+1} \psi_n$$

$$= \sum_n \mathcal{L}(n)$$

where the fermion Lagrangian is given by

$$\mathcal{L}(n) = -r\bar{\psi}_n \psi_n + 2K\bar{\psi}_{n+1}\psi_n$$

$$= 2K[\bar{\psi}_{n+1} - \bar{\psi}_n]\psi_n + (2K - r)\bar{\psi}_n \psi_n \qquad (6.7.1)$$

The path integral is given by

$$Z_P = \prod_n \int d\bar{\psi}_n d\psi_n e^S \qquad (6.7.2)$$

The path integral is written in the fermion coherent state basis, similar to the coherent basis path integral for bosons (real and complex degrees of freedom).

Note that the dual fermion variable ψ_n at time n propagates to variable $\bar{\psi}_{n+1}$ at time $n+1$. The partition function is written by repeatedly using the completeness equation given in Eq. 6.3.4 and yields

$$Z_P = \int D\bar{\psi}D\psi \langle \bar{\psi}_{n+1}|e^{-\epsilon H}|\psi_n\rangle e^{-r\bar{\psi}_n\psi_n} \langle \bar{\psi}_n|e^{-\epsilon H}|\psi_{n-1}\rangle e^{-r\bar{\psi}_{n-1}\psi_{n-1}} \cdots$$

Hence, from the definition of fermionic operators given in Eq. 6.8.4

$$e^{2K\bar{\psi}_{n+1}\psi_n} = \langle \bar{\psi}_{n+1}|e^{-\epsilon H}|\psi_n\rangle \qquad (6.7.3)$$

$$\simeq e^{-\epsilon H(\bar{\psi}_{n+1},\psi_n)} \langle \bar{\psi}_{n+1}|\psi_n\rangle = \frac{1}{r}e^{-\epsilon H(\bar{\psi}_{n+1},\psi_n)} e^{r\bar{\psi}_{n+1}\psi_n}$$

Dropping the index of time on the fermion variables yields, up to an irrelevant constant, the particle Hamiltonian

$$-\epsilon H_P(\bar{\psi}, \psi) \simeq 2K\bar{\psi}\psi - r\bar{\psi}\psi$$

$$\Rightarrow H_P(\bar{\psi}, \psi) \simeq \left(\frac{r - 2K}{\epsilon}\right)\bar{\psi}\psi = \langle \bar{\psi}|H|\psi\rangle$$

Note that the Hamiltonian is automatically normal ordered.

The limit of continuous time is taken by defining the following continuum fermions $\bar{\psi}(t), \psi(t)$ and Hamiltonian

$$\bar{\psi}_t = \sqrt{2K}\bar{\psi}_n \;\; ; \;\; \psi_t = \sqrt{2K}\psi_n \;\; ; \;\; t = n\epsilon$$

$$H_P(\bar{\psi}, \psi) = \left(\frac{r - 2K}{2K\epsilon}\right)\bar{\psi}_t\psi_t = \omega\bar{\psi}_t\frac{\delta}{\delta\bar{\psi}_t} \;\; ; \;\; \omega = \frac{r - 2K}{2K\epsilon} \qquad (6.7.4)$$

The continuum Lagrangian $\mathcal{L}(t)$ and action, from Eqs. 6.7.1 and 6.7.4, are given by

$$\mathcal{L}(n) = -\epsilon \left[\frac{\partial \bar{\psi}_t}{\partial t} \psi_t + \frac{r - 2K}{2\epsilon K} \bar{\psi}_t \psi_t \right] \equiv \epsilon \mathcal{L}(t)$$

$$S_P = \int dt \mathcal{L}(t) \quad ; \quad \mathcal{L}(t) = - \left[\frac{\partial \bar{\psi}_t}{\partial t} \psi_t + \omega \bar{\psi}_t \psi_t \right]$$

S_P is the action for a particle propagating forward in time. Consider another action that describes the time evolution of *antiparticles*, namely

$$S_A = -\sum_n \bar{\chi}_n \chi_n + 2K \sum_n \bar{\chi}_n \chi_{n+1}$$

with

$$Z_A = \int D\bar{\chi} D\chi e^{S_A} \tag{6.7.5}$$

The antifermion dual variable $\bar{\chi}_n$ propagates from time n to the variable χ_{n+1} at time $n+1$. If one thinks of the variable χ_{n+1} as representing a particle, then one can think of the antiparticle as being equivalent to a particle propagating **backwards** in time; although this way of thinking is not required, it does help to develop some physical intuition about the peculiarities of antiparticles.

Given the nature of the state space of the antiparticles, Z_A can be decomposed as follows

$$Z_A = \int D\bar{\chi} D\chi \ldots \langle \chi_{n+1} | e^{-\epsilon H_A} | \bar{\chi}_n \rangle e^{-r \bar{\chi}_n \chi_n} \langle \chi_n | e^{-\epsilon H_A} | \bar{\chi}_{n-1} \rangle \ldots \tag{6.7.6}$$

and by inspection

$$\langle \chi_{n+1} | e^{-\epsilon H_A} | \bar{\chi}_n \rangle = \langle \chi_{n+1} | \bar{\chi}_n \rangle e^{2K \bar{\chi}_n \chi_{n+1}}$$

$$= \frac{1}{r} e^{r \bar{\chi}_n \chi_{n+1}} e^{-\epsilon H_A (\bar{\chi}_n \chi_{n+1})}$$

$$\Rightarrow H_A = \left(\frac{1 - 2K}{\epsilon} \right) \bar{\chi} \chi = \langle \chi | H | \bar{\chi} \rangle$$

Note the order of the matrix element of the antifermions, namely $\langle \chi | H | \bar{\chi} \rangle$, is the reverse of the fermion case. Define continuum fermion variables by

$$\bar{\chi}_t = \sqrt{2K} \bar{\chi}_n \quad ; \quad \chi_t = \sqrt{2K} \chi_n \quad ; \quad \omega = \left(\frac{r - 2K}{2K\epsilon} \right) \quad ; \quad t = n\epsilon$$

Anticommuting the fermionic variables and ignoring an irrelevant constant yields, similar to Eq. 6.7.4, the following continuum antifermion Hamiltonian

$$H_A = -\omega \chi_t \bar{\chi}_t \tag{6.7.7}$$

After normal ordering, interpreting $\hat{H}_A$ as an operator yields, using $\bar{\chi} = -\delta/\delta\chi$, the following

$$H_A = \omega \chi_t \frac{\delta}{\delta \chi_t}$$

6.8   Fermionic Operators

Operators for fermions, such as the Hamiltonian and momentum have a representation in terms of fermionic variables. Given that conjugate of the fermionic coordinate $\bar{\psi}$ is given by ψ, all operators for fermions are expressed in terms of both the fermion coordinate and its dual. In this sense, the operators of the fermion degree of freedom are defined analogous to bosonic operators that are defined on phase space [Baaquie (2014)].

The particle coordinate operator and its conjugate are given by

$$\bar{\psi} \; ; \; \psi = \frac{1}{r}\frac{\delta}{\delta\bar{\psi}} \quad \Rightarrow \quad \{\bar{\psi},\psi\} = \frac{1}{r} \tag{6.8.1}$$

For the antiparticle, one has

$$\chi \; ; \; \bar{\chi} = -\frac{1}{r}\frac{\delta}{\delta\chi} \quad \Rightarrow \quad \{\bar{\chi},\chi\} = -\frac{1}{r}$$

Note that for both the particle and antiparticle operators, the metric on state space given by r enters the anticommutation equation.

Consider a fermionic operator $\mathcal{O}$; the operator maps from $V_{\bar{\psi}}$ to itself and hence, $\mathcal{O}$ is an element of the tensor product space $V_{\bar{\psi}} \otimes V_{\psi}$.

$$\mathcal{O} \in \mathcal{V} \otimes \mathcal{V}_D \equiv V_{\bar{\psi}} \otimes V_{\psi}$$

The matrix elements of operator $\mathcal{O}$ are $\mathcal{O}(\bar{\psi},\psi)$ and given by

$$\langle\bar{\psi}|\mathcal{O}|\psi\rangle = \langle\bar{\psi}|\psi\rangle\mathcal{O}(\bar{\psi},\psi) = \frac{1}{r}e^{r\bar{\psi}\psi}\mathcal{O}(\bar{\psi},\psi) \tag{6.8.2}$$

In particular, the Hamiltonian operator H is defined by

$$\langle\bar{\psi}|H|\psi\rangle = H(\bar{\psi},\psi)\langle\bar{\psi}|\psi\rangle = \frac{1}{r}e^{r\bar{\psi}\psi}H(\bar{\psi},\psi) \tag{6.8.3}$$

and yields, from Eq. 6.8.3, the following

$$\langle\bar{\psi}|e^{-\epsilon H}|\psi\rangle = \langle\bar{\psi}|[1-\epsilon H]\psi\rangle = \frac{1}{r}[1-\epsilon H(\bar{\psi},\psi)]e^{r\bar{\psi}\psi} + O(\epsilon^2)$$

$$\simeq \frac{1}{r}e^{-\epsilon H(\bar{\psi},\psi)}e^{r\bar{\psi}\psi} + O(\epsilon^2) \tag{6.8.4}$$

Hermitian conjugation for an operator $\mathcal{O}$, denoted by $\mathcal{O}^\dagger$, is given by

$$\mathcal{O}^\dagger(\bar{\psi},\psi) = \mathcal{O}^*(\bar{\psi}^c,\psi^c)$$

where conjugation of the basis states is given by Eq. 6.5.2 and all fermionic variables need to be reversed in taking the conjugation. An operator is Hermitian if

$$\mathcal{O}^\dagger(\bar{\psi},\psi) = \mathcal{O}^*(\bar{\psi}^c,\psi^c) = \mathcal{O}(\bar{\psi},\psi) \tag{6.8.5}$$

6.9 Fermion-Antifermion Hamiltonians

Consider fermionic Hamiltonian operators based on a state space with metric given by r.

From Eq. 6.7.4, a Hamiltonian for fermions is given by

$$H_P = \omega\bar{\psi}\psi = \frac{\omega}{r}\bar{\psi}\frac{\delta}{\delta\bar{\psi}} \tag{6.9.1}$$

The eigenstates and eigenenergies of the the Hamiltonian are given by

$$H_P\Phi_n = E_n\Phi_n$$
$$\Phi_0 = 1 \quad E_0 = 0$$
$$\Phi_1 = \bar{\psi} \quad E_1 = \omega/r$$

From Eq. 6.7.7, a typical antifermion Hamiltonian is given by

$$H_A = \tilde{\omega}\chi\frac{\delta}{r\delta\chi} = -\tilde{\omega}\chi\bar{\chi} \tag{6.9.2}$$

with eigenstates and eigenenergies

$$\Phi_0^A = 1 \quad , \quad E_0 = 0$$
$$\Phi_1^A = \chi \quad , \quad E_1 = \omega/r$$

Consider a fermion and antifermion system. The Hilbert state space is four-dimensional since there are four possible states for the system, as enumerated in Eq. 6.5.1. A Hermitian Hamiltonian for the fermion-antifermion system is given by

$$H = \omega\bar{\psi}\frac{\delta}{r\delta\bar{\psi}} + \omega'\chi\frac{\delta}{r\delta\chi} + \lambda\bar{\psi}\chi\frac{\delta^2}{r^2\delta\bar{\psi}\delta\chi}$$
$$= \omega\bar{\psi}\psi + \omega'\bar{\chi}\chi - \lambda\bar{\psi}\chi\psi\bar{\chi}$$

Hermitian conjugation, explicitly shown below, shows that, for ω, ω', λ real, the Hamiltonian is Hermitian.

$$H^\dagger = \omega^*(\bar{\psi}\psi)^c + (\omega')^*(\bar{\chi}\chi)^c - \lambda^*\bar{\chi}^c\psi^c\chi^c\bar{\psi}^c$$
$$= \omega\bar{\psi}\psi + \omega'\bar{\chi}\chi - \lambda(-\chi)\bar{\psi}(-\bar{\chi})\psi$$
$$= \omega\bar{\psi}\psi + \omega'\bar{\chi}\chi - \lambda\chi\bar{\psi}\bar{\chi}\psi$$
$$= \omega\bar{\psi}\psi + \omega'\bar{\chi}\chi - \lambda\bar{\psi}\chi\psi\bar{\chi}$$
$$= H$$

The eigenfunctions can be read off by inspection and are the following

$$\Phi_0 = 1/r \quad ; \quad E_0 = 0$$
$$\Phi_1 = \bar{\psi}/\sqrt{r} \quad ; \quad E_1 = \omega/r$$
$$\Phi_2 = \chi/\sqrt{r} \quad ; \quad E_2 = \omega'/r$$
$$\Phi_3 = \bar{\psi}\chi \quad ; \quad E_3 = \omega/r + \omega'/r - \lambda/r^2$$

6.10 A Quadratic Hamiltonian

Consider the following Hermitian Hamiltonian

$$H = \omega\bar{\psi}\psi + \omega'\chi\bar{\chi} + i\lambda(\bar{\psi}\chi + \bar{\chi}\psi)$$
$$= \omega\bar{\psi}\frac{\delta}{r\delta\bar{\psi}} + \omega'\chi\frac{\delta}{r\delta\chi} + i\lambda(\bar{\psi}\chi + \bar{\chi}\psi) \qquad (6.10.1)$$

Two of the four orthogonal eigenstates can be obtained by inspection and yields

$$\Phi_1 = \chi/\sqrt{r} \quad ; \quad E_1 = \omega'/r$$
$$\Phi_2 = \bar{\psi}/\sqrt{r} \quad ; \quad E_2 = \omega/r$$

For the other two eigenfunctions consider the ansatz

$$\Phi = N(\bar{\psi}\chi + ic)$$

where N is a normalization constant. Applying the Hamiltonian given in Eq. 6.10.1 on Φ yields

$$H\Phi = \left[\frac{\omega}{r} + \frac{\omega'}{r} - c\lambda\right]\left(\bar{\psi}\chi - i\frac{\lambda}{r^2}\frac{1}{\frac{\omega}{r} + \frac{\omega'}{r} - c\lambda}\right)$$

One obtains the eigenvalue condition

$$c = -\frac{\lambda}{r^2}\frac{1}{\frac{\omega}{r} + \frac{\omega'}{r} - c\lambda} \quad \Rightarrow \quad cr = -\frac{\lambda}{\omega + \omega' - cr\lambda}$$

which has two solutions given by

$$c_\pm = \frac{1}{2r\lambda}\left(\omega + \omega' \pm \sqrt{(\omega + \omega')^2 + 4\lambda^2}\right) \quad ; \quad c_+c_- = -\frac{1}{r^2} \qquad (6.10.2)$$

Hence, the remaining two eigenfunctions and eigenvalues are given by

$$\Phi_3 = \frac{1}{\sqrt{1 + r^2c_+^2}}[\bar{\psi}\chi + irc_+] \quad ; \quad E_3 = \frac{1}{r}(\omega + \omega' - r\lambda c_+)$$

$$\Phi_4 = \frac{1}{\sqrt{1 + r^2c_-^2}}[\bar{\psi}\chi + irc_-] \quad ; \quad E_4 = \frac{1}{r}(\omega + \omega' - r\lambda c_-) \qquad (6.10.3)$$

The interpretation of the state Φ_3 is that $r^2|c_+|^2/\sqrt{1 + r^2|c_+|^2}$ is the likelihood that the system has no particles and $1/\sqrt{1 + r^2|c_+|^2}$ is the likelihood of having a fermion-antifermion pair; with a similar interpretation for Φ_4.

6.10.1 *Orthogonality and completeness*

To illustrate the workings of fermion calculus, the orthogonality of the states Φ_3 , Φ_4 is explicitly verified. Using the rules for forming the conjugate state function yields

$$\langle\Phi_3|\Phi_4\rangle = \int d\bar{\psi}d\psi d\bar{\chi}d\chi\langle\Phi_3|\psi;\bar{\chi}\rangle e^{-r\bar{\psi}\psi - r\bar{\chi}\chi}\langle\bar{\psi};\chi|\Phi_4\rangle$$

$$= \int d\bar{\psi}d\psi d\bar{\chi}d\chi\Phi_3^\dagger[\psi;\bar{\chi}\,]\Phi_4[\bar{\psi};\chi]e^{-r\bar{\psi}\psi - r\bar{\chi}\chi}$$

$$= \int d\bar{\psi}d\psi d\bar{\chi}d\chi\left(-\bar{\chi}\psi - irc_+\right)\left(\bar{\psi}\chi + irc_-\right)e^{-r\bar{\psi}\psi - r\bar{\chi}\chi}$$

$$= (1 + r^2 c_+ c_-) \int d\bar{\psi} d\psi d\bar{\chi} d\chi \ \bar{\psi}\psi\bar{\chi}\chi$$

$$= (1 + r^2 c_+ c_-) = 0 \ \text{since} \ c_+ c_- = -1/r^2$$

The completeness equation can be expressed in terms of the eigenfunctions by the following

$$\mathbb{I} = \sum_{i=1}^{4} |\Phi_i\rangle\langle\Phi_i| \tag{6.10.4}$$

To verify Eq. 6.10.4, one needs to prove that

$$\sum_{i=1}^{4} \langle\bar{\psi}, \chi|\Phi_i\rangle\langle\Phi_i|\psi, \bar{\chi}\rangle = \frac{1}{r^2} \exp\{r\bar{\psi}\psi + r\bar{\chi}\chi\}$$

Using the explicit form of the state functions derived above yields

$$\sum_{i=1}^{4} \langle\bar{\psi}, \chi|\Phi_i\rangle\langle\Phi_i|\psi, \bar{\chi}\rangle = \frac{1}{r}(\bar{\chi}\chi + \bar{\psi}\psi) + N_+[\bar{\psi}\chi + irc_+][-\bar{\chi}\psi - irc_+]$$

$$+ N_-[\bar{\psi}\chi + irc_-][-\bar{\chi}\psi - irc_-]$$

$$= \frac{1}{r}(\bar{\chi}\chi + \bar{\psi}\psi) + [r^2 c_+^2 N_+ + r^2 c_-^2 N_-] + [N_+ + N_-]\bar{\psi}\psi\bar{\chi}\chi$$

$$- i[rc_+ N_+ + rc_- N_-][\bar{\psi}\chi + \bar{\chi}\psi]$$

where

$$N_+ = \frac{1}{1 + r^2 c_+^2} \ ; \ N_- = \frac{1}{1 + r^2 c_-^2}$$

From Eq. 6.10.2 it follows that

$$r^2 c_+^2 N_+ + r^2 c_-^2 N_- = 1 = N_+ + N_-$$

$$rc_+ N_+ + rc_- N_- = 0$$

Hence

$$\sum_{i=1}^{4} \langle\bar{\psi}, \chi|\Phi_i\rangle\langle\Phi_i|\psi, \bar{\chi}\rangle = \frac{1}{r^2} + \frac{1}{r}(\bar{\chi}\chi + \bar{\psi}\psi) + \bar{\psi}\psi\bar{\chi}\chi = \frac{1}{r^2} \exp\{r\bar{\psi}\psi + r\bar{\chi}\chi\}$$

thus verifying Eq. 6.10.4.

6.11 Fermion-Antifermion Lagrangian

Consider the propagation of a fermion-antifermion system given by the Lagrangian

$$\epsilon\mathcal{L}_n = -r\bar{\psi}_n\psi_n - \bar{\chi}_n\chi_n + 2K(\bar{\psi}_{n+1}\psi_n + \bar{\chi}_n\chi_{n+1}) \tag{6.11.1}$$

$\mathcal{L}_n$ consists of a fermion propagating forward in time and its antifermion (since K is the same for both) propagating 'backward' in time.

Define two-component spinor

$$\Psi = \begin{pmatrix} \psi \\ \chi \end{pmatrix} \qquad \bar{\Psi} = (\bar{\psi}\ \bar{\chi}) \ ; \ \gamma_0 = \begin{pmatrix} 1 & 0 \\ 0 & -1 \end{pmatrix} \qquad (6.11.2)$$

Then, from Eqs. 6.11.1 and 6.11.2

$$\mathcal{L}_n = -r\bar{\Psi}_n \Psi_n + K\left\{\bar{\Psi}_{n+1}(1+\gamma_0)\Psi_n + \bar{\Psi}_n(1-\gamma_0)\Psi_{n+1}\right\} \qquad (6.11.3)$$

The fermion action is

$$S = \epsilon \sum_n \mathcal{L}_n$$

Consider the continuum limit by defining $t = n\epsilon$ and take the limit of $\epsilon \to 0$ yields

$$K\bar{\psi}_{n+1}\psi = K(\bar{\psi}_{n+1} - \bar{\psi}_n)\psi_n + K\bar{\psi}_n\psi_n$$
$$\cong K\epsilon\partial_0\bar{\psi}_n\psi_n + K\bar{\psi}_n\psi_n$$

Similarly

$$K\bar{\chi}_n\chi_{n+1} = \epsilon K\bar{\chi}_n\partial_0\chi + K\bar{\chi}_n\chi_n$$

Hence

$$\epsilon\mathcal{L}_n = (-r + 2K)[\bar{\psi}_n\psi_n + \bar{\chi}_n\chi_n] + 2K\epsilon[\partial_0\bar{\psi}_n\psi_n + \bar{\chi}_n\partial_0\chi_n]$$

Define continuum fermionic variables by

$$\bar{\psi}_t = (2K)^{1/2}\bar{\psi}_n \ ; \ \psi_t = (2K)^{1/2}\psi_n$$
$$\bar{\chi}_t = (2K)^{1/2}\bar{\chi}_n \ ; \ \chi_t = (2K)^{1/2}\chi_n$$

The continuum Lagrangian is given

$$\epsilon\mathcal{L}(t) = \left(\frac{2K-r}{2K}\right)[\bar{\psi}_t\psi_t + \bar{\chi}_t\chi_t] + \epsilon[\bar{\psi}_t(-\partial_0)\psi_t + \bar{\chi}_t\partial_0\chi_t]$$

Define

$$m\epsilon = -\left[\frac{2K-r}{2K}\right] \ \Rightarrow \ 2K = \frac{r}{1+m\epsilon}$$

Hence, the continuum action and Lagrangian are given by

$$S = \int_{-\infty}^{+\infty} dt\mathcal{L}$$
$$\mathcal{L}(t) = -\bar{\Psi}(\gamma_0\partial_0 + m)\Psi$$
$$: \text{One-dimensional Dirac Lagrangian}$$

For Minkowski spacetime, both Euclidean time and γ_0 have to be continued to Minkowski time in the following manner

$$t_E \to t_M = -it_E \ ; \ \gamma_0 \to \gamma_0^M = \gamma_0$$

This yields the Minkowski Dirac equation in one spacetime dimension as

$$\mathcal{L}_M = -\bar{\Psi}\left(-i\gamma_0^M\frac{\partial}{\partial t_M} + m\right)\Psi \qquad (6.11.4)$$

6.12 Fermionic Transition Probability Amplitude

The derivation done earlier for finding the eigenstates and eigenenergies of the fermionic Hamiltonian can be obtained directly working with the fermion transition probability amplitude. Recall for a fermion particle

$$\langle \bar{\psi} | e^{-\epsilon H} | \psi \rangle = e^{2K\bar{\psi}\psi}$$

The eigenvalue and eigenstates that are given by $\Phi_0 \sim 1$ and $\Phi_1 \sim \bar{\psi}$ can be directly obtained by performing fermion integration in the following manner

$$\langle \bar{\psi} | e^{-\epsilon H} | \Phi_0 \rangle = \int d\varsigma d\bar{\varsigma} \langle \bar{\psi} | e^{-\epsilon H} | \varsigma \rangle e^{-\bar{\varsigma}\varsigma} \langle \bar{\varsigma} | \Phi_0 \rangle = \int d\varsigma d\bar{\varsigma} e^{2K\bar{\psi}\varsigma} e^{-\bar{\varsigma}\varsigma}$$

$$= \int d\varsigma d\bar{\varsigma} (1 + 2K\bar{\psi}\varsigma)(1 - \bar{\varsigma}\varsigma) = 1$$

$$\Rightarrow \Phi_0(\bar{\psi}) = \langle \bar{\psi} | \Phi_0 \rangle = 1 \quad ; \quad E_0 = 0$$

For the eigenstate $\Phi_1 \sim \bar{\psi}$ consider the following calculation

$$\langle \bar{\psi} | e^{-\epsilon H} | \Phi_1 \rangle = \int d\varsigma d\bar{\varsigma} e^{2K\bar{\psi}\varsigma} e^{-\bar{\varsigma}\varsigma} \bar{\varsigma}$$

$$= 2K\bar{\psi} \int d\varsigma d\bar{\varsigma} \varsigma \bar{\varsigma} = 2K\bar{\psi} = e^{-\epsilon E_1} \bar{\psi}$$

$$\Rightarrow \langle \psi | \Phi_1 \rangle = \bar{\psi}$$

yields the expected answer. The eigenenergy is given by

$$\Rightarrow E_1 = -\frac{1}{\epsilon} \ln(2K) = -\frac{1}{\epsilon} \ln\left(\frac{1}{1 + m\epsilon}\right) \simeq m + O(\epsilon)$$

Similarly for the antifermion recall

$$\langle \chi | e^{-\epsilon H} | \bar{\chi} \rangle = e^{2K\bar{\chi}\chi}$$

Hence for the antifermion ground state $e^{-\epsilon H} | \Phi_0 \rangle = | \Phi_0 \rangle$ since $\Phi_0 \sim 1$.

For $\langle \chi | \Phi_1 \rangle = \chi$ one has

$$\langle \chi | e^{-\epsilon H} | \Phi_1 \rangle = \int \langle \chi | e^{-\epsilon H} | \bar{\varsigma} \rangle e^{-\bar{\varsigma}\varsigma} \langle \varsigma | \Phi_1 \rangle = \int_{\bar{\varsigma}\varsigma} e^{2K\bar{\varsigma}\chi} e^{-\bar{\varsigma}\varsigma} \varsigma$$

$$= \int_{\bar{\varsigma}\varsigma} (1 + 2K\bar{\varsigma}\chi)(1 - \bar{\varsigma}\varsigma)\varsigma = 2K\chi \int_{\bar{\varsigma}\varsigma} (-\bar{\varsigma}\varsigma) = 2K\chi$$

Note the transition amplitude automatically yields a normal ordered Hamiltonian with the energy given by

$$E_1 = -\frac{1}{\epsilon} \ln(2K) \simeq m$$

Until now, fermions and the antifermions have equal mass but have not been coupled, and hence their contrasting properties have not come into play. One can couple them by nonlinear terms in the Lagrangian such as $\lambda \bar{\psi}\psi\bar{\chi}\chi$ as well as by their coupling to gauge fields that is now briefly explored.

6.13   Quark Confinement

Consider one-dimensional toy-model of quarks (fermions) and antiquarks (antifermions) given by

$$\mathcal{L}_n = -\bar{\psi}_n \psi_n + 2K(\bar{\psi}_{n+1}\psi_n + \bar{\chi}_n \chi_{n+1}) - \bar{\chi}_n \chi_n$$

A gauge transformation on the fermions is defined by

$$\psi_n \to e^{i\phi_n}\psi_n \ , \ \ \bar{\psi}_n \to \bar{\psi}_n e^{-i\phi_n}$$

$$\chi_n \to e^{i\phi_n}\chi_n \ , \ \ \bar{\chi}_n \to \bar{\chi}_n e^{-i\phi_n}$$

To leave $\mathcal{L}_n$ invariant we need to fix up the nearest neighbor term. Introduce gauge field e^{iB_n} and modify $\mathcal{L}_n$ to

$$\mathcal{L}_n = -\bar{\psi}_n \psi_n - \bar{\chi}_n \chi_n + 2K(\bar{\psi}_{n+1}e^{-iB_n}\psi_n + \bar{\chi}_n e^{iB_n}\chi_{n+1})$$

Under a gauge transformation, let B_n have the transformation

$$e^{iB_n} \to e^{i\phi_n}e^{iB_n}e^{-i\phi_{n+1}}$$

and hence the combined transformations on the fermions and gauge field leave the Lagrangian $\mathcal{L}$ invariant.

Note B_n is an angular variable taking values in $[-\pi, +\pi]$. The partition function is

$$Z = \prod_n \int_{-\pi}^{\pi} \frac{dB_n}{2\pi} d\bar{\psi}_n d\psi_n d\bar{\chi}_n d\chi_n e^S$$

Since e^{iB_n} couples nearest neighboring instants of time, one can derive as before that the transition amplitude for the quark-antiquark system, namely

$$\langle \bar{\psi}_{n+1}, \chi_{n+1}|e^{-\epsilon H}|\psi_n, \bar{\chi}_n \rangle$$

by generalizing Eq. 6.7.3. One can choose to include the integration over the gauge field in the definition of the transition probability amplitude for the fermions and this yields

$$\langle \bar{\psi}_{n+1}, \chi_{n+1}|e^{-\epsilon H}|\psi_n, \bar{\chi}_n \rangle = \int_{-\pi}^{\pi} \frac{dB_n}{2\pi} e^{\mathcal{L}_n}$$

Dropping the subscripts from $\mathcal{L}_n$ yields

$$\langle \bar{\psi}, \chi|e^{-\epsilon H}|\psi, \bar{\chi} \rangle = \int_{-\pi}^{\pi} \frac{dB}{2\pi} \left[1 + 2K\bar{\psi}e^{-iB}\psi\right]\left[1 + 2K\bar{\chi}e^{iB}\chi\right]$$

$$= 1 + (2K)^2 \bar{\psi}\psi\bar{\chi}\chi$$

The transition probability amplitude acting on a general state function $|\Phi\rangle$ yields[3]

$$\langle \bar{\psi}, \chi|\Phi \rangle = \int d\bar{\zeta}d\varsigma d\bar{\xi}d\xi \langle \bar{\psi}, \chi|e^{-\epsilon H}|\varsigma, \bar{\xi} \rangle e^{-\bar{\zeta}\varsigma - \bar{\xi}\xi} \langle \bar{\zeta}, \xi|\Phi \rangle$$

$$= \int d\bar{\zeta}d\varsigma d\bar{\xi}d\xi \left[1 + (2K)^2 \bar{\psi}\varsigma\bar{\xi}\chi\right] e^{-\bar{\zeta}\varsigma - \bar{\xi}\xi} \Phi(\bar{\zeta}, \xi) \qquad (6.13.1)$$

---

[3]The metric r is set to 1 as it does not affect the conclusions.

The only non-zero integral is

$$\int d\bar{\zeta} d\zeta d\bar{\xi} d\xi \; \bar{\zeta}\zeta\bar{\xi}\xi = 1$$

Solving the eigenvalue equation of the quark-antiquark Hamiltonian

$$e^{-\epsilon H}|\Phi_n\rangle = e^{-\epsilon E_n}|\Phi_n\rangle$$

using Eq. 6.13.1 yields the following eigenfunctions and eigenvalues.

$\Phi_n(\bar{\zeta}, \xi)$	E_n	State
1	0	vacuum
$\bar{\psi}$	∞	one quark
χ	∞	one antiquark
$\bar{\psi}\chi$	$2\left[-\frac{\ln(2K)}{\epsilon}\right]$	quark+antiquark

Note that single quark $\bar{\psi}$ and antiquark χ states are confined since they have infinite energy and cannot propagate in time; they are fixed at whatever moment in time that they are created. Only the paired states of quark-antiquark, namely $\bar{\psi}\chi$, are the finite energy eigenstates and hence can propagate in time.

6.14 Summary

The fermion variable takes only two values and is fundamentally different from the real variable. The fermion degree of freedom describes a system that is essentially discrete and at the same time quite distinct from the Ising variable, that belongs to the category of real variables.

A differential and integral calculus was developed for the fermion variable and the concepts applicable to a real variable were generalized to the fermion case. Gaussian integration was defined for both real and complex fermions and the results are similar to the real variable case, but with a few significant differences.

Fermion and antifermion variables emerge naturally based on the manner in which the fermion and antifermion degrees of freedom are defined as well as the conjugation of the state vectors. The state space and Hamiltonian for fermions and antifermions was derived and the state space was shown to behave in the manner that one intuitively expects for a discrete system. A few simple models of the fermion and antifermion path integral, Hamiltonian and state space were discussed. A one-dimensional toy model based on fermion and antifermion degrees of freedom coupled to a gauge field was shown to exhibit quark confinement.

PART 2

Lattice Gauge Field

Chapter 7

Non-Abelian Lattice Gauge Field

7.1 Introduction

Lattice gauge fields are defined on a lattice spacetime and is a quantum field theory in which all the ultra-violet divergences are regularized and rendered finite [Wilson (1974)]. The two defining features of the lattice theory are the following: firstly, the gauge degrees of freedom are compact group elements of the gauge group and secondly, that spacetime lattice has a finite lattice spacing that plays the role of an ultra-violet cutoff for the quantum field theory. The continuum limit of the lattice gauge field exists only if the theory is renormalizable.

The fundamental feature of gauge-invariance dominates the entire analysis, narrative and discourse of gauge field theories. The following are a number of very general properties of gauge fields:

- In all spacetime dimensions, only vector or higher order tensor fields can be used to define gauge invariant theories.
- Gauge-invariance implies that all the components of the gauge vector field are not dynamical. There is a redundancy in describing a gauge field with a vector field.
- The irreducible information carried by the gauge field is given by all *gauge-invariant* functions – local and non-local – of the gauge field. *Wilson loops*, which are exponential line integrals (path ordered for non-Abelian gauge fields) of the gauge field taken over a closed loop encode all the information of a gauge field.
- Wilson loops contain global information about the underlying topology of the manifold on which the gauge field theory is defined. The topological aspect of the Wilson loops is clearly brought out in the derivation of Jones polynomials by Witten (1989).
- Renormalization theory shows that, in $d = 4$ spacetime dimensions, only vector quantum fields that are defined by a gauge-invariant Lagrangian or Hamiltonian can be renormalized.

Lattice gauge fields are Yang-Mills non-Abelian gauge fields defined on a lattice. To preserve exact gauge invariance on the lattice, the gauge field is defined using the finite group element of the gauge group – unlike the case of continuum Yang-Mills that is defined by the Lie algebra of the gauge group. The lattice gauge field is quantized on a discrete lattice embedded in a four-dimensional Euclidean spacetime. The following are some of the advantages of going to a lattice.

- Since the lattice provides an ultraviolet cutoff there are no ultraviolet divergences in the theory.
- Using the lattice as a cutoff allows one to formulate the cutoff lattice gauge field theory to have *exact* local gauge invariance.
- The gauge field is defined on a finite dimensional lattice, with the Feynman path integral being reduced to a finite dimensional multiple integral – opening the way to the numerical and algorithmic study of the gauge field.

Consider gauge group G, which for definiteness is taken to be $SU(\mathcal{N})$, with the Lie algebra given by

$$[X^a, X^b] = iC^{abc}X^c \; ; \; \mathrm{Tr}(X^a X^b) = \delta^{ab}/s^2$$

where $\{X^a\}$ are the generators of $SU(\mathcal{N})$. In the fundamental representation, the generators are given by $\mathcal{N} \times \mathcal{N}$ Hermitian traceless matrices with $s^2 = 2$. The structure constants C^{abc} for a compact Lie group are completely anti-symmetric in the group indices a, b, c. The group index a has the range of $1, 2, ..., \mathcal{N}^2 - 1$ for $SU(\mathcal{N})$.

The Yang-Mills action is defined by [Zinn-Justin (1993)]

$$\mathcal{S} = \int d^4x \mathcal{L}(x)$$

with the Yang-Mills gauge field A_μ^a, field tensor $F_{\mu\nu}^a$ and Lagrangian given by

$$\mathcal{L}_{YM}(x) = -\frac{1}{4}\sum_{\mu\nu;a}(F_{\mu\nu}^a)^2 \; ; \; F_{\mu\nu}^a = \partial_\mu A_\nu^a - \partial_\nu A_\mu^a - g_0 s C^{abc} A_\mu^b A_\nu^c$$

where g_0, is the *dimensionless* bare coupling constant.

To define the continuum gauge transformation, define the finite group element

$$\Phi(x) = e^{i\phi^a(x)X^a}$$

The non-Abelian gauge transformation for the continuum is given by

$$A_\mu^a(x)^a \rightarrow A_\mu^a(x)X^a(\Phi) = \Phi(x)A_\mu^a(x)X^a\Phi^\dagger(x) - i\frac{1}{sg_0}\Phi(x)\frac{\partial \Phi^\dagger(x)}{\partial x_\mu}$$

and which yields

$$F_{\mu\nu}^a(x) \rightarrow F_{\mu\nu}^a(x)X^a(\Phi) = \Phi(x)F_{\mu\nu}^a(x)X^a\Phi^\dagger(x)$$

and leaves the Yang-Mills Lagrangian invariant

$$\mathcal{L}(x) \rightarrow \mathcal{L}^{(\Phi)}(x) = \mathcal{L}(x)$$

The degrees of freedom of the Yang-Mills gauge field A_μ^a are element of the *Lie algebra* of gauge group G. The main insight of Wilson (1974) was to identify that it is the *finite group elements* of the gauge group G that are the correct non-Abelian degrees of freedom, and which entails exponentiating the Lie algebra. All quantum field theories are defined using a cutoff, and Wilson introduced a lattice cutoff to introduce the finite group elements as the degrees of freedom for the non-Abelian gauge field.

Consider an asymmetric four-dimensional Euclidean lattice with N^4 lattice sites: let the lattice spacing in the time direction be denoted by ϵ and in the space direction by a. The Lagrangian on the finite lattice is written down below; it is shown later, in Section 7.2, that the lattice Lagrangian produces the Yang-Mills continuum Lagrangian on taking the classical limit of $\epsilon, a \to 0$.

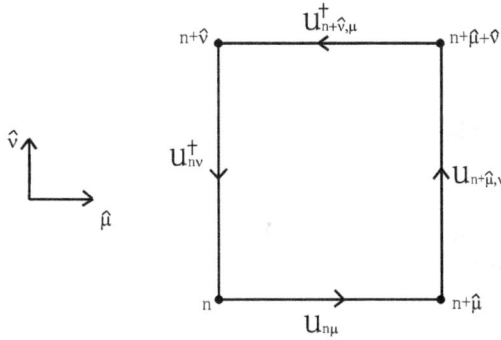

Fig. 7.1   The Wilson plaquette $W_{n,\mu\nu}$.

Let n specify the lattice site and $\hat{\mu}$ the directions on the lattice. The local vector gauge field degrees of freedom $U_{n\mu}$ are finite group elements belonging to the gauge group G, which for definiteness is taken to be $SU(\mathcal{N})$. Define the degrees of freedom $U_{n\mu}$, lattice gauge field tensor $W_{n\mu\nu}$ and gauge transformation by the finite group element as follows

$$W_{n\mu\nu} = e^{if^a_{n\mu\nu}X^a} \quad ; \quad U_{n\mu} = e^{iB^a_{n\mu}X^a} \quad ; \quad \Phi_n = e^{i\phi^a_n X^a} \tag{7.1.1}$$

The Wilson plaquette $W_{n\mu\nu}$, shown in Figure 7.1, is defined by

$$W_{n\mu\nu} = U_{n\mu}U_{n+\hat{\mu},\nu}U^\dagger_{n+\hat{\nu},\mu}U^\dagger_{n\nu} \quad ; \quad \mu,\nu = 0,1,2,3 \tag{7.1.2}$$

The Wilson plaquette is the generalization of the Yang-Mills field tensor. The asymmetric lattice Lagrangian is given by (Tr signifies trace) [Wilson (1974)]

$$\mathcal{L}(n) = \frac{1}{4g_0^2}\frac{a}{\epsilon}\sum_{i=1}^{3}\text{Tr}(W_{n,0i} + W^\dagger_{n,0i}) + \frac{1}{2g_0^2}\frac{\epsilon}{a}\sum_{i\neq j=1}^{3}\text{Tr}(W_{n,ij}) \quad ; \quad i,j = 1,2,3 \tag{7.1.3}$$

and the asymmetric lattice action is given by

$$\mathcal{S} = \sum_n \mathcal{L}(n) \tag{7.1.4}$$

For a symmetric lattice, with $\epsilon = a$, the gauge field Wilson action functional is defined by

$$\mathcal{S} = \frac{1}{2g_0^2}\sum_n\sum_{\mu\neq\nu}\text{Tr}(W_{n\mu\nu}) \tag{7.1.5}$$

Gauge transformations for the lattice are defined by

$$U_{n\mu} \to \Phi_n U_{n\mu} \Phi_{n+\hat{\mu}}^{\dagger} \tag{7.1.6}$$

where Φ_n, is also an element of the gauge group, and yields

$$W_{n,\mu\nu} \to \Phi_n W_{n,\mu\nu} \Phi_n^{\dagger}$$

Note that $\mathcal{S}$ is invariant under local gauge transformations. The gauge field theory is quantized by integrating $e^{\mathcal{S}}$ over all possible values for $U_{n\mu}$

$$Z(g_0^2) = \prod_n \prod_\mu \int dU_{n\mu} e^{\mathcal{S}} = \int DU e^{\mathcal{S}} < \infty \tag{7.1.7}$$

where $dU_{n\mu}$ is the invariant Haar measure on the $SU(\mathcal{N})$ group space defined in Section 2.5.

The partition function is perfectly finite since each group integration $dU_{n\mu}$ is over a compact space. There is no need to fix a gauge to make the theory finite. In the strong coupling regime for which $g_0 \gg 1$, one can expand $\exp\{\mathcal{S}\}$ as a power series in the action $\mathcal{S}$; the compact integrations $\int DU$ then exhibit the area law for the Wilson loops and lead to the confinement of quarks.

7.2   The Weak Coupling Approximation

The Lagrangian of the lattice gauge theory is studied for its weak coupling behavior, namely in the limit of $g_0 \to 0$ [Baaquie (1977b)]. This limit is of central importance since it is well known that non-Abelian gauge fields are asymptotically free and hence have $g_0 \to 0$ at short distances. As one separates the quarks, the coupling constant grows continuously and the gauge field becomes strongly coupled at large distances. One needs to show that the theory goes over smoothly from the weak coupling to the strong coupling sectors, without a phase transition separating these two limiting cases.

Recall from Eq. 7.1.1 $B_{n\mu}^a$ be the local lattice spacetime gauge field, ϕ_n^a be a local scalar field, and let $f_{n\mu\nu}^a$ be the local gauge field tensor. Then

$$W_{n\mu\nu} = e^{if_{n\mu\nu}^a X^a} \quad ; \quad U_{n\mu} = e^{iB_{n\mu}^a X^a} \quad ; \quad \Phi_n = e^{i\phi_n^a X^a}$$

$B_{n\mu}^a$ and $f_{n\mu\nu}^a$ are compact (bounded) variables that take values in the compact parameter space of $SU(\mathcal{N})$. We consider the case when $B_{n\mu}^a \simeq 0$. Using the CBH-equation $e^A e^B = e^{A+B+[A,B]/2+\cdots}$ yields [Baaquie (1977b)]

$$f_{n\mu\nu}^a = \Delta_\mu B_{n\mu}^a - \Delta_\nu B_{n\mu}^a - \frac{1}{2} C^{abc} \Big(B_{n+\hat{\nu},\mu}^b B_{n+\hat{\mu},\nu}^c + B_{n+\hat{\nu},\mu}^b B_{n,\nu}^c + B_{n\mu}^b B_{n+\hat{\mu},\nu}^c$$

$$- B_{n\mu}^b B_{n\nu}^c - B_{n\mu}^b B_{n+\hat{\nu},\mu}^c - B_{n+\hat{\mu},\nu}^b B_{n,\nu}^c \Big) + O(B^3) = -f_{n\nu\mu}^a \tag{7.2.1}$$

where repeated *group indices* are summed over and

$$\Delta_\mu h_n \equiv h_{n+\hat{\mu}} - h_n$$

is the finite lattice derivative. In general, the lattice field tensor $f_{n\mu\nu}^a$ is an infinite power series of the $\{B_{n\mu}^a, B_{n\hat{\mu}\nu}^a, B_{n+\hat{\nu}\mu}^a, B_{n\nu}^a\}$ variables and is an analytic function of these variables as a consequence of the group multiplication law.

From Eq. 7.1.6, the gauge transformation on the $B_{n\mu}^a$ variables is given by

$$e^{iB_{n\mu}^a(\phi)X^a} = e^{i\phi_n^a X^a} e^{iB_{n\mu}^a X^a} e^{-i\phi_{n+\hat{\mu}}^a X^a} \tag{7.2.2}$$

Using the equation

$$\exp(A)\exp(B) = \exp\left\{A + B + \frac{1}{2}[A,B] + \frac{1}{12}[[A,B],B] - \frac{1}{12}[[A,B],A] + \cdots\right\}$$

yields

$$B_{n\mu}^a(\phi) = B_{n\mu}^a - \Delta_\mu \phi_n^a - \frac{1}{2}C^{abc}(\phi_n^b + \phi_{n+\hat{\mu}}^b)B_{n\mu}^c + \frac{1}{2}C^{abc}\phi_n^b\phi_{n+\hat{\mu}}^c$$
$$+ \frac{1}{12}C^{abe}C^{cde}B_{n\mu}^b B_{n\mu}^c(\phi_n^d - \phi_{n+\hat{\mu}}^d) + O(\phi^3, B^3\phi) \tag{7.2.3}$$

For the four-dimensional asymmetric lattice with time lattice ϵ and space lattice a, the continuum limit is obtained by defining the continuum gauge field A_μ as follows

$$x = na \;\; ; \;\; B_{n0}^a = sg_0\epsilon A_0^a(x) \;\; ; \;\; B_{ni}^a = sg_0 a A_i^a(x) \;\; i = 1,2,3 \tag{7.2.4}$$

From Eq. 7.2.5, taking the limit of $\epsilon, a \to 0$ yields

$$f_{nij}^a = sg_0 a^2\left(\partial_i A_j^a - \partial_j A_i^a - g_0 s C^{abc} A_i^b A_j^c\right) = sg_0 a^2 F_{ij}^a(x) \tag{7.2.5}$$

$$f_{n,0i}^a = sg_0 a\epsilon\left(\partial_0 A_i^a - \partial_i A_0^a - g_0 s C^{abc} A_0^b A_i^c\right) = sg_0 a\epsilon F_{0i}^a(x) \tag{7.2.6}$$

From Eq. 7.1.3, expanding the lattice Lagrangian to $O(B^4)$ yields

$$\mathcal{L}(n) = \frac{1}{4g_0^2}\frac{a}{\epsilon}\sum_{i=1}^{3} 2\frac{i^2}{2}\cdot\mathrm{Tr}(f_{n,0i}^a X^a)^2 + \frac{1}{2g_0^2}\frac{\epsilon}{a}\sum_{i\neq j=1}^{3}\frac{i^2}{2}\mathrm{Tr}(f_{nij}^a X^a)^2 + O(B^5)$$

$$= -\frac{1}{4g_0^2}\frac{a}{\epsilon}(g_0 a\epsilon)^2\sum_{a,i=1}^{3}(F_{0i}^a(x))^2 - \frac{1}{4g_0^2}\frac{\epsilon}{a}(g_0 a^2)^2\sum_{i\neq j=1}^{3}(F_{ij}^a(x))^2$$

$$= -\frac{1}{4}a^3\epsilon\sum_{\mu\nu;a}(F_{\mu\nu}^a)^2 \;\; ; \;\; F_{\mu\nu}^a = \partial_\mu A_\nu^a - \partial_\nu A_\mu^a - g_0 s C^{abc} A_\mu^b A_\nu^c \tag{7.2.7}$$

Eqs. 7.2.7 and 7.1.4 yield the continuum Yang-Mills action

$$S = \sum_n \mathcal{L}(n) = -\frac{1}{4}a^3\epsilon\sum_n\sum_{\mu\nu;a}\int d^4x (F_{\mu\nu}^a)^2 = -\frac{1}{4}\sum_{\mu\nu;a}\int d^4x (F_{\mu\nu}^a)^2$$

To define the continuum gauge transformation, define

$$\phi_n^a \to \phi^a(x) \;\; \Rightarrow \;\; \Phi_n \to \Phi(x) = e^{i\phi^a(x)X^a}$$

From Eqs. 7.2.2 and 7.2.4, gauge transformation for the continuum is given by

$$\mathbb{I} + iasg_0 A_\mu^a(\Phi) X^a = \Phi(x)[\mathbb{I} + iasg_0 A_\mu^a X^a]\Phi^\dagger(x + \hat{\mu})$$

$$\Rightarrow \quad A_\mu^a(\phi) X^a = \Phi(x) A_\mu^a X^a \Phi^\dagger(x) - i\frac{1}{sg_0}\Phi(x)\frac{\partial \Phi^\dagger(x)}{\partial x_\mu}$$

Using the definition of vielbien and adjoint representation given in Eqs. 2.3.3 and 2.2.6, respectively yields

$$A_\mu^a(\phi) = \rho_{ab}(\phi) A_\mu^b + \frac{1}{sg_0} e_{ab}(\phi)\frac{\partial \phi^b(x)}{\partial x_\mu}$$

with, in matrix notation

$$\rho(\phi) = e^{-\omega} \quad ; \quad e = \frac{1}{\omega}(1 - e^{-\omega}) \quad ; \quad \omega_{ab} = C^{a\alpha b}\phi^\alpha(x)$$

7.3   Gauge-Fixing the Lagrangian

For concreteness, consider the effect of the gauge transformation on the path integral of the action functional. The action is gauge invariant under gauge transformations

$$\mathcal{S}[W(\Phi)] = \mathcal{S}[W] \quad : \text{ independent of } \Phi_n$$

Although the path integral is well defined, it is not suited for Feynman perturbation expansion since, due to gauge invariance, all the gauge field variables $B_{n\mu}^a$ are not constrained to be small as $g_0 \to 0$. The gauge-fixing term is introduced to constrain all variables to be small. This means, in terms of the original variables $B_{n\mu}^a$, that the action has added to it a term that necessarily breaks gauge invariance.

To leave the gauge-invariant sector unchanged by the gauge-fixing procedure a counter-term, also called a ghost action, needs to be added to the action. The counter-term is a gauge-invariant functional of the gauge field and is evaluated from the gauge-fixing term via a path integral.

Let $\mathcal{S}_\alpha$ be the gauge-fixing term and $\mathcal{S}_c$ be the counter-term. The modified action is defined as

$$\mathcal{S}' = \mathcal{S} + \mathcal{S}_\alpha + \mathcal{S}_c$$

The actions $\mathcal{S}$ and $\mathcal{S}'$ give the same gauge-invariant expectation values.

7.4   Zero Mode

In spite of gauge-fixing the lattice gauge field, note for any finite lattice, there is a *single* degree of freedom that is still unconstrained by the gauge fixing. This is because on the lattice, there are only $N^4 - 1$ independent gauge transformations and hence even after gauge-fixing there is *one* degree of freedom that is left unconstrained.

Consider a four-dimensional symmetric periodic lattice of period N. Choosing a gauge for the finite lattice requires one to partition the lattice into domains that consist of the bulk and boundary lattice sites. The various domains are shown explicitly for $d = 3$ in Figure 8.3, and the construction generalizes to any d dimensions. To keep track of the single variable that is not constrained by the gauge-fixing, define the *single* lattice site $\vec{N} \equiv (N, N, N, N)$.[1] On gauge-fixing the gauge field, all the field variables can be approximated by letting the range of each degree of freedom $B^a_{n\mu}$ take values in the range of $[-\infty, +\infty]$, and hence allowing one to use Gaussian integration for developing a Feynman perturbation expansion about the free field. *However*, the *single* degree of freedom variable $B^a_{\vec{N}\mu}$ – with $U_{\vec{N}\mu} = e^{iB^a_{\vec{N}\mu}X^a}$ continues to take values in the compact group space, that is $B^a_{\vec{N}\mu} \in G$; in particular note that $B^a_{\vec{N}\mu} \notin [-\infty, +\infty]$.

In the case of the Abelian Maxwell gauge field, both for the lattice and continuum, the zero mode of the gauge field decouples from the theory (in the absence of fermions) since the Abelian theory is linear.

Although this single variable may seem unimportant and an artifact of the lattice this is far from true. The single variable $B^a_{N\mu}$ will later be mapped, via a discrete Fourier transform, to the zero momentum degree of freedom, namely the zero mode. It is a remarkable feature of the lattice gauge field on a finite lattice, due to the discrete Fourier transform, that the zero mode can be separated out from the other degrees of freedom. The fact that the zero mode can be isolated is the fundamental basis for a program of studying mass renormalization of the lattice gauge field for a finite lattice, and is further discussed in Section 9.

7.5 Gauge-Fixed Path Integral

One has a wide choice as to what functional of the field variables $\mathcal{S}_\alpha$ should be. The only necessary condition is that

$$Z = \prod_{n \neq N} \prod_{\mu, a} \int_{-\infty}^{\infty} dB^a_{n\mu} \mu(B^a_{n\mu}) \int_G \prod_\mu dU_{N\mu} e^{\mathcal{S} + \mathcal{S}_\alpha + \mathcal{S}_c} < \infty \qquad (7.5.1)$$

where $\mu(B^a_{n\mu})$ is the Haar measure and $N \equiv (N, N.N, N)$.

On the finite lattice, since there are only $N^4 - 1$ independent gauge transformations, choose $\Phi_N = \mathbb{I}$. To define $\mathcal{S}_c$ introduce the following notation

$$U^{(\phi)}_{n\mu} = \Phi_n U_{n\mu} \Phi^\dagger_{n+\hat{\mu}} \quad ; \quad \Phi_N = \mathbb{I}$$

$$D\Phi = \prod_{n \neq N} d\Phi_n \quad ; \quad DU = \prod_n \prod_\mu dU_{n\mu}$$

Define $\mathcal{S}_c$ by

$$e^{\mathcal{S}_c[U]} = 1 / \int D\Phi e^{\mathcal{S}_\alpha[U^{(\phi)}]} \quad : \quad \text{gauge invariant} \qquad (7.5.2)$$

Note the identity

$$1 = \int D\Phi e^{\mathcal{S}_\alpha[U^{(\phi)}]} / \int D\Phi' e^{\mathcal{S}_\alpha[U^{(\phi')}]}$$

---

[1]For notational convenience $\vec{N} \equiv (N, N, N, N)$ will be denoted by N unless necessary.

Let $K[U]$ be an arbitrary *gauge-invariant* function. Perform the *inverse* gauge transformation on $\{U_{n\mu}\}$ variables, namely

$$U'_{n\mu} = \Phi_n^\dagger U_{n\mu} \Phi_{n+\hat{\mu}} \quad \Rightarrow \quad DU' = DU$$

$$e^{\mathcal{S}_c[U^{(\phi)}]} = e^{\mathcal{S}_c[U']} \quad \Rightarrow \quad K[U] e^{\mathcal{S}_c[U^{(\phi)}]} = K[U'] e^{\mathcal{S}_c[U']}$$

We hence have the following

$$\int DU K[U] = \left(\int D\Phi \right) \left\{ \int DU' K[U'] e^{\mathcal{S}_\alpha [U'] + \mathcal{S}_c[U']} \right\}$$

$$= \int DU K[U] e^{\mathcal{S}_\alpha[U] + \mathcal{S}_c[U]}$$

using the gauge group is compact to obtain

$$\int D\Phi = 1$$

We thus see that $e^{\mathcal{S}_\alpha + \mathcal{S}_c}$ leaves the gauge-invariant sector *unchanged*. Hence, in particular, for $K[U] = \exp\{\mathcal{S}\}$

$$Z(g_0) = \int DU e^{\mathcal{S}[U]} = \int DU e^{\mathcal{S} + \mathcal{S}_\alpha + \mathcal{S}_c} < \infty \qquad (7.5.3)$$

Result Eq. 7.5.3 is valid exactly for the lattice theory.

Lattice gauge theory, in summary, defines a completely finite quantum field theory ($Z(g_0) < \infty$) due to a) the finite lattice rendering the degrees of freedom to be finite and b) the divergences due to gauge invariance are rendered finite due to the compactness of the gauge group. The lattice formulation reduces to the continuum case in the weak-coupling approximation.

7.6 The Faddeev-Popov Counter-Term

Choose a specific $\mathcal{S}_\alpha$ and calculate $\mathcal{S}_c$ for it. Consider the following gauge-fixing term

$$e^{\mathcal{S}_\alpha[B]} = \prod_{n,a}' \delta(s_n^a - t_n^a) \quad ; \quad \prod_n' = \prod_{n \neq N}$$

where $\{t_n^a\}$ are fixed numbers and

$$s_n^a = \Delta_\mu B_{n-\hat{\mu},\mu}^a = \Delta_\mu (B_{n,\mu}^a - B_{n-\hat{\mu},\mu}^a) \quad \Rightarrow \quad \sum_n s_n^a = 0 \qquad (7.6.1)$$

Hence, for the s_n^a, there are $(N^4 - 1)$-independent variables as well.

Recall that $B_{n,\mu}^a(\phi)$ is defined by

$$\exp[i B_{n,\mu}^a(\phi) X^a] = \Phi_n U_{n\mu} \Phi_{n+\hat{\mu}}^\dagger$$

and

$$s_n^a(\phi) = \sum_\mu \Delta_\mu (B_{n-\hat{\mu},\mu}^a(\phi)) \quad \Rightarrow \quad \sum_n s_n^a(\phi) = 0 \qquad (7.6.2)$$

Hence, as expected, there are only $N^4 - 1$ independent variables for $s_n^a(\phi)$.

From Eq. 7.5.2, using the δ-functions in the numerator yields

$$e^{\mathcal{S}_\alpha + \mathcal{S}_c} = \frac{\prod'_{n,a} \delta(s_n^a - t_n^a)}{\int D\phi \prod'_n \delta(s_n^a(\phi) - t_n^a)} = \frac{\prod'_{n,a} \delta(s_n^a - t_n^a)}{\int D\phi \prod'_n \delta(s_n^a(\phi) - s_n^a)} \tag{7.6.3}$$

The combined effect of $e^{\mathcal{S}_\alpha + \mathcal{S}_c}$ is to leave the gauge-invariant sector unchanged and yields

$$Z(g_0) = \int DU e^{\mathcal{S} + \mathcal{S}_\alpha + \mathcal{S}_c} = \int DU e^{\mathcal{S}}$$

Since $Z(g_0)$ is independent of $\{t_n^a\}$ and it follows that

$$Z(g_0) = (\text{constant}) \prod_{n,a}' \int_{-\infty}^{+\infty} Z(g_0) \exp\left\{ -\frac{\alpha}{2} \sum_{n,a}' (t_n^a)^2 \right\}$$

$$= \int DU e^{\mathcal{S} + \mathcal{S}_c + \mathcal{S}_\alpha} \equiv \int DU e^{\mathcal{S}'}$$

where

$$\mathcal{S}_\alpha = -\frac{\alpha}{2} \sum_{n,a}' (s_n^a)^2 \quad ; \quad \sum_n' = \sum_{n \neq N}$$

and from Eq. 7.6.3

$$e^{-\mathcal{S}_c} = \int D\phi \prod_{n,a}' \delta(s_n^a(\phi) - s_n^a) \tag{7.6.4}$$

$\mathcal{S}_\alpha + \mathcal{S}_c$ is not gauge invariant, but it has a lower symmetry, which is the BRST symmetry. Lattice BRST is discussed in Section 7.9.

In summary

$$\mathcal{S}_{GF} = \mathcal{S} + \mathcal{S}_\alpha + \mathcal{S}_c$$

Let $\alpha = O(1/g_0^2)$; then the modified action $\mathcal{S}_{GF} = \mathcal{S} + \mathcal{S}_\alpha + \mathcal{S}_c$, restricts *all* the variables (except $B_{N\mu}^a$) to be $O(g_0)$ and is suited for studying the weak coupling sector using Feynman diagrams.

7.7 Abelian Gauge-Fixed Path Integral

Before analyzing the non-Abelian case, the results for the Abelian gauge field are summarized. The gauge field action for the symmetric lattice, from Eq. 7.1.5, is given by

$$S = \frac{1}{2g_0^2} \sum_n \sum_{\mu \neq \nu} e^{if_{n\mu\nu}} = \frac{1}{2g_0^2} \sum_n \sum_{\mu \neq \nu} \cos\{f_{n\mu\nu}\}$$

$$f_{n\mu\nu} = \Delta_\mu B_{n\mu} - \Delta_\nu B_{n\mu} \tag{7.7.1}$$

The weak coupling limit of the action, from Eq. 7.7.1, is given by

$$S \simeq -\frac{1}{4g_0^2}\sum_n\sum_{\mu\neq\nu}(f_{n\mu\nu})^2\}$$ (7.7.2)

From Eq. 7.2.2, gauge transformations are given by

$$B_{n,\mu}(\phi) = B_{n,\mu} - \phi_{n+\hat{\mu}} + \phi_n$$

A covariant gauge-fixing action is given by

$$S_\alpha = -\frac{\alpha}{2}\sum_n\sum_\mu(\Delta_\mu B_{n-\hat{\mu},\mu})^2$$

From Eq. 7.7.2

$$S \simeq -\frac{1}{2g_0^2}\sum_n\sum_{\mu,\nu}\left((\Delta_\mu B_{n\nu})^2 - \Delta_\mu B_{n\nu}\Delta_\nu B_{n\mu}\right)$$ (7.7.3)

Note that

$$\sum_{\mu,\nu}\Delta_\mu B_{n\nu}\Delta_\nu B_{n\mu} = \left(\sum_\mu\Delta_\mu B_{n-\hat{\mu},\mu}\right)^2$$

and hence

$$S = -\frac{1}{2g_0^2}\sum_n\left[\sum_{\mu,\nu}\left(\Delta_\mu B_{n\nu}\right)^2 - \left(\sum_\mu\Delta_\mu B_{n-\hat{\mu},\mu}\right)^2\right]$$ (7.7.4)

The covariant gauge-fixing action is chosen to offset the second term in the action. The Feynman gauge consists of choosing $\alpha = 1/g_0^2$ and yields the gauge-fixed action

$$S + S_\alpha = -\frac{1}{2g_0^2}\sum_n\sum_{\mu,\nu}(\Delta_\mu B_{n\nu})^2$$ (7.7.5)

In the Feynman gauge, the Abelian gauge field consists of *four decoupled* scalar quantum fields.

The Faddeev-Popov counter-term, from Eq. 7.6.4, is given by

$$e^{-S_c} = \int D\phi\prod_{n,a}^{\prime}\delta(s_n^a(\phi) - s_n^a)$$

For the Abelian case

$$s_n(\phi) - s_n = \sum_\mu\Delta_\mu B_{n-\hat{\mu},\mu}(\phi) - \sum_\mu\Delta_\mu B_{n-\hat{\mu},\mu} = -\Delta^2\phi_n$$

where the lattice Laplacian is given by

$$\Delta^2 h_n = \sum_\mu(h_{n+\hat{\mu}} - 2h_n + h_{n-\hat{\mu}})$$

Hence

$$e^{-S_c} = \int D\phi\prod_n\delta(-\Delta^2\phi_n) = \int D\phi\frac{1}{\det(\Delta^2)}\prod_n\delta(\phi_n)$$

$$\Rightarrow\quad e^{S_c} = \det(\Delta^2)$$

The Faddeev-Popov counter-term for the Abelian case does not depend on the gauge field. Fermion integration can be used to represent the determinant. The ghost fields $c, \bar{c}$ are complex, and from Eq. 6.6.9 the counter-term can be written as

$$e^{\mathcal{S}_c} = \int Dc D\bar{c} \exp\left\{ -\sum_n \sum_\mu \bar{c}_n \Delta^2 c_n \right\} = \int Dc D\bar{c} \exp\left\{ \sum_n \sum_\mu \Delta_\mu \bar{c}_n \Delta_\mu c_n \right\}$$

The fields $\bar{c}, c$ are complex scalar fermions and do not obey the spin-statistic rule that requires that all fermions must have half-integer spin, given by $1/2, 3/2, ...$; for this reason $\bar{c}, c$ are called Faddeev-Popov *ghost fields*.

The fermionic ghost field has spin zero – dictated by the rules of fermion integration – and hence does not represent a physically realizable quantum field with its own particles and so on. Nevertheless, ghost fields play a key role in the study of gauge field theories – and superstring theory as well – and their full significance is mysterious and has not yet been fully understood.

The gauge fixed lattice path integral for the Abelian field is

$$Z = \prod_{n,\mu} \int dB_{n\mu} d\bar{c}_n dc_n e^{\mathcal{S} + \mathcal{S}_\alpha + \mathcal{S}_c}$$

Consider the continuum limit of the Abelian gauge field. From Eq. 7.2.7, for the Lagrangian for the Abelian case

$$\mathcal{L}(n) = -\frac{1}{4} a^3 \epsilon \sum_{\mu\nu} (F_{\mu\nu})^2 \quad ; \quad F_{\mu\nu} = \partial_\mu A_\nu - \partial_\nu A_\mu \tag{7.7.6}$$

The continuum action is given by

$$\mathcal{S} = \sum_n \mathcal{L}(n) = -\frac{1}{4} a^3 \epsilon \sum_n \sum_{\mu\nu} \int d^4 x (F_{\mu\nu})^2 = -\frac{1}{4} \sum_{\mu\nu} \int d^4 x (F_{\mu\nu})^2$$

In summary, the gauge-fixed continuum action for the Abelian gauge field, with a covariant gauge-fixing term, is given by

$$\mathcal{S}_{GF} = -\frac{1}{4} \int d^4 x \sum_{\mu\nu} (F_{\mu\nu})^2 - \frac{\alpha}{2} \int d^4 x \left(\sum_\mu \partial_\mu A_\mu \right)^2 + \int d^4 x \partial_\mu \bar{c} \partial_\mu c$$
$$= \mathcal{S} + \mathcal{S}_\alpha + \mathcal{S}_c \tag{7.7.7}$$

with the path integral given by

$$Z = \int DA_\mu D\bar{c} Dc \, e^{\mathcal{S}_{GF}}$$

7.8 Lattice Faddeev-Popov Non-Abelian Ghost Action

The ghost action, for a finite lattice, is given by

$$e^{-\mathcal{S}_c} = \int D\Phi \prod_{n,a}' \delta(s_n^a(\phi) - s_n^a) \tag{7.8.1}$$

where, from Eq. 7.6.1

$$s_n^a = \sum_\mu \Delta_\mu B_{n+\hat{\mu},\mu}^a$$

For $B_{n,\mu}^a(\phi)$ defined by

$$\exp[iB_{n,\mu}^a(\phi)X^a] = \Phi_n U_{n\mu}\Phi_{n-\hat{\mu}}^\dagger$$

Eq. 7.6.2 yields

$$s_n^a(\phi) = \sum_\mu \Delta_\mu B_{n-\hat{\mu},\mu}^a(\phi)$$

Define $u_n^a = u_n^a(\phi)$ by

$$u_n^a \equiv s_n^a(\phi) - s_n^a$$

where $s_n^a = s_n^a(\phi)|_{\phi=0}$. Make a change of variable from $\{\phi_n^a\}$ to $\{u_n^a\}$ to evaluate Eq. 7.8.1. The δ functions make $u_n^a(\phi) = 0$; this in turn implies $\phi_n^a = 0$ as the unique solution for which $u_n^a = 0$ (as long as $B_{n\mu}^a \simeq 0$). Then from Eq. 7.8.1

$$e^{-\mathcal{S}_c[B]} = \int D\Phi \prod_{n,a}' \delta(s_n^a(\phi) - s_n^a) = \int D\Phi \prod_{n,a}' \delta(u_n^a) \tag{7.8.2}$$

Noteworthy 7.1: Momentum space

To define the degrees of freedom in momentum space, consider the Fourier transform of the variables of h_n, which is an arbitrary function of n. The torus structure of the lattice yields that $h_{n+N} = h_n$: periodic in all the coordinates, with period N. Hence h_n can be expanded in terms of the basis functions

$$\{e^{ik_\mu n_\mu}\}, k_\mu = 0, 2\pi/N, \ldots 2\pi(N-1)/N \ ; \ \delta_{k,q} \equiv N^4 \prod_{i=0}^{2} \delta_{k_i+q_i}$$

That is

$$h_n = \sum_k e^{ikn} h_k \equiv \frac{1}{N^4}\left(\prod_\mu \sum_{k_\mu=0}^{2\pi(N-1)/N} e^{ik_\mu n_\mu}\right) h_k \tag{7.8.3}$$

$$h_k = \sum_n e^{-ikn} h_n = \left(\prod_\mu \sum_{n_\mu=0}^{N} e^{ik_\mu n_\mu}\right) h_n \ : \ k_\mu \equiv 2\pi n_\mu/N$$

Define the Fourier transforms

$$u_n^a = \sum_k e^{ikn} u_k^a \ ; \ B_{n\mu}^a = \sum_k e^{ikn} B_{k\mu}^a \ ; \ \phi_n^a = \sum_k e^{ikn} \phi_k^a$$

The variable $u_n^a = u_n^a(\phi)$ is analyzed. From Eqs. 7.6.2 and 7.2.3

$$u_k^a = \sum_k e^{-ikn} u_n^a$$

$$= \sum_\mu |1 - e^{ik_\mu}|^2 \phi_k^a + \frac{1}{2}C^{abc}\sum_{k,\mu}(1 - e^{-ik_\mu})(1 + e^{-iq_\mu})B_{k-q,\mu}^b \phi_q^c \tag{7.8.4}$$

$$+ \frac{1}{12}C^{abc}C^{cde}\sum_{p,q}\sum_\mu (1 - e^{-ik_\mu})(1 - e^{-iq_\mu})B_{k-p-q,\mu}^b B_{p\mu}^c \phi_q^d + O(B^3\phi, B^2\phi^2)$$

Note from above equation that $u^a_{k=0} = 0$; in particular, it is not coupled to the ϕ^a_n. Hence $u^a_{k=0}$ can be redefined to be

$$u^a_{k=0} = \phi^a_{k=0}$$

Define

$$d_k = 1 \ \text{if} \ k = 0 \ ; \ d_k = \sum_\mu |1 - e^{ik_\mu}|^2 \ \text{if} \ k \neq 0,$$

Then, from Eq. 7.8.4

$$u^a_k = d_k \phi^a_k + \sum_q (M^{ab}(k,q) + L^{ab}(k,q))\phi^b_q$$

$$\equiv \sum_q T^{ab}(k,q)\phi^b_q \tag{7.8.5}$$

where

$$M^{ad}(k,q) = \frac{1}{2}C^{abc}\sum_{k,\mu}(1 - e^{-ik_\mu})(1 + e^{-iq_\mu})B^b_{k-q,\mu} \tag{7.8.6}$$

and

$$L^{ad}(k,q) = \frac{1}{12}C^{abe}C^{cde}\sum_{p,q}\sum_\mu(1 - e^{-ik_\mu})(1 - e^{-iq_\mu})B^b_{k-p-q,\mu}B^c_{p\mu}$$

Eq. 7.8.5 yields

$$Du = \det(T)D\phi$$

Due to the delta-functions in Eq. 7.8.2, $D\Phi = D\phi$. Hence the counter-term, from Eq. 7.8.2, is given by

$$e^{\mathcal{S}_c[B]} = \det(T)$$

Using property $\det(AB) = \det A \det B$ and that d_k is independent of the gauge field $\{B^a_{n\mu}\}$ yields, up to an overall constant is independent of the gauge field, the following

$$e^{\mathcal{S}_c[B]} = \det(d)\det\left(1 + \frac{1}{d}M + \frac{1}{d}L\right) \simeq \det(d)\det\left(1 + \frac{1}{d}M\right)\det\left(\frac{1}{d}L\right)$$

$$= \det(d + M)\exp\left[\text{Tr}(\frac{1}{d}L)\right] \tag{7.8.7}$$

Consider

$$\text{Tr}\left(\frac{1}{d}L\right) = \sum_a \sum_k \frac{1}{d_k}L^{aa}(k,k)$$

$$= -\frac{\mathcal{N}}{12}\left\{\sum_k \frac{1}{d_k}\sum_\mu |1 - e^{ik_\mu}|^2\right\} \times \sum_{q,a,\nu} B^a_{-q\nu}B^a_{q\nu}$$

Using the identities

$$\sum_k \frac{1}{d_k}|1 - e^{ik_\mu}|^2 = \frac{1}{4} \ \text{and} \ C^{abc}C^{abc} = \mathcal{N}\delta^{aa}$$

yields

$$\mathrm{Tr}\left(\frac{1}{d}L\right) = -\frac{\mathcal{N}}{48}\sum_k\sum_{a,\mu} B^a_{-k\mu}B^a_{k\mu} = -\frac{\mathcal{N}}{48}\sum_n\sum_{a,\mu}(B^a_{n\mu})^2$$

Note that $\mathrm{Tr}[(1/d)L]$ is completely local. Hence, from Eq. 7.8.7

$$e^{\mathcal{S}_c[B]} = \det(d_k\delta^{ab}\delta_{k,q} + M^{ad}(k,q)) \times \exp\left[-\frac{\mathcal{N}}{48}\sum_{n\mu a}(B^a_{n\mu})^2\right] + O(B^3) \qquad (7.8.8)$$

Eq. 7.8.8 is the Faddeev-Popov ghost action (counter term), accurate to $O(B^2)$. The determinant in the expression for $\mathcal{S}_c[B]$ can be represented by fermion integration variables as given in Eq. 6.6.9. Let the complex *ghost fermion fields* be $c^a_n, c^{a\dagger}_n$; the Faddeev-Popov ghost action is given by

$$e^{\mathcal{S}_c[B]} = \exp\left[-\frac{\mathcal{N}}{48}\sum_{n\mu a}(B^a_{n\mu})^2\right] \times \prod_{n,a}\int dc^a_n dc^{a\dagger}_n \exp\{-c^\dagger(d+M)c\} \qquad (7.8.9)$$

$$(d+M)^{ab}(k,q) = d_k\delta^{ab}\delta_{k,q} + M^{ad}(k,q) + O(B^3)$$

Unlike the continuum case, for the ghost action the lattice gauge field yields an infinite order polynomial in the lattice gauge field $B^a_{n\mu}$. In particular, the extra local term $\sum_{n\mu a}(B^a_{n\mu})^2$ is absent in the continuum formulation. This term is *quadratically divergent* and plays an important role in ensuring that there is no mass renormalization necessary for the lattice gauge field.

7.9   Lattice BRST Symmetry

BRST symmetry for the Abelian case, and in continuum spacetime, was studied in great detail in Baaquie (2018). For the Abelian case, the theory is linear and hence all the consequences of BRST symmetry can be worked out explicitly. In general, BRST symmetry yields a state space is defined by BRST cohomology. It was shown in Baaquie (2018) that for the Abelian case, the BRST state space is given by the Gupta-Bleuler analysis for a covariant gauge.

For the non-Abelian gauge field, such an explicit analysis is not possible due to the nonlinearities of the gauge field. However, the important result that BRST cohomology yields the correct state space continues to hold for non-Abelian gauge fields [Zinn-Justin (1993)]. The BRST transformation is shown to naturally extend to the lattice gauge theory. The main focus of the BRST symmetry for the lattice case is to show that zero mass renormalization, to all orders in perturbation theory, holds for the lattice gauge field.

Recall that in the last section it was shown that

$$Z = \int DU e^{\mathcal{S}} = \int DU e^{\mathcal{S}+\mathcal{S}_\alpha+\mathcal{S}_c}$$

Note that $\mathcal{S}_\alpha$ of necessity breaks gauge invariance. However, the term $\mathcal{S}_\alpha + \mathcal{S}_c$, is invariant under the BRST transformation. This invariance is more restricted than gauge invariance, but its usefulness lies in that it holds for the gauge theory in the presence of gauge fixing.

To define the BRST transformation, e^{S_c} is first rewritten in a more formal way. For an infinite-size lattice, from Eq. 7.6.4

$$e^{-S_c} = \int D\phi \prod_{n,a} \delta(s_n^a(\phi) - s_n^a)$$

The value of $\{\phi_n^a\}$ for which the δ functions are satisfied is $\phi_n^a = 0$. For an infinite lattice, up to an irrelevant constant

$$D\phi = \prod_{n,a} d\phi_n^a \mu(\phi_n) \to \prod_n \mu(\phi_n = 0) \prod_{n,a} d\phi_n^a = \prod_{n,a} d\phi_n^a$$

Therefore, up to an irrelevant constant

$$e^{-S_c} = \prod_{n,a} \int d\phi_n^a \delta(s_n^a(\phi) - s_n^a)$$

Make the following change of variable, from $\{\phi_n^a\}$ to $\{\zeta_n^a\}$

$$\zeta_n^a = s_n^a(\phi) - s_n^a$$

Define the notation

$$\frac{\partial h_n^a}{\partial \phi_m^b} \equiv \frac{\partial h_n^a(\phi)}{\partial \phi_m^b}\bigg|_{\phi=0} \qquad (7.9.1)$$

Noteworthy 7.2: Commutation equation

The definition of differentiation given in Eq. 7.9.1 obeys the commutation equations of the underlying Lie algebra. To illustrate this, let B be a constant matrix and consider the transformation given by[2]

$$B(\phi) = UBU^\dagger \; ; \; U = \exp\{i\phi_\beta X^\beta\}$$

Then, from Eq. 7.9.1

$$\frac{\partial B}{\partial \phi_a} = \frac{\partial B(\phi)}{\partial \phi_a}\bigg|_{\phi=0} = \frac{\partial}{\partial \phi_a}\left(B + i[\phi_\beta X^\beta, B]\right)\bigg|_{\phi=0} = i[X^a, B] \qquad (7.9.2)$$

To evaluate the second differentiation, from Eq. 7.9.2

$$\frac{\partial^2 B}{\partial \phi_b \partial \phi_a} = \left[\frac{\partial}{\partial \phi_b}\left(\frac{\partial B}{\partial \phi_a}\right)(\phi)\right]\bigg|_{\phi=0} = \left[\frac{\partial}{\partial \phi_b}\left(i[X^a, B(\phi)]\right)\right]\bigg|_{\phi=0} = i^2[X^a, [X^b, B]]$$

Hence from above equation

$$\left[\frac{\partial}{\partial \phi_a}, \frac{\partial}{\partial \phi_b}\right]B = \left(\frac{\partial^2}{\partial \phi_a \partial \phi_b} - \frac{\partial^2}{\partial \phi_b \partial \phi_a}\right)B$$
$$= [X^a, [X^b, B]] - [X^b, [X^a, B]] \qquad (7.9.3)$$

---

[2]U can be a constant gauge transformation.

The right hand side of above equation Eq. 7.9.3, using the Jacobi identity given in Eq. 2.2.2, yields

$$[[B, X^a], X^b] + [X^a, [B, X^b]] = [B, [X^a, X^b]]$$

$$= iC^{abc}[B, X^c] = -iC^{abc}[X^c, B] = -C^{abc}\frac{\partial B}{\partial \phi_c} \qquad (7.9.4)$$

Hence, from Eqs. 7.9.3 and 7.9.4

$$\left[\frac{\partial}{\partial \phi_a}, \frac{\partial}{\partial \phi_b}\right] = C^{abc}\frac{\partial}{\partial \phi_c} \qquad (7.9.5)$$

It follows from Eqs. 7.9.5 that for the field ϕ_n

$$\left[\frac{\partial}{\partial \phi_n^a}, \frac{\partial}{\partial \phi_m^b}\right] = \delta_{n,m} C^{abc}\frac{\partial}{\partial \phi_n^c} \qquad (7.9.6)$$

Due to the δ functions the Jacobian of the transformation is evaluated for $\phi_n^a = 0$ and yields

$$d\zeta_n^a = \left.\frac{\partial s_n^a(\phi)}{\partial \phi_m^b}\right|_{\phi=0} d\phi_m^b = \frac{\partial s_n^a}{\partial \phi_m^b} d\phi_m^b \qquad (7.9.7)$$

with all repeated indices being summed over. Hence

$$\prod_{n,a} d\zeta_n^a = \det\left(\frac{\partial s_n^a}{\partial \phi_m^b}\right) \prod_{n,a} d\phi_n^a \qquad (7.9.8)$$

and

$$e^{\mathcal{S}_c} = 1/\prod_{n,a}\int d\phi_n^a \delta(s_n^a(\phi) - s_n^a) = \det\left(\frac{\partial s_n^a}{\partial \phi_m^b}\right)/\left\{\prod_{n,a}\int d\zeta_n^a \delta\left(\zeta_n^a\right)\right\}$$

$$= \det\left(\frac{\partial s_n^a}{\partial \phi_m^b}\right) \qquad (7.9.9)$$

To define the BRST transformation, the determinant $e^{\mathcal{S}_c}$ needs to be represented using fermion integration. For $c_n^a, c_n^{\dagger a}$ complex scalar fermion fields, from Eq. 6.6.9

$$e^{\mathcal{S}_c} = \det\left(\frac{\partial s_n^a}{\partial \phi_m^b}\right) = \prod_{n,a}\int dc_n^d dc_n^{\dagger a} \exp\left(\sum_{mn;ab} c_n^{\dagger a}\frac{\partial s_m^b}{\partial \phi_n^a} c_m^b\right) \qquad (7.9.10)$$

Hence, summing over all repeated indices

$$\mathcal{S}_\alpha = -\frac{1}{2}\alpha s_n^\alpha s_n^\alpha \quad ; \quad \mathcal{S}_c = c_n^{\dagger a}\frac{\partial s_m^b}{\partial \phi_n^a} c_m^b \qquad (7.9.11)$$

Let λ be a spacetime-independent *fermionic variable* that anticommutes with all other fermion variables and commutes with bosons. The lattice BRST transformation is defined on the gauge field by the following gauge transformation

$$\exp\{i\phi_n^a X^a\} = \exp\{i\lambda c_n^{\dagger a} X^a\} \quad \Rightarrow \quad \phi_n^a = \lambda c_n^{\dagger a} \qquad (7.9.12)$$

The fermionic variable λ picks out the infinitesimal neighborhood of the gauge transformation.

Define the BRST transformation by

$$\delta_B \mathcal{O} = \mathcal{O}_{BRST} - \mathcal{O}$$

The Abelian case has been discussed in detail in Baaquie (2018). For both the Abelian and non-Abelian case, the BRST transformation is exactly nilpotent since it can be shown that [Zinn-Justin (1993)]

$$\delta_B^2 = 0$$

The Slavnov-Taylor identity, discussed in Section 9.9, is obtained from the BRST symmetry and is an on-shell identity for the two-point correlation function. The lattice Slavnov-Taylor identity is sufficient to show that there is no mass renormalization for the non-Abelian lattice gauge field is an exact result.

The BRST transformation on the ghost fields is defined by[3]

$$\delta_B c_n^\alpha = -\alpha \lambda s_n^a \quad ; \quad \delta_B c_n^{\dagger a} = \frac{1}{2} \lambda C^{abc} c_n^{\dagger b} c_n^{\dagger c} \tag{7.9.13}$$

The BRST transformation for the gauge field is given by

$$\delta_B B_{n\mu}^a = [B_{n\mu}^a(\phi + \lambda c^\dagger) - B_{n\mu}^a(\phi)]\Big|_{\phi=0} = \lambda \frac{\partial B_{n\mu}^a}{\partial \phi_m^b} c_m^{\dagger b} \tag{7.9.14}$$

Since the BRST transformation in Eq. 7.9.14 is a special case of a linearized gauge transformation, it leaves the gauge field action $\mathcal{S}$ invariant and hence

$$\delta_B \mathcal{S} = 0 \quad ; \quad \text{BRST invariant}$$

The BRST variation of the gauge-fixing term yields, from Eq. 7.9.12

$$\delta_B s_n^a = [s_n^a(\phi + \lambda c^\dagger) - s_n^a(\phi)]\Big|_{\phi=0}$$

$$\Rightarrow \quad \delta_B s_n^a = \lambda \frac{\partial s_n^a}{\partial \phi_m^b} c_m^{\dagger b} \quad ; \quad \delta_B \left(\frac{\partial s_n^a}{\partial \phi_m^b} \right) = \lambda \frac{\partial^2 s_n^a}{\partial \phi_m^b \partial \phi_i^\gamma} c_i^{\dagger \gamma}$$

The BRST transformation of the gauge-fixing $\mathcal{S}_\alpha$ yields

$$\mathcal{S}_\alpha \to -\frac{\alpha}{2} \left(s_n^a + \lambda \frac{\partial s_n^a}{\partial \phi_m^b} c_m^{\dagger b} \right)^2 \quad \Rightarrow \quad \delta_B \mathcal{S}_\alpha = -\alpha \lambda s_n^a \frac{\partial s_n^a}{\partial \phi_m^b} c_m^{\dagger b} \tag{7.9.15}$$

and for the ghost counter-term

$$\mathcal{S}_c \to \left(c_n^{\dagger a} + \frac{\lambda}{2} c^{\alpha\beta\gamma} c_n^{\dagger \beta} c_n^{\dagger \gamma} \right) \left(\frac{\partial s_m^b}{\partial \phi_n^a} + \lambda \frac{\partial^2 s_m^b}{\partial \phi_n^d \partial \phi_i^\gamma} c_i^{\dagger \gamma} \right) (c_m^b - \alpha \lambda s_m^b)$$

Simplifications using anticommutation of fermion variables yields

$$\delta_B \mathcal{S}_c = \alpha \lambda c_n^{\dagger a} \frac{\partial s_m^b}{\partial \phi_n^a} s_m^b + \frac{\lambda}{2} C^{\alpha\beta\gamma} \frac{\partial s_m^b}{\partial \phi_n^a} c_n^{\dagger \beta} c_n^{\dagger c} c_m^b - \lambda \frac{\partial^2 s_m^b}{\partial \phi_n^d \partial \phi_i^c} c_n^{\dagger d} c_i^{\dagger c} c_m^b \tag{7.9.16}$$

---

[3]Unlike the Abelian case, for non-Abelian case one can define a transformation that in the Abelian limit consists of $\delta \bar{c} = 0$ and with $\delta c \neq 0$. This gives the same BRST charge in the Abelian limit.

Therefore, Eqs. 7.9.15 and 7.9.16 yield

$$\delta_B(\mathcal{S}_\alpha + \mathcal{S}_c) = \lambda\left(\frac{1}{2}C^{\alpha\beta\gamma}\frac{\partial s_m^b}{\partial\phi_n^\alpha}c_n^{\dagger\beta}c_n^{\dagger\gamma}c_m^b - \frac{\partial^2 s_m^b}{\partial\phi_n^a\partial\phi_i^c}c_n^{\dagger a}c_i^{\dagger c}c_m^b\right) \tag{7.9.17}$$

Consider the following

$$\frac{\partial^2 s_m^b}{\partial\phi_n^d\partial\phi_i^c}c_n^{\dagger d}c_i^{\dagger c}c_m^b = \frac{1}{2}\left\{\frac{\partial^2 s_m^b}{\partial\phi_n^a\partial\phi_i^c}c_n^{\dagger a}c_i^{\dagger c} + \frac{\partial^2 s_m^b}{\partial\phi_n^c\partial\phi_i^a}c_i^{\dagger c}c_n^{\dagger a}\right\}c_m^b$$

$$= \frac{1}{2}\left\{\frac{\partial^2}{\partial\phi_n^a\partial\phi_i^c} - \frac{\partial^2}{\partial\phi_n^c\partial\phi_i^a}\right\}s_m^b c_i^{\dagger a}c_n^{\dagger c}c_m^b = \frac{1}{2}C^{\gamma ac}\left\{\frac{\partial s_m^b}{\partial\phi_n^\gamma}\right\}c_n^{\dagger a}c_n^{\dagger c}c_m^b \tag{7.9.18}$$

since, from Eq. 7.9.6

$$\left[\frac{\partial}{\partial\phi_n^a}, \frac{\partial}{\partial\phi_n^b}\right] = \delta_{n,m}C^{abc}\frac{\partial}{\partial\phi_n^c} \tag{7.9.19}$$

Therefore, the terms in parentheses of Eq. 7.9.17 cancel and yields

$$\delta_B(\mathcal{S}_\alpha + \mathcal{S}_c) = 0 \quad : \text{ BRST invariant} \tag{7.9.20}$$

Hence we have proved that $\mathcal{S} + \mathcal{S}_\alpha + \mathcal{S}_c$, is invariant under the BRST transformation, namely

$$\delta_B(\mathcal{S} + \mathcal{S}_\alpha + \mathcal{S}_c) = 0 \quad : \text{ BRST invariant} \tag{7.9.21}$$

All the steps taken for proving the BRST invariance of the lattice gauge-fixed action can be seen to carry over unchanged to the continuum case. The only expression where the lattice structure enters into lattice BRST transformation is via the gauge-fixing term s_n^a, as given in Eq. 7.9.13.

7.10   Summary

The lattice gauge field was defined on a four-dimensional Euclidean spacetime lattice using the finite group elements as the degrees of freedom. The Wilson plaquette provides a generalization of the Yang-Mills field tensor and leads to a quantized theory that, in principle, does not need any gauge-fixing. To study the weak coupling limit of the lattice gauge field, gauge-fixing is required.

Gauge-fixing leads to the introduction of the Faddeev-Popov ghost field to keep the gauge invariant sector of the theory unchanged under gauge-fixing. The Abelian lattice gauge field was studied to have a better idea of gauge-fixing. The gauge-fixed lattice gauge theory is invariant under a generalized lattice BRST transformation.

Chapter 8

Abelian Lattice Gauge Field in $d = 3$

8.1 Introduction

Lattice gauge theory is motivated by the requirement of exact local gauge invariance for the gauge field theory with a lattice cutoff. This requirement leads to a redefinition of the conventional continuum action, such that the lattice action is made of function of the *finite* group elements of the gauge group, as discussed in Chapter 7.

The analysis of the Abelian gauge field is a warm up for the more complex non-Abelian case. A modified form of the Wilson action is studied for the Abelian field in three-dimensional Euclidean spacetime. The system is shown to be identical to a discrete Gaussian quantum field that is a generalization of the Ising model. In the weak-coupling limit, the system factorizes into the ordinary Coulomb interaction and a new set of interactions that are a direct reflection of the compactness of the Abelian lattice gauge field. If expanded about the naive vacuum, the Abelian gauge field is divergent for $d \leq 4$. An exact calculation shows the existence of at least two phases. A mean-field calculation shows the system to be in the same phase for all coupling g^2, except for the $g^2 = 0$. This Chapter is based largely on Baaquie (1977a).

Consider a finite d-dimensional Euclidean lattice which is periodic and consists of a lattice of N^d lattice points. Let $B_{n\mu}$ be the local spacetime Abelian degree of freedom at the lattice site n. $B_{n\mu} = a^{(d-2)/2} g A_\mu(x = na)$, where a is the lattice spacing, g is the bare coupling constant, and $A_\mu(x)$ is the continuum field. Let lattice field tensor be

$$f_{n\mu\nu} = B_{n\mu} + B_{n+\hat{\mu},\nu} - B_{n+\hat{\nu},\mu} - B_{n\nu}$$

The continuum Maxwell action is defined by

$$e^{\mathcal{S}_c} = \prod_{n\mu\nu}{}' \exp\left(-\frac{1}{4g^2} f_{n\mu\nu}^2\right)$$

where the prime denotes $\mu \neq \nu$. The Wilson action for the Abelian case, from Eq. 7.1.5, is defined by

$$e^{\mathcal{S}_c} \to e^{\mathcal{S}} = \prod_{n\mu\nu}{}' \exp\left[\frac{1}{2g^2}\left(e^{if_{n\mu\nu}} - 1\right)\right] = \prod_{n\mu\nu}{}' \exp\left(\frac{1}{4g^2}\left(\cos f_{n\mu\nu} - 1\right)\right)$$

Note that since the action is now only function of $e^{iB_{n\mu}}$, the Feynman path integral is defined on the compact space of $-\pi \leq B_{n\mu} \leq +\pi$ for all n, μ and hence leading to a theory that needs no gauge fixing. The modified Wilson action in $d = 3$ is defined by the following prescription:

$$e^{\mathcal{S}_c} = \prod_{n\mu\nu}{}' \exp\left(-\frac{1}{4g^2} f_{n\mu\nu}^2\right) = \prod_{ni} \exp\left[-\frac{1}{2g^2}\left(\frac{1}{2}\epsilon_{ijk} f_{njk}\right)^2\right]$$

$$e^{\mathcal{S}_c} \to e^{\mathcal{S}} = \prod_{ni} \sum_{i_{ni}=-\infty}^{+\infty} \exp\left[-\frac{1}{2g^2}\left(\epsilon_{ijk}\delta_j B_{nk} - 2\pi\, l_{ni}\right)^2\right] \qquad (8.1.1)$$

Using the identity

$$\sum_{n=-\infty}^{+\infty} \delta(x - n) = \sum_{i=-\infty}^{+\infty} e^{2\pi i l x} \qquad (8.1.2)$$

yields (dropping an overall constant)

$$e^{\mathcal{S}} = \sum_{\{l\}} \exp\left(-\frac{g^2}{2}\sum_{ni} l_{ni}^2 + i\sum_n \epsilon_{ijk} l_{ni}\delta_j B_{nk}\right) \; ; \; \sum_{\{l\}} = \prod_{ni} \sum_{l_{ni}=-\infty}^{+\infty} \qquad (8.1.3)$$

where

$$\delta_i f_n \equiv f_n - f_{n-\hat{i}}$$

is the finite lattice derivative.

Similar systems have been studied in statistical mechanics. Note that e^A is a function of $e^{i f_{nij}}$ and hence is a periodic in the B_{ni} variables with period 2π. In fact, it is to have the periodicity in the f_{nij} that the construction in Eq. 8.1.1 was carried out. The strong-coupling limit ($g^2 \gg 1$) yields

$$e^{\mathcal{S}} \simeq \prod_{ni}\left[1 + \exp(e^{-g^2/2})(e^{i\epsilon_{ijk} f_{njk}/2} + e^{-i\epsilon_{ijk} f_{njk}/2}) + O(e^{-2g^2})\right]$$

$$\simeq \exp\left(e^{-g^2/2}\sum_{nij} e^{i f_{nij}}\right) \qquad (8.1.4)$$

Eq. 8.1.4 yields the (strong-coupling) Wilson action, but with a coupling-constant redefinition. See Appendix A for further discussion on this action.

8.2   Strong Coupling Representation

Since the action is a function of $e^{iB_{ni}}$, all possible unique values are covered by letting $-\pi \leq B_{ni} \leq \pi$. Hence the Feynman path integral is defined by

$$Z = \prod_{ni} \int_{-\pi}^{+\pi} dB_{ni}\, e^{\mathcal{S}[B]} \equiv \int DB\, e^{\mathcal{S}[B]} \qquad (8.2.1)$$

The gauge transformation on the lattice for the Abelian case, from Eq. 7.1.6, is defined by

$$e^{iB_{ni}} \to e^{i\phi_n} e^{iB_{ni}} e^{-i\phi_{n+\hat{i}}} \quad \Rightarrow \quad B_{ni} \to B_{ni} - \delta_i\phi_{n+\hat{i}} + 2\pi h_{ni} \qquad (8.2.2)$$

where h_{ni} are integers and ϕ_n is a (compact) scalar quantum field. The most general integer-valued external current that is gauge invariantly coupled to B_{ni} is given by

$$\exp\left(-i\epsilon_{ijk}\sum_n j_{ni}\delta_j B_{n+\hat{j},k}\right) = \exp\left(i\epsilon_{ijk}\sum_n \delta_i j_{nj} B_{nk}\right)$$

where $\{j_{ni}\}$ is an arbitrary collection of integers.

The moment generating function, using Eq. 8.1.3, is given by

$$Z[j] = \prod_{ni}\int_{-\pi}^{+\pi} dB_{ni}\exp\left(i\epsilon_{ijk}\sum_n \delta_i j_{nj} B_{nk}\right)e^{\mathcal{S}}$$

$$= \sum_l \exp\left(-\frac{g^2}{2}\sum_{ni} l_{ni}^2\right)\prod_{ni}\int_{-\pi}^{+\pi} dB_{ni}e^{i\epsilon_{ijk}\delta_i(l_{nj}+j_{nj})B_{nk}}$$

Displacing $l_{nj} \rightarrow l_{nj} - j_{nj}$ yields

$$Z[j] = \sum_l \exp\left(-\frac{g^2}{2}\sum_{ni}(l_{ni}-j_{ni})^2\right)\prod_{nk}\left[\int_{-\pi}^{+\pi} dB_{nk}e^{i\epsilon_{ijk}(\delta_i l_{nj})B_{nk}}\right]$$

$$= \sum_l \exp\left(-\frac{g^2}{2}\sum_{ni}(l_{ni}-j_{ni})^2\right)\left[\prod_{nk}\delta(\epsilon_{ijk}\delta_i l_{nj})\right] \qquad (8.2.3)$$

Noteworthy 8.1: Gauge invariance

No gauge fixing was required to obtain Eq. 8.2.3 due the gauge field being a compact degree of freedom. To make this result more explicit consider from Eq. 8.2.2, the change of variables (up to a normalization) yields

$$B_{ni} = C_{ni} - \delta_i\phi_{n+\hat{i}} \;;\; C_{n0} = 0 \;:\; B_{n0} = -\delta_0\phi_{n+\hat{0}} \;\Rightarrow\; \prod_{i=1}^{3} dB_i = dC_1 dC_2 d\phi$$

Hence the change of variables yields, from Eq. 8.2.1

$$Z = \prod_{ni}\int_{-\pi}^{+\pi} dB_{ni}e^{\mathcal{S}[B]} = \left[\prod_n\int_{-\pi}^{+\pi} d\phi_n\right]\prod_n\int_{-\pi}^{+\pi} dC_{n1}dC_{n2}e^{\mathcal{S}[C]}$$

Since

$$\prod_n\int_{-\pi}^{+\pi}\frac{d\phi_n}{2\pi} = 1$$

The component of B_{n0}, given by ϕ_n, does not couple to the gauge-invariant sector and its integration yields a finite constant. Hence, the compact gauge field does not require gauge-fixing.

Note that the Kronecker δ functions in Eq. 8.2.3 imply that

$$\epsilon_{ijk}\delta_j l_{nk} = 0 \;\forall\, n, i \qquad (8.2.4)$$

It is shown in Appendix A that Eq. 8.2.4 is uniquely solved by

$$l_{ni} \equiv \delta_i l_n$$

Note l_n is an integer-valued scalar quantum field; the functional sum

$$\prod_{ni} \sum_{l_{ni}=-\infty}^{+\infty}$$

is restricted by Eq. 8.2.4 to become

$$\prod_{n} \sum_{l_{n}=-\infty}^{+\infty}$$

Hence, from Eq. 8.2.3, the generating function is given by

$$Z[j] = \prod_{n} \sum_{l_n=-\infty}^{+\infty} \exp\left[-\frac{g^2}{2}\sum_{ni}(\delta_i l_n - j_{ni})^2\right] \tag{8.2.5}$$

Eq. 8.2.5 is a *discrete Gaussian quantum field theory* and is a generalized Ising model, with the infinite-spin field l_n at the lattice site n. Eq. 8.2.5 is the *strong coupling representation* of $Z[j]$.

The theory is well defined, on a finite lattice ($N < \infty$), for all $d \geq 1$. The limit of $N \to \infty$ seems to exist only for $d > 2$ (see Section 8.6). $Z[j]$ reduces, for the case of $l_n = \pm 1$, to the ordinary Ising model in the presence of a magnetic field.

8.3   Weak Coupling Representation

Equation 8.2.5 is suitable for the strong-coupling expansion, but unsuited for the weak-coupling sector. Eq. 8.2.5 is rewritten using the identity of Eq. 8.1.2. To do so, consider a periodic lattice with total number lattice points N^d. Since

$$l_n = l_{n+N}$$

define the following Fourier transform

$$l_n = \frac{1}{N^d}\left(\prod_{i=1}^{d}\sum_{k_i=0}^{2\pi(N-1)/N} e^{ik_i n_i}\right) l_k \equiv \sum_{k} e^{ikn} l_k$$

The following notation is henceforth employed

$$\delta_{k,q} = N^d \prod_{i=1}^{d}\delta_{k_i,q_i} \ , \quad ; \quad r_{qi} = 1 - e^{-iq_i} \quad ; \quad d_q = \sum_{i}|r_{qi}|^2 \quad ; \quad D_n = \sum_{q}\frac{e^{iqn}}{d_q}$$

Using the identity of Eq. 8.1.2 yields

$$Z[j] = \sum_{l}\prod_{n}\int_{-\infty}^{+\infty} dx_n \exp(2\pi i \sum_{n} x_n l_n) \exp\left[-\frac{g^2}{2}\sum_{ni}(\delta_i x_n - j_{ni})^2\right]$$

$$= \exp\left(-\frac{g^2}{2}\sum_{ni}j_{ni}^2\right)\sum_{l}\prod_{n}\int_{-\infty}^{+\infty} dx_n \exp\left[\sum_{n}x_n(2\pi i l_n - g^2\delta_i j_{n+\hat{i},i})\right]$$

$$\times \exp\left[-\frac{g^2}{2}\sum_{ni}(\delta_i x_n)^2\right]$$

$$= e^{F_c[j]}\bar{Z}[j] \tag{8.3.1}$$

where

$$\bar{Z}[j] = \sum_l \exp\left(-\frac{2\pi^2}{g^2}\sum_{nn'} l_n D_{n-n'} l_{n'}\right)\exp\left(i\sum_n l_n \theta_n\right) \quad (8.3.2)$$

and

$$\theta_m = 2\pi\sum_{ni} D_{m-n}(j_{n+i,i} - j_{ni}) \quad (8.3.3)$$

The partition function $\bar{Z}[j]$ is due to the compactness (periodicity) of the field variables $e^{iB_{ni}}$. Note that $\bar{Z}[j]$ is nonanalytic about $g^2 = 0$, owing to the discreteness of the degrees of freedom $\{l_n\}$.

The Coulomb component of the partition function is given by

$$F_c = -\frac{g^2}{2}\sum_{ni} j_{ni}^2 + \frac{g^2}{2}\sum_{nm} \delta_i j_{n+\hat{i},i} D_{n-m} \delta_k j_{m+\hat{k},k} \quad (8.3.4)$$

For $d = 3$, F_c can be rewritten as

$$F_c = -\frac{g^2}{2}\sum_{nn'} \epsilon_{ijk}\delta_i j_{nk}\epsilon_{iab}\delta_a j_{nb} D_{n-n'}$$

Eq. 8.3.1 is the *weak-coupling representation* for $Z[j]$, suitable for studying the theory for $g \simeq 0$. The results obtained are exact, and note that no gauge fixing was involved in obtaining the result.

8.4   Gauge Invariance

Consider gauge invariance for the system in the $\{j_{ni}\}$ basis. The gauge-invariant coupling of $\{j_{ni}\}$ to the Abelian field implies invariance under the transformation

$$j_{ni} \to j_{ni} + \delta_i h_n \quad (8.4.1)$$

where $\{h_n\}$ is an arbitrary collection of integers. Eq. 8.4.1 is the gauge transformation on the external current j_{ni}. Eq. 8.4.1 yields

$$Z[j] = Z[j - \delta h] \quad (8.4.2)$$

The invariance is realized quite differently for the strong-weak-coupling representations of $Z[j]$. For the strong-coupling representation, Eq. 8.2.5 yields

$$Z[j - \delta h] = \sum_l \exp\left[-\frac{g^2}{2}\sum_{ni}(\delta_i l_n - j_{ni} + \delta_i h_n)^2\right]$$

$$= \sum_l \exp\left[-\frac{g^2}{2}\sum_{ni}(\delta_i l_n - j_{ni})^2\right] = Z[j] \quad (8.4.3)$$

where l_n has been displaced to $l_n - h_n$ to obtain Eq. 8.4.3. Any cutoff on l_n, say $-H \geq l_n \geq H$ (H = integer), will destroy invariance in Eq. 8.4.2. (Note that if $e^{iB_{nu}}$ is made discrete, then $\{l_n\}$ becomes bounded.) For a strong coupling expansion, it seems easier to use the $e^{iB_{ni}}$ variables.

Gauge invariance of $Z[j]$ for the weak-coupling representation is realized as follows. From Eqs. 8.3.3 and 8.3.4

$$F_c = -\frac{g^2}{2}\sum_k j_{-ki}j_{ki} + \frac{g^2}{2}\sum_k j_{-ki}\frac{r_{ki}r_{ki}^*}{d_k}j_{kl}$$

and

$$\theta_n = -2\pi\sum_k e^{ikn}\frac{r_{ki}^*j_{ki}}{d_k} \tag{8.4.4}$$

From Eq. 8.4.1, consider the gauge transformation

$$j'_{ki} \to j'_{ki} + r_{ki}h_k \tag{8.4.5}$$

Therefore, for the transformation given in Eq. 8.4.5

$$F_c[j - \delta h]$$

$$= -\frac{g^2}{2}\sum_k (j_{-ki} + r_{ki}^*h_{-k})(j_{ki} + r_{ki}h_k) + \frac{g^2}{2}(j_{-ki} + r_{ki}^*h_{-k})\frac{r_{ki}r_{ki}^*}{d_k}(j_{kj}r_{kj}h_k)$$

$$= F_c[j] - \frac{g^2}{2}\sum_k d_k h_{-k}h_k - g^2\sum_k j_{ki}r_{ki}h_k + \frac{g^2}{2}\sum_k \frac{d_k^2}{d_k}h_{-k}h_k + g^2\sum_k j_{-ik}r_{ki}\frac{d_k}{d_k}h_k$$

Due to cancellation of terms above

$$F_c[j - \delta h] = F_c[j]$$

and from Eq. 8.4.4

$$\theta_n[j - \delta h] = -2\pi\sum_k e^{ikn}\frac{r^*(j_{ki} + r_{ki}h_k)}{d_k} = \theta_n[j] - 2\pi h_n$$

Therefore, from Eq. 8.3.2

$$Z[j - \delta h] = e^{F_c[j]}\sum_l \exp\left(-\frac{2\pi^2}{g^2}\sum_{nn'}\bar{l}_n D_{n-n'}\bar{l}_{n'}\right)$$

$$\times \exp\left\{i\sum_n \bar{l}_n\theta_n[j]\right\}\exp\left(-2\pi i\sum_n \bar{l}_n h_k\right)\exp\left(-2\pi i\sum_n \bar{l}_n h_k\right)$$

$$= e^{F_c[j]}\sum_l \exp\left(-\frac{2\pi^2}{g^2}\sum_{nn'}\bar{l}_n D_{n-n'}\bar{l}_{n'} + i\sum_n \bar{l}_n\theta_n[j]\right) = Z[j] \tag{8.4.6}$$

Hence, the partition function is gauge-invariant

$$Z[j - \delta h] = Z[j]$$

Note that Eq. 8.4.6 is gauge invariant, *order by order*, in $\bar{l}_n$; that is, one can obtain an *exact* gauge-invariant weak-coupling expansion by imposing a cutoff on $\bar{l}_n$.

8.5 Wilson Loop

F_c is simply the ordinary Coulomb interaction in $2 + 1$ dimensions. To show this, the Wilson loop – given by a closed gauge field loop – is computed. An appropriate choice of $\{j_{ni}\}$, given in Appendix B, yields the Coulomb contribution to the Wilson loop. Consider the contour Γ that is a planar square of a side L, as shown in Figure 8.1. From Eq. 8.3.1

$$E[e^{i\sum_\Gamma B_{m\lambda}}] \equiv \frac{1}{Z} \int DB e^S e^{i\sum_\Gamma B_{m\lambda}} = e^{F_c[j]}\bar{Z}[j] \simeq e^{F_c[j]} + O(e^{-1/g^2})$$

The sum along the closed contour Γ is denoted by $\sum_\Gamma$. As shown in Appendix B, for large L the Coulomb contribution to the expectation value yields

$$E[e^{i\sum_\Gamma B_{m\lambda}}]_c = e^{F_c[j]} \simeq e^{-(g^2/\pi)L \ln L} + \text{lower order}$$

Let $E(L)$ be the energy for the state. Then [Wilson (1974)]

$$E\left[\exp\left(i\sum_\Gamma B_{m\lambda}\right)\right]_c = e^{-LE_c(L)}$$

and hence the Coulomb part of the energy is

$$E_c(L) = \frac{1}{\pi}g^2 \ln L + O(1)$$

Note that $E_c(L)$ is the expected logarithmically growing Coulomb potential for $d = 3$.

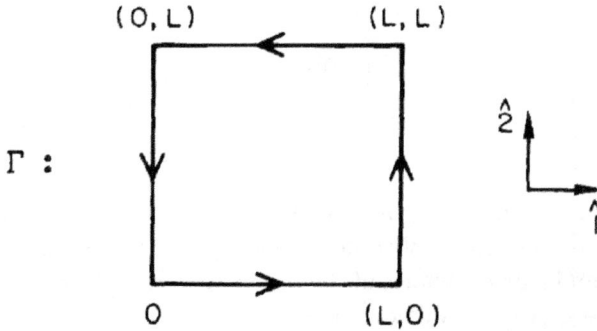

Fig. 8.1   Contour product of the gauge field around the contour, which is denoted by Γ.

8.6 Phase Transition

In the limit of $N \to \infty$, the phase diagram of the partition function is analyzed.

Consider the weak-coupling representation given in Eq. 8.3.1. Set $j_{ni} = 0$, as this does not change the bulk properties (phase diagram) of the system. A weak-coupling expansion about the gauge-invariant naive vacuum is given by $\bar{l}_n = 0$, and yields (for d dimensions)

$$Z = \prod_n \sum_{l_n=-\infty}^{+\infty} \exp\left(-\frac{2\pi^2}{g^2}\sum_{nn'} l_n D_{n-n'} l_{n'}\right) \tag{8.6.1}$$

$$\simeq 1 + 2N^d e^{-(2\pi^2/g^2)D_0} 4\sum_{n,n'} \cosh D_{n-n'} + O(e^{-6\pi^2 D_0/g^2})$$

Hence

$$\frac{1}{N^d}\ln Z \simeq 2e^{-(2\pi^2/g^2)D_0} + 2e^{-(4\pi^2/g^2)D_0} N^d \sum_n (\cosh D_n - 1) + O(e^{-9\pi^2 D_0/g^2})$$

$$= 2e^{-(2\pi^2/g^2)D_0} + 2e^{-(4\pi^2/g^2)D_0} N^d \sum_n D_n^2 \tag{8.6.2}$$

Note that for $|n| \gg 1$, we have $(d > 2)$

$$D_n = \sum_q \frac{e^{iqn}}{d_q} \sim \frac{1}{n^{d-2}}$$

and yields

$$\sum_{n \neq 0} (\cosh D_n - 1) \geq \sum_{|n|>R} (\cosh D_n - 1) \simeq \sum_{|n|>R} D_n^2$$

Note that

$$\sum_{|n|>1} D_n^2 \simeq \int_1^N \frac{d^d n}{n^{2d-4}} \simeq \int_1^N n^{d-1}\frac{dn}{n^{2d-4}} \simeq N^{4-d}$$

Hence, for $N \to \infty$

$$\frac{1}{N^d}\ln Z \sim \begin{cases} N^{4-d}, & 2 < d < 4 \\ \ln N, & d = 4 \\ \text{constant}, & d > 4 \end{cases}$$

The result above yields that $\ln Z/N^d$ is divergent (as $N \to \infty$) for $d \leq 4$. The divergence can arise because a) this theory undergoes a phase transition for $g^2 < g_c^2$, or b) because of an incorrect choice of the native vacuum. Hence, the phase diagram for the system needs to be studied.

Consider the limit of $g^2 \to 0$. Note that for $g^2 \simeq 0$ the gauge field becomes a massless scalar field. To see this, let $\phi_n = gl_n$. From Eq. 8.2.5 (with $j_{ni} \simeq 0$)

$$Z = \lim_{g^2 \to 0}\left(\frac{1}{g}\right)^{N^d} \prod_n g\sum_{l_n} \exp\left[-\frac{1}{2}\sum_{ni}[\delta_i(gl_n)]^2\right]$$

$$= (\text{constant}) \prod_n \int_{-\infty}^{+\infty} d\phi_n \exp\left[-\frac{1}{2}\sum_{ni}(\delta_i\phi_n)^2\right]$$

ϕ_n becomes a *continuous variable* – ranging from $-\infty$ to $+\infty$ – in the limit of $g^2 = 0$. Note that the massless free field is a lattice system at the critical point and has infinite correlation length. The weak-coupling expansion of Eq. 8.3.1 is generated by making all the discrete variables $\{l_n\}$ continuous in Eq. 8.3.1, *except* for a finite number of l_n's, which are kept discrete.

The $g^2 \simeq \infty$ phase is well defined. From Eq. 8.6.1, defining $\phi_n = l_n/g$, yields (up to an irrelevant constant)

$$Z(g^2 = \infty) = \prod_n \int_{-\infty}^{+\infty} d\phi_n \exp\left(-2\pi^2 \sum_{n,n'} \phi_n D_{n-n'} \phi_n\right)$$

The propagator for the above system is nonzero only between nearest neighbors, giving a correlation length of unity. The correlation length for the system at $g^2 = 0$ is infinity. Hence, the existence of (at least) two distinct phases is an exact result.

8.6.1 *Mean field approximation*

The phase diagram needs to be studied to ascertain whether the phase transition takes place at $g^2 = 0$, or at some finite $g_c^2 > 0$. The crudest first approximation for the phase diagram is given by mean field theory. In this approximation, one replaces the entire field theory – consisting of N^d degrees of freedom – by a system involving only one degree of freedom.

Recall from Eq. 8.2.5 that

$$A = -\frac{g^2}{2} \sum_{ni} (l_n - l_{n-\hat{i}})^2$$

Consider a particular lattice site n; replace the interaction of l_n with a constant background field of strength M. Then for this one site

$$A \rightarrow g^2 d(l - M)^2$$

and

$$Z_{MF} = \sum_{l=-\infty}^{+\infty} e^S = \sum_{l=-\infty}^{+\infty} e^{-g^2 d(l-M)^2}$$

Let

$$f(M) = \frac{1}{Z_{MF}} \sum_{l=-\infty}^{+\infty} l e^{-g^2 d(l-M)^2} \tag{8.6.3}$$

Since M is the background field that is self consistently created by all the $\{l_n\}$'s, one has the following mean-field equation for M

$$M = f(M) \tag{8.6.4}$$

Eq. 8.6.4 fixes M as a function of g^2 and yields a solution given by $M = M(g^2)$. Note that for $n = $ integer, $f(M + n) = f(M) + n$, so M is only analyzed for $M \in (0, 1)$. The solution of Eq. 8.6.4, for $g^2 \neq 0$, is rather trivial, since from Eq. 8.6.3

$$f\left(\frac{1}{2}\right) = 1 - f\left(\frac{1}{2}\right) \quad \Rightarrow \quad f\left(\frac{1}{2}\right) = \frac{1}{2} \tag{8.6.5}$$

Hence, from Eqs. 8.6.4 and 8.6.5

$$M = \frac{1}{2}$$

for all $g^2 \neq 0$.

To investigate the $g^2 = 0$ behavior, note that for $g^2 \simeq 0$

$$f(M) = M + \frac{2\pi}{g^2 d} e^{-\pi^2/g^2 d} \sin 2\pi M + O(e^{-4\pi^2/g^2 d}) \qquad (8.6.6)$$

Therefore, for $g^2 \simeq 0$, we see that $F(M) = M$; this gives *all* $M \in (0,1)$ as solutions of the mean-field equation Eq. 8.6.4. Hence, there is a singularity (instability) of the system at $g^2 = 0$. The singularity of the system at $g^2 = 0$ indicates the phase transition to the free massless scalar field.

A numerical calculation shows that Eq. 8.6.6 is a good approximation for $g^2 d <$ 0.4. Hence, the system is exponentially close to the $g^2 = 0$ massless free field phase for $0 < g^2 d < 0.4$. The crossover from strong- to weak-coupling is for $0.4 < g^2 d < 0.6$. A schematic plot $\xi =$ of correlation length as a function of g is given in Figure 8.2. The mean-field calculation in general is not reliable for quantitative information, although it is usually good enough to describe, qualitatively, the phase diagram of the system.

A tentative conclusion can be drawn about the divergence found in perturbation theory: for the $g^2 = 0$ naive vacuum there is phase transition of the system at $g^2 = 0$; and that defining a perturbation about the correct vacuum will give a convergent expansion of $(1/N^d) \ln Z$ for all $g^2 > 0, d > 4$.

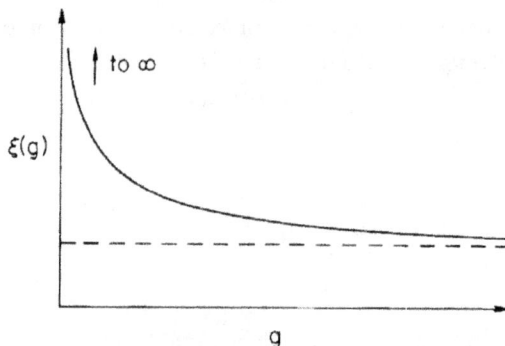

Fig. 8.2   Schematic plot of the mean-field result for the correlation length ξ as a function of g.

8.7   Summary

Gauge fields are compact degrees of freedom, and this aspect does not come out clearly in the continuum formulation. The $d = 3$ Abelian lattice gauge field is seen to be dual to a discrete Gaussian degree of freedom. The path integral over the Abelian gauge field can be carried out without any gauge-fixing using the compactness of the degrees of freedom. The redundant degrees of freedom of the Abelian gauge field do not lead to a divergence and factor out from the path integral.

The Wilson loop for $d = 3$ exhibits the expected Coulomb potential. The Abelian gauge field undergoes a phase transition in $d = 3$. If one studies the discrete Gaussian theory in all dimensions, a weak coupling expansion shows that the theory has a phase transition for $2 < d \leq 4$.

8.8   Appendix A

Define $\vec{N} = (N, N, N)$. It is shown that

$$Z[j] = \prod_{n \neq \vec{N}} \sum_{l_n=-\infty}^{+\infty} \exp\left[-\frac{g^2}{2} \sum_{n \neq \vec{N}} \sum_i (\delta_i l_n - j_{ni})^2 \right]$$

For a precise definition of the modified action, note that for a periodic lattice

$$\sum_n f_{nij} = 0 \quad \Rightarrow \quad f_{\vec{N}ij} = - \sum_{n \neq \vec{N}} f_{nij}$$

Hence, eliminating $f_{\vec{N}ij}$ from the action gives

$$e^A = \left(\prod_{n \neq \vec{N}} \prod_i \sum_{l_{ni}=-\infty}^{+\infty} \right) \exp\left[-\frac{1}{2g^2} \left(\frac{1}{2} \epsilon_{ijk} f_{njk} - 2\pi l_{ni} \right)^2 \right]$$

From Eq. 8.2.3, the partition function $Z[j]$ is given by

$$Z[j] = \sum_{\vec{l}}{}' \exp\left[-\frac{1}{2} g^2 \sum_{ni}{}' (l_{ni} - j_{ni})^2 \right] \left[\prod_{nk} \delta(\epsilon_{ijk} \delta_i l_{nj}) \right]$$

where the primes on $\sum$ means that $n \neq \vec{N}$.

Make a change of variables so as to saturate the constraint imposed by Kronecker δ functions. Choose the generalized *axial gauge* that is defined for a finite periodic lattice; for $d = 3$, factorize the lattice into disjoint domains $D^{(i)}$ defined in Figure 8.3. Consider the following generalization of the axial gauge:

$$n \in D^{(i)} \; ; \; l_{nj} \to \tilde{l}_{nj} \; : \; \begin{cases} l_{nj} = \tilde{l}_{nj} + \delta_j l_n \; (j \neq i) \\ l_{ni} = \delta_i l_n \, , \tilde{l}_{ni} \equiv 0 \end{cases}$$

In other words, the generalized axial gauge yields the following

$$l_{ni} = \tilde{l}_{ni} + \delta_i l_n \quad \Rightarrow \quad \tilde{l}_{ni} = 0 \; ; \; n \in D^{(i)}$$

In the new variables, the Kronecker δ functions reduce to the restriction that all $\{\tilde{l}_{nj}\}$ be zero; hence, dropping the superfluous Kronecker δ's, yields

$$Z[j] = \prod_n{}' \sum_{l=-\infty}^{+\infty} \exp\left[-\frac{1}{2} g^2 \sum_{ni}{}' (\delta_i l_n - j_{ni})^2 \right]$$

The special behavior of the lattice site (N, N, N) is not required for any of the results.

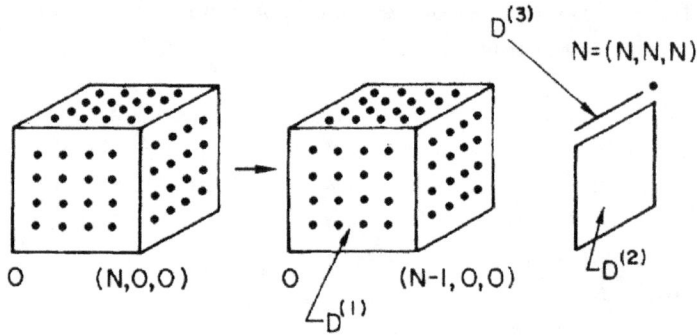

Fig. 8.3   Partition of the lattice points into disjoint sets denoted by $n \in D^{(i)}, i = 1, 2, 3$. Lattice site $\vec{N} = (N, N, N)$ is a domain by itself.

8.9   Appendix B

To obtain the Coulomb force, the expectation value of the Wilson loop

$$E\left[\exp\left(i\sum_{\Gamma} B_{m\lambda}\right)\right]$$

is computed. The Coulomb force is given by $F_c[j]$, with the remaining portion yielding the quantum corrections.

Γ is a planar contour restricted to the 12-plane (see Figure 8.1). In $d = 3$, the planar loop can be represented (up to a gauge transformation) by the following choice of $\{j_{ni}\}$:

$$j_{ni} = \delta_{i3} \sum_{l_1,l_2=0}^{L-1} \delta_{n,l_1\hat{1}+l_2\hat{2}} \tag{8.9.1}$$

Note that $\{j_{ni}\}$ is nonzero on the planar grid of points enclosed by the contour Γ; for $\{j_{ni}\}$ given in Eq. 8.9.1, it follows that

$$E\left[\exp\left(i\sum_{\Gamma} B_{m\lambda}\right)\right] = E\left[\exp\left(i\sum_{n} \epsilon_{ijk}\delta_i j_{nj} B_{nk}\right)\right]$$

The Coulomb part is given by

$$E\left[\exp\left(i\sum_{\Gamma} B_{m\lambda}\right)\right] = e^{F_c[j]} + O(e^{-1/g^2})$$

using $\{j_{ni}\}$ given by Eq. 8.9.1.

Chapter 9

Lattice Gauge Field Mass Renormalization

9.1 Introduction

The computation of mass renormalization to one-loop is summarized. Following the derivation given in Baaquie (1977b), the one-loop calculation is carried out explicitly to provide an exemplar of lattice perturbation theory as well as to illustrate the nonlinearities of the Yang-Mills gauge fields.

This computation is based on weak coupling limit of the gauge field and some of the specific feature of the lattice regularization of the gauge field are necessary. The one-loop computation shows that there is no mass renormalization for the lattice gauge field, and hence the gauge field remains massless in spite of the nonlinear interactions. Zero mass renormalization is essential for a well defined regularization of the gauge quantum field, and the lattice provides the desired result.

The absence of mass renormalization for the lattice theory is due to exact gauge invariance on the lattice that forbids the appearance of a mass term. Since the lattice gauge field breaks Euclidean invariance (and in turn Lorentz invariance), the one-loop calculation shows that gauge-invariance is more important than Euclidean invariance for renormalizing the gauge field.

Recall an important feature of the lattice regularization is that there are N^4 degrees of freedom (integration variables) and hence one can isolate every single momentum mode of the lattice gauge field. In particular, the zero momentum mode of the gauge field directly contains information about the mass term of the field: if the gauge field does not have a bare mass and its renormalized mass is zero (massless), then the zero mode is unconstrained. If the gauge field has a quadratic mass divergence, then this implies that the theory is not renormalizable, and hence inconsistent.

The zero mode $B_{k=0,\mu}^a$ is the result of the value of the gauge field having a constant value, say c_μ^a at every point of spacetime. This can be seen from the following

$$B_{k,\mu}^a = \sum_n e^{ikn} B_{n,\mu}^a \rightarrow \sum_n e^{ikn} c_\mu^a = c_\mu^a \delta_{k,0}$$

The energy needed to displace the field simultaneous at every point of spacetime is a global property of the gauge field and depends crucially on its long distance properties. In particular, the zero mode of the gauge field carries the long distance information about the gauge field. For example, the Abelian photon field, because it is massless, can propagate out to infinitely large distances.

Integrating out all the degrees of freedom *except* for $B^a_{k=0,\mu}$ is the extreme case of thinning out the degrees of freedom of the gauge field. The sum total of the effects of all the higher momentum modes are summed up in the properties of the zero mode. In particular, if the higher momentum modes each contribute to the mass of the gauge field, then this would lead to a divergent mass-term for the gauge field. Since a mass term is forbidden on the lattice due to gauge invariance, the divergent mass term would render the entire theory non-renormalizable. Hence, obtaining zero mass contributions from the higher momentum modes, and in particular, that the gauge field does not require mass renormalization, is essential for the consistency of the (lattice) gauge field.

The non-Abelian gauge field, unlike the Abelian case, can generate a spectrum of states for which there is a mass gap due to the nonlinearities of the gauge field. In fact, the massive states of the gauge field are low energy *collective bound states* of the gauge field and the mass of these states does not arise from the higher momentum modes in the action, but rather arise from strong coupling effects that reflect the *long distance* behavior of the gauge field.

9.2 The Propagator and Mass Renormalization

The main features of the calculation are discussed before going into the details. Define the (global) color-singlet propagator

$$D_{n\mu\nu} = \frac{1}{Z} \int DU B^a_{n\mu} B^a_{0\nu} e^{S+S_\alpha+S_c} \qquad (9.2.1)$$

$$D_{k\mu\nu} = \sum_n e^{ikn} D_{n\mu\nu} \qquad (9.2.2)$$

Using translational invariance (due to the periodic lattice) and using Eq. 7.8.3 gives

$$D_{k\mu\nu} = \frac{1}{N^4} \frac{1}{Z} \int dU B^a_{-k\mu} B^a_{k\nu} e^{S'} / Z \qquad (9.2.3)$$

Let $D^0_{n\mu\nu}$ be the bare propagator defined by the quadratic part of A' of the gauge-fixed action, and let $\prod_{\mu\nu}(k)$ be the proper self-energy. Then, in matrix notation, Dyson's equation states that

$$D_k = D^0_k + D^0_k \Pi(k) D_k \qquad (9.2.4)$$

Recall from Eq. 7.2.4 that $B^\alpha_{n\mu} = a g_o s A^\alpha_{n\mu}$ is dimensionless, making the lattice self-energy $\Pi(k)$ in Eq. 9.2.4 dimensionless. Hence, the continuum self-energy, denoted by $\Pi^{phy}(p)$ – which has the dimension of mass [Zinn-Justin (1993)], is given by dimensional analysis. The only dimensional quantity in the lattice theory is the lattice spacing a, and yields

$$\Pi^{\text{phy}}(p) = \frac{1}{a^2}\Pi(k = pa) \tag{9.2.5}$$

$$= \frac{1}{a^2}\{\Pi(0) + [\Pi(pa) - \Pi(0)]\} \tag{9.2.6}$$

It can be shown using perturbation theory that $(p \neq 0)$

$$\lim_{a \to 0} \frac{1}{a^2}[\Pi(pa) - \Pi(0)] \sim \text{logarithmic divergences in } a \tag{9.2.7}$$

The $a \to 0$ limit yields

$$\Pi^{\text{phy}}(p) = \frac{1}{a^2}\Pi(0) + \text{logarithmic divergences in } a \tag{9.2.8}$$

The logarithmic divergences in a are removed by wave-function renormalization.

For there to be no mass renormalization, the quadratic divergence $\frac{1}{a^2}\Pi(0)$ must be zero, and hence

$$\Pi(0) = 0 \tag{9.2.9}$$

$g_0 \to 0$ yields the expansion

$$\Pi(0) = \Pi_0 + \Pi_1 g_0 + \Pi_2 g_0^2 + \cdots \tag{9.2.10}$$

In the lowest-order calculation done explicitly in this Chapter, it will be shown that $\Pi_0 = 0$. From Eq. 9.2.4

$$D_k = \frac{1}{D_k^{(0)-1} - \Pi(k)} \tag{9.2.11}$$

For $N \to \infty, k \to 0, D_k^{(0)-1} \to 0$; hence

$$D_{k=0} = -\frac{1}{\Pi(0)} \tag{9.2.12}$$

In order to obtain $\Pi(0)$, the propagator is evaluated

$$D \equiv D_{k=0,\mu\mu} = \frac{1}{N^4}\frac{1}{Z}\int dU B_{k=0,\mu}^a B_{k=0,\mu}^a e^{S'}/Z \tag{9.2.13}$$

To calculate D, in the path integral the integrations over *all* $\{B_{k\mu}^a, k \neq 0\}$ is performed; this yields an effective action involving only the $B_{k=0,\mu}^a$ variables, and provides the result for $\Pi(0)$. In the following Sections, the partition function $Z(g_0)$ is analyzed and yields $\Pi(0)$. The general result that $\Pi(0) = 0$ to all orders in perturbation theory is proven by using the Slavnov-Taylor identity, and is discussed in Section 9.9.

9.3　The Computational Scheme

Recall from Eq. 7.5.1 that with a change of notation $B_{n\mu}^a \rightarrow \tilde{B}_{n\mu}^a$

$$Z = \prod_{n\neq N} \prod_{\mu\neq a} \int_{-\infty}^{\infty} d\tilde{B}_{n\mu}^a \mu(\tilde{B}_{n\mu}^a) \prod_\alpha d\bar{U}_{N\alpha} e^{\mathcal{S}+\mathcal{S}_\alpha+\mathcal{S}_c} \tag{9.3.1}$$

The zero mode cannot be treated like the other variables since there is no Gaussian factor for it in the integrand. In the Fourier transform of $\{\tilde{B}_{n\mu}^a\}$ the role of $U_{N\alpha}$ is taken by the variable $B_{k=0,\mu}^a$ since there is no Gaussian factor for it either. (This can be easily seen later.) The zeroth mode in the action needs to be isolated for the computation of mass renormalization and, as mentioned, entails integrating out all the other non-zero modes.

An efficient method for computing the quadratic mass divergence is to isolate the zero mode and then integrate out *all* the non-zero momentum degrees of freedom. The net effect of this procedure, which in effect yields the effective action, is to provide the requisite mass renormalization.

To carry out this computation, consider the following Fourier expansion. The original variables are denoted by

$$\tilde{B}_{n\mu}^a = \sum_k e^{ik\mu} B_{k\mu}^a$$

The zero mode is isolated by the following expansion

$$\tilde{B}_{n\mu}^a = B_{k=0,\mu}^a/N^4 + \sum_k{}' e^{ik\mu} B_{k\mu}^a \;\; ; \;\; \sum_k{}' \equiv \sum_{k\neq 0}$$

$$\tilde{B}_{n\mu}^a = \theta_\mu + B_{n\mu}$$

where the zero and non-zero modes are defined by

$$\theta_\mu^a = B_{k=0,\mu}^a/N^4 \;\; ; \;\; B_{n\mu}^a = \sum_k{}' e^{ik\mu} B_{k\mu}^a \tag{9.3.2}$$

Gauge-fixing with $\alpha = 1/2g_0^2$ in $\mathcal{S}_\alpha + \mathcal{S}_c$ yields

$$B_{n\mu}^a = O(g_0) \;\; ; \;\; \theta_\mu^a = O(1) \tag{9.3.3}$$

The path integral yields the following expression for $W(\theta)$, the zero mode effective action

$$Z = \prod_{\mu,a} \int_G d\theta_\mu^a e^{W(\theta)} \;\; ; \;\; \int DB = \prod_{\mu,a} \prod_n \int_{-\infty}^{+\infty} dB_{n\mu}^a$$

$$e^{W(\theta)} = \int DB \mu[\theta + B] e^{\mathcal{S}[\theta+B]+\mathcal{S}_c[\theta+B]+\mathcal{S}_\alpha[\theta+B]}$$

The effective action $W(\theta)$ is being evaluated using the background field method – with the role of the background field being played by the zero mode – and this method can be executed directly in the continuum. The result, if one uses dimensional regularization, can be read off directly from the action; since every term that yields a quadratic divergence is rendered zero by dimensional regularization, the computation turns out to be essentially trivial.

The case for the lattice is quite different from continuum dimensional regularization since all the quadratic divergences – which are all set to zero by dimensional regularization – have been regulated by the lattice cutoff and yield finite values; hence one needs to compute all the quadratically divergent terms to verify that their sum indeed cancels exactly.

The effective action $W(\theta)$ is given by the following expansion in powers of θ (the pre-factor of $\mathcal{N}$ from $SU(\mathcal{N})$ and the volume factor N^4 have been factored out for future convenience)

$$W(\theta) = \delta m \; \cdot \; \frac{1}{4}\mathcal{N}N^4 \Big(\sum_{\mu,a} \theta^a_{n\mu}\Big)^2 + O(\theta^3) \tag{9.3.4}$$

δm is given by a perturbation expansion in g_0 and only the leading term needs to be computed, namely

$$\delta m = O(1) + O(g_0^2)$$

Note that δm is quadratically divergent and has contributions from the following sources

- From the measure term $\mu[\theta + B] : m_\mu$.
- From the gauge field action $\mathcal{S}[\theta + B] : m_A$.
- From the gauge-fixing term $\mathcal{S}_\alpha[\theta + B] : m_\alpha$.
- From the Faddeev-Popov ghost counter-term $\mathcal{S}_c[\theta + B] : m_c$.

Hence

$$\delta m = m_\mu + m_A + m_c + m_\alpha + O(g_0^2) \tag{9.3.5}$$

9.4 Determination of m_α and m_μ

The gauge-fixing term $\mathcal{S}_\alpha$ chosen in Eq. 7.6.2 is

$$\mathcal{S}_\alpha = -\frac{1}{4g_0^2}\sum_{n\mu}(\Delta_\mu \tilde{B}^a_{n-\hat{\mu},\mu})^2$$

and makes it independent of θ. Hence

$$\mathcal{S}_\alpha[\tilde{B}] = \mathcal{S}_\alpha[\theta + B] = \mathcal{S}_\alpha[B] \quad : \quad \text{independent of } \theta$$

and

$$m_\alpha = 0$$

The measure term for $SU(\mathcal{N})$, shown in Figure 9.1(a), yields from Eq. 2.5.3

$$\mu[\tilde{B}] = \exp\Big\{-\frac{\mathcal{N}}{24}\sum_{n,\mu,a}(\tilde{B}^a_{n\mu})^2\Big\} = \exp\Big\{-\frac{1}{6}\cdot\frac{\mathcal{N}}{4}N^4\Big(\sum_{\mu,a}\theta^a_{n\mu}\Big)^2\Big\} + O(g_0^2)$$

and hence, from Eqs. 9.3.4 and 9.3.5

$$m_\mu = -\frac{1}{6} \tag{9.4.1}$$

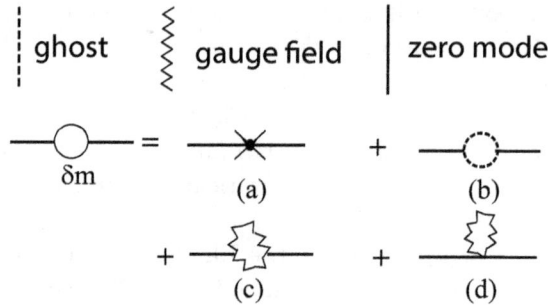

Fig. 9.1   Feynman diagrams for the computation of the lowest-order mass renormalization of the gauge field quantum. The external lines correspond to the zero mode θ^a. (a) is the contribution m_μ from the measure term μ. (b) shows the contribution m_c from the Faddeev-Popov ghost action $\mathcal{S}_c$. (c) and (d) are the contributions that yield m_a from the nonlinear gauge field action $\mathcal{S}$.

9.5   Determination of m_c for $\mathcal{S}_c[\theta + B]$

From Eq. 7.8.8 we have

$$e^{\mathcal{S}_c[B]} = \det(d_k \delta^{ab} \delta_{k,q} + M^{ad}(k,q)) \times \exp\left[-\frac{\mathcal{N}}{48} \sum_{n,\mu a}(B^a_{n\mu})^2 \right] + O(g_0^3)$$

where, from Eq. 7.8.6

$$M^{ad}(k,q) = \frac{1}{2} C^{abc} \sum_\mu (1 - e^{-ik_\mu})(1 + e^{-iq_\mu}) B^b_{k-q,\mu}$$

To extract the mass renormalization term m_c, note that the matrix $M^{ad}(k,q)$ depends linearly on B and yields

$$\mathcal{S}_c[\tilde{B}] = \mathcal{S}_c[\theta + B] = \mathcal{S}_c[\theta] + O(B) : \quad \text{independent of } B$$

Hence

$$M_{ab}(k,q) = \frac{1}{2} C^{abc} \sum_\mu (1 - e^{ik_\mu})(1 + e^{iq_\mu}) B^c_{k-q,\mu}$$

$$= -\frac{1}{2} C^{abc} \sum_\mu (e^{ik_\mu} - e^{-ik_\mu}) \theta^c_\mu \delta_{k,q} + O(g_0)$$

Let

$$M_{ab}(k,q) = n^{ab}(k) \delta_{k,q} + O(g_0)$$

Therefore, up to an irrelevant constant

$$e^{\mathcal{S}_c} = \exp\left[-\frac{\mathcal{N}N^4}{48} \sum_{\mu a}(\theta^a_\mu)^2 \right] \times \det\left(1 + \frac{1}{d}n\right) + O(g_0) \tag{9.5.1}$$

$$\simeq \exp\left(-\frac{1}{12}\phi \right) \exp\left[-\frac{1}{2} \text{Tr}\left(\frac{1}{d}n\frac{1}{d}n \right) \right] \tag{9.5.2}$$

where

$$\phi = \mathcal{N} N^{-4} \frac{1}{4} \sum_{\mu} \theta_{\mu}^2 \tag{9.5.3}$$

Hence, the leading $O(1)$ contribution of the counter-term $\mathcal{S}_c$ to mass renormalization *decouples* from the functional integral $\int DB$ and depends only on the ghost fields $c, c^{\dagger}$ and is shown in the Feynman diagrams given in Figure 9.1(b). The trace yields

$$\text{Tr}\left(\frac{1}{d}n\frac{1}{d}n\right) \to \phi I \tag{9.5.4}$$

where the lattice constant I is given by

$$I = \sum_{k \neq 0} \frac{|e^{ik_1} - e^{-ik_1}|^2}{d_k^2} \tag{9.5.5}$$

The lattice constant I arises from Figure 9.1(b): the two propagators yield the denominator d_k^2 whereas the ghost vertex yields the factor $|e^{ik_1} - e^{-ik_1}|^2$.

The contribution of the Faddeev-Popov ghost counter-term to mass renormalization yields

$$e^{\mathcal{S}_c[\theta]} = \exp\left[-\frac{\mathcal{N} N^4}{8} I \sum_{\mu a} (\theta_{\mu}^a)^2\right] \times \exp\left[-\frac{\mathcal{N} N^4}{48} \sum_{\mu a} (\theta_{\mu}^a)^2\right] + O(g_0^3)$$

$$= \exp\left[-m_c \frac{\mathcal{N} N^4}{4} \sum_{\mu a} (\theta_{\mu}^a)\right]^2 = \exp[-m_c \phi]$$

where

$$m_c = -\left(\frac{1}{12} + \frac{1}{2}I\right) \tag{9.5.6}$$

9.6 Expansion of the Gauge Field Action

For the gauge group is $SU(\mathcal{N})$, the following normalization is chosen

$$\text{Tr}(X^a X^b) = \frac{1}{2}\delta^{ab} \quad ; \quad \text{Tr}(\mathbb{I}) = \mathcal{N} \tag{9.6.1}$$

The lattice gauge field action is expanded as a polynomial in θ_{μ}^a and $B_{n\mu}^a$ and only terms that are at most of $O(\theta^2 B^2)$ will be retained. The following notation is used

$$a \cdot b = a^{\alpha} b^{\alpha}, (a \times b)^{\alpha} = c^{\alpha \beta \gamma} a^{\beta} b^{\gamma}$$

The action as a polynomial in θ_{μ}^a, is written as

$$S = A_0 + A_I^{(1)} + A_I^{(2)} + O(\theta^3) \tag{9.6.2}$$

The expansion in Eq. 9.6.2 has the generic form

$$S = B^2 + \theta B^2 + \theta^2 B^2 + \cdots$$

After considerable simplification [Baaquie (1977b)]

$$A_0 = -\frac{1}{4g_0^2} \sum_n \sum_{\mu\nu} [(\Delta_\nu B_{n\mu})^2 - \Delta_\mu B_{n\nu} \Delta_\nu B_{n\mu}], \qquad (9.6.3)$$

$$A_I^{(1)} = \frac{C^{abc}}{4g_0^2} \sum_n \sum_{\mu\nu} \theta_\nu^a (\Delta_\mu B_{n\nu}^b - \Delta_\nu B_{n\mu}^b)(B_{n+\hat{\nu},\mu}^c + B_{n\mu}^c) \qquad (9.6.4)$$

Henceforth, the $\sum'_{\mu\nu}$ is replaced by $\sum_{\mu\nu}$ using antisymmetry of the summand. Note that the general structure of $A_I^{(2)}$ is (the matrix M defined below is different from martrix as defined earlier in Eq. 7.8.6)

$$A_I^{(2)} = \frac{1}{4g_0^2} \sum_{mn} \sum_{\mu\nu} \sum_{ab} B_{n\mu}^a M_{\mu\nu}^{ab} B_{m\nu}^b \qquad (9.6.5)$$

In the final calculation only terms of $O(\theta^2)$ are kept in performing the $B_{n\mu}^a$ integrations; since $M_{\mu\nu}^{ab} = O(\theta^2)$, owing to integration over the $B_{n\mu}^a$, only the trace of $M_{\mu\nu}^{ab}$ is evaluated; hence one needs only the diagonal elements $M_{\mu\mu}^{aa}$ of the matrix$M_{\mu\nu}^{ab}$. The diagonal terms yield

$$A_I^{(2)} = \frac{1}{2g_0^2} \sum_n \sum_{\mu\nu}' \left[\frac{1}{12} (\theta_\mu \times B_{n\mu})^2 - \frac{1}{12} (B_{n\mu} \times \theta_\mu) \cdot (B_{n+\hat{\nu},\mu} \times \theta_\mu) \right.$$
$$\left. - \frac{1}{2} (B_{n\mu} \times \theta_\mu) \cdot (B_{n+\hat{\nu},\mu} \times \theta_\nu) \right] + \text{ off diagonal terms in} (ab, \mu\nu)$$

Choose the Feynman gauge given by

$$S_\alpha = -\frac{1}{4g_0^2} \sum_n \left(\sum_\mu \Delta_\mu B_{n+\hat{\mu},\mu} \right)^2 \qquad (9.6.6)$$

Using the definition of $B_{n\mu}^a = \sum'_k B_{k\mu}^a$ yields the quadratic term for the action

$$A_0 + S_\alpha = -\frac{1}{4g_0^2} \sum_k' \sum_{\mu\nu} \left(\sum_\sigma |1 - e^{ik_\sigma}|^2 \right) B_{-k\mu}^a B_{k\nu}^a$$
$$= -\frac{1}{4g_0^2} \sum_k' \sum_\mu d_k B_{-k\mu}^a B_{k\nu}^a$$

For the nonlinear terms

$$A_I^{(1)} = -\frac{C^{abc}}{4g_0^2} \frac{1}{2} \sum_k' \sum_{\mu\nu} \left[\theta_\nu^a (e^{ik_\mu} - 1)(e^{ik_\nu} + 1) - \theta_\mu^a (e^{-ik_\nu} - 1)(e^{ik_\mu} + 1) \right.$$
$$\left. - 4i\delta_{\mu\nu} \sum_\sigma \theta_\sigma^a \sin k_\sigma \right] B_{-k\mu}^b B_{k\nu}^c$$

$$\Rightarrow A_I^{(1)} = -\frac{1}{4g_0^2} \sum_k' \sum_{\mu\nu} B_{-k\mu}^b N_{\mu\nu}^{bc} B_{k\nu}^c \qquad (9.6.7)$$

and

$$A_I^{(2)} = \frac{C^{abc}C^{a\beta\gamma}}{2g_0^2} {\sum_k}' \sum_\mu \left[\frac{1}{4}\theta_\mu^b \theta_\mu^\beta - \frac{1}{24}\theta_\mu^b \theta_\mu^\beta \sum_{\nu \neq \mu}(e^{ik_\nu} + e^{-ik_\nu}) \right.$$

$$\left. - \frac{1}{4} \sum_{\nu \neq \mu} \theta_\nu^b \theta_\nu^\beta (e^{ik_\nu} + e^{-ik_\nu}) \right] B_{-k\mu}^c B_{k\nu}^\gamma$$

$$\Rightarrow A_I^{(2)} = -\frac{1}{4g_0^2} \sum_k \sum_{\mu,\nu} B_{-k\mu}^c M_{\mu\nu}^{cd} B_{k\nu}^d \qquad (9.6.8)$$

Collecting Eqs. 9.6.6, 9.6.7 and 9.6.8 gives

$$S + S_\alpha = -\frac{1}{4g_0^2} {\sum_k}' \sum_{\mu\nu} d_k \left(\delta_{\mu\nu}^{ab} + \frac{1}{d_k} N_{\mu\nu}^{ab} + \frac{1}{d_k} M_{\mu\nu}^{ab} \right) B_{-k\mu}^a B_{k\nu}^b \qquad (9.6.9)$$

Note the important fact that both $N_{\mu\nu}^{ab}$ and $M_{\mu\nu}^{ab}$ have been made *explicitly* Hermitian; this is necessary to carry out the integration over the momentum modes of the gauge field given by $B_{k\mu}^a$. Let

$$L = 1 + \frac{1}{d}N + \frac{1}{d}M \qquad (9.6.10)$$

To determine the last piece of mass renormalization, namely m_a, we collect all the results that have been obtained. The effective action is given by

$$e^{W(\theta)} = \int DB \mu[\theta + B] e^{S[\theta+B] + S_c[\theta+B] + S_\alpha[\theta+B]}$$

$$= \mu[\theta] e^{S_c[\theta]} \int DB e^{S[\theta+B] + S_\alpha[B]}$$

9.7 Determination of m_A for $\mathcal{S}[\theta + B] + \mathcal{S}_\alpha[B]$

Collecting all the results yields

$$Z = \prod_{a,\mu} \int_G d\theta_\mu^a e^{S_c(\theta)} \prod_{k \neq 0} \prod_{\mu,a} \int_{-\infty}^{\infty} dB_{k\mu}^a \mu(\theta) e^{S + S_\alpha}$$

$$= \prod_{a,\mu} \int_G d\theta_\mu^a e^{S_c(\theta)} \frac{\mu(\theta)}{\sqrt{\det L}} \qquad (9.7.1)$$

The gauge field action, to leading order in g_0, can be represented schematically as follows

$$e^{S[\theta+B]} = e^{S_0[B]} \{ 1 + (\theta N B)^2 + (\theta)^2 B^2 M + O(g_0^2) \}$$

and where $S_0[B] = O(B^2)$ is the quadratic part of the gauge field action $S[\theta + B]$.

Since θ_μ^a is being evaluated to $O(\theta^2)$, the exponential functions of θ_μ^a is expanded in a power series. The following identity is used

$$\prod_{a,\mu} \int_G d\theta_\mu^a \theta_\tau \theta_\sigma = \delta_{\tau\sigma} \prod_{a,\mu} \int_G d\theta_\mu^a \theta_\tau^2 = \delta_{\tau\sigma} \times (\text{constant}) \qquad (9.7.2)$$

The use of Eq. 9.7.2 is signified by an arrow ($\rightarrow$). Note that for $SU(\mathcal{N})$

$$C^{aba}C^{ab\beta} = \mathcal{N}\delta^{\alpha\beta}$$

Note that matrices $N \simeq O(\theta), M \simeq O(\theta^2)$ and that $\mathrm{Tr}(N) = 0$. Feynman perturbation expansion is based on the quadratic action $\mathcal{S}_0[B] + \mathcal{S}_\alpha[B]$. Performing the functional integral $\int DB$ yields two terms, shown by in Figures 9.1(c) and (d) given below

$$\det L = \det \left(1 + \frac{1}{d}N + \frac{1}{d}M\right) = \exp\left[\mathrm{Tr}\left(\frac{1}{d}N + \frac{1}{d}M\right) - \frac{1}{2}\mathrm{Tr}\left(\frac{1}{d}N\frac{1}{d}N\right)\right]$$

$$= \exp\left[\mathrm{Tr}\left(\frac{1}{d}M\right) - \frac{1}{2}\mathrm{Tr}\left(\frac{1}{d}N\frac{1}{d}N\right)\right] + O(\theta^3)$$

From Eq. 9.5.3

$$\phi = \mathcal{N}N^{-4}\frac{1}{4}\sum_\mu \theta_\mu^2$$

Then

$$\mathrm{Tr}\left(\frac{1}{d}M\right) \rightarrow \phi\sum_k{}' \frac{4}{d_k}\left(-3 + \frac{7}{16}d_k\right) \tag{9.7.3}$$

The term above is represented by the Feynman diagrams given in Figure 9.1(d). Define lattice constants for the four-dimensional hyper-cubic lattice

$$J = \sum_k{}' \frac{1}{d_k} \ ; \quad K = \sum_k{}' \frac{1 - \cos k_1 \cos k_2}{d_k^2} \tag{9.7.4}$$

and yields

$$\mathrm{Tr}\left(\frac{1}{d}M\right) = 12\phi J - \frac{7}{4}\phi \tag{9.7.5}$$

The lattice constant J arises from the tadpole diagram given in Figure 9.1(d). Also

$$\mathrm{Tr}\left(\frac{1}{d}N\frac{1}{d}N\right) \rightarrow (+\phi)12\sum_k{}' \left(\frac{|e^{ik_1} - e^{-ik_1}|^2}{d_k^2} + 2\frac{1 - \cos k_1 \cos k_2}{d_k^2}\right)$$

$$\Rightarrow \mathrm{Tr}\left(\frac{1}{d}N\frac{1}{d}N\right) = 12\phi(I + 2K)$$

The term above is represented by the Feynman diagrams given in Figure 9.1(c). The lattice constant K arises from Figure 9.1(c), and two propagators yield the denominator d_k^2 whereas the non-Abelian vertices yield the factor $1 - \cos k_1 \cos k_2$.

The result of performing the $\int DB$ functional integral for the gauge fixed action yields

$$\det L \simeq e^{(12J - 7/4 - 6I - 12K)\phi} \quad \Rightarrow e^{m_A\phi} = \frac{1}{(\det L)^{1/2}} \tag{9.7.6}$$

We hence have

$$\int DBe^{\mathcal{S}[\theta + B] + \mathcal{S}_\alpha[B]} = \exp\left\{-\frac{1}{2}\left(12J - \frac{7}{4} - 6I - 12K\right)\frac{\mathcal{N}N^4}{4}\sum_{\mu a}(\theta_\mu^a)^2\right\}$$

and

$$m_A = -\frac{1}{2}\left(12J - \frac{7}{4} - 6I - 12K\right) \tag{9.7.7}$$

9.8   One-Loop Mass Renormalization

The final result is given by

$$Z = \prod_{a,\mu} \int_G d\theta_\mu^a e^{S_c} \frac{\mu(\theta)}{(\det L)^{1/2}} = \prod_{a,\mu} \int_G d\theta_\mu^a e^{\delta m \phi}$$

The renormalized mass, from Eq. 9.3.5, is given by

$$\delta m = m_\mu + m_c + m_A$$

From Eqs. 9.4.1 and 9.5.6 yield

$$m_\mu = -\frac{1}{6} \quad ; \quad m_c = -\frac{1}{12} - \frac{1}{2} I$$

Equations above and Eq. 9.7.7 yield, to order $O(g_0^2)$

$$\delta m = \frac{5}{8} + \frac{5}{2} I + 6K - 6J \tag{9.8.1}$$

Using the four-dimensional hyper-cubic symmetry for the summation

$$1 = \sum_k{}' \frac{d_k^2}{d_k^2} = \sum_k{}' \frac{1}{d_k^2} [16(\cos^2(k_1) - 1) - 48(1 - \cos k_1 \cos k_2) + 16 d_k]$$

gives the following

$$1 = -4I - 48K + 16J \quad \Rightarrow \quad I = 4J - 12K - \frac{1}{4} \tag{9.8.2}$$

Hence Eqs. 9.8.1 and 9.8.2 yield

$$\delta m = 4(6K - J)$$

It can be shown that the lattice constants obey the identity [Baaquie (1976)]

$$J = 6K + O(e^{-N^4})$$

and yield the final result

$$\delta m = 0 \quad : \quad \text{zero mass renormalization}$$

The one-loop computation shows that if there was a quadratic mass divergence, then perturbation theory would have consistently evaluated it.

The absence of a quadratic mass divergence in the lattice theory is expected to follow from exact gauge invariance for the lattice: the gauge invariant action does not generate any terms that are not gauge invariant, and that includes a mass term that breaks gauge invariance. It is worthwhile to note that the absence of the quadratic mass divergence is realized in a nontrivial manner by a remarkable cancellation of lattice constants.

9.9   Slavnov-Taylor Identity

The Slavnov-Taylor identity has been discussed for Yang-Mills fields by Ryder (2001), Das (2006) and others. The continuum results are readily generalized to the case of lattice gauge fields.

The Slavnov-Taylor identity provides a general proof that mass renormalization is zero to all order in perturbation theory [Baaquie (1977b)]. An identity is obtained for $D_{k\mu\nu}^{ab}$, defined in Eq. 9.2.3, and this will give the desired result.

The Faddeev-Popov ghost action and the gauge-fixing term is given by

$$
\mathcal{S}_c = c_n^{\dagger a} \frac{\partial s_m^b}{\partial \phi_n^a} c_m^b \;\; ; \;\; s_n^a = \sum_\mu \Delta_\mu B_{n-\hat{\mu},\mu}^a \tag{9.9.1}
$$

Recall from Eq. 7.9.13 that

$$
\delta_B c_n^a = -\alpha\lambda s_n^a \;\; ; \;\; \delta_B s_n^a = \lambda \frac{\partial s_n^a}{\partial \phi_m^b} c_m^{\dagger b} = -\lambda \frac{\delta \mathcal{S}_c}{\delta c_n^a} \tag{9.9.2}
$$

Note that $\delta/\delta c_n^a$ anticommutes with all fermion variables. We have the following expression for the BRST transformation of s_n^a (re-writing Eq. 9.9.2)

$$
\mathcal{S}_c = c_n^{\dagger a} \frac{\partial s_m^b}{\partial \phi_n^a} c_m^b \Rightarrow s_n^a \to s_n^a - \lambda e^{-\mathcal{S}_c} \frac{\delta e^{\mathcal{S}_c}}{\delta c_n^b} \tag{9.9.3}
$$

Consider the expression

$$
c_I^a \Delta_\sigma B_{n-\hat{\sigma},\sigma}^a e^{\mathcal{S}_\alpha + \mathcal{S}_c}
$$

Performing the BRST transformation, and using the invariance of the gauge-fixed action as given in Eq. 7.9.21, yields the following

$$
c_I^a \Delta_\sigma B_{n-\hat{\sigma},\sigma}^a e^{\mathcal{S}_\alpha + \mathcal{S}_c} \to \Big(c_I^a \Delta_\sigma B_{n-\hat{\sigma},\sigma}^a + e^{-\mathcal{S}_c} \lambda c_I^a \frac{\delta e^{\mathcal{S}_c}}{\delta c_n^b}
$$
$$
- \alpha\lambda \Delta_\mu B_{I-\hat{\mu},\mu}^a \Delta_\sigma B_{n-\hat{\sigma},\sigma}^b \Big) e^{\mathcal{S}_\alpha + \mathcal{S}_c} \tag{9.9.4}
$$

Let

$$
E[\mathcal{O}] \equiv \prod_n \prod_{\mu,a} \int dU_{n\mu} dc_n^a dc_n^{\dagger a} e^{\mathcal{S}} \; \mathcal{O}
$$

BRST invariance and Eq. 9.9.4 yields

$$
E[c_I^a \Delta_\sigma B_{n-\hat{\sigma},\sigma}^a e^{\mathcal{S}_\alpha + \mathcal{S}_c}] = E[c_I^a \Delta_\sigma B_{n-\hat{\sigma},\sigma}^a e^{\mathcal{S}_\alpha + \mathcal{S}_c}] + \lambda E\Big[c_I^a \frac{\delta e^{\mathcal{S}_c}}{\delta c_n^b} e^{\mathcal{S}_\alpha} \Big]
$$
$$
- \alpha\lambda E\Big[\Delta_\mu B_{I-\hat{\mu},\mu}^a \Delta_\sigma B_{n-\hat{\sigma},\sigma}^b e^{\mathcal{S}_\alpha + \mathcal{S}_c} \Big]
$$
$$
\Rightarrow E[\Delta_\mu B_{I-\hat{\mu},\mu}^a \Delta_\sigma B_{n-\hat{\sigma},\sigma}^b e^{\mathcal{S}_\alpha + \mathcal{S}_c}] = \frac{1}{\alpha} E[e^{\mathcal{S}_\alpha} c_I^a \frac{\delta e^{\mathcal{S}_c}}{\delta c_n^b}] \tag{9.9.5}
$$

To perform integration by parts for the fermion variables, note that

$$
E\Big[e^{\mathcal{S}_\alpha} \frac{\delta}{\delta c_n^b} (c_I^a e^{\mathcal{S}_c}) \Big] = 0 = \delta^{ab} \delta_{n-I} E[e^{\mathcal{S}_c + \mathcal{S}_\alpha}] - E\Big[e^{\mathcal{S}_\alpha} c_I^a \frac{\delta e^{\mathcal{S}_c}}{\delta c_n^b} \Big] \tag{9.9.6}
$$

Therefore, from Eqs. 9.9.5 and 9.9.6

$$\frac{1}{Z}E[\Delta_\mu B^a_{I-\hat\mu,\mu}\Delta_\sigma B^b_{I-\hat\sigma,\sigma}e^{\mathcal{S}_\alpha+\mathcal{S}_c}] = \frac{1}{\alpha}\delta^{ab}\delta_{n-I} \ ; \ \ Z = E[e^{\mathcal{S}_c+\mathcal{S}_\alpha}]$$

Fourier transforming the above equation and using translational invariance yields the Slavnov-Taylor identity

$$(1 - e^{ik_\mu})(1 - e^{ik_\nu})D^{ab}_{k\mu\nu} = \frac{1}{\alpha}\delta^{ab} \tag{9.9.7}$$

From above we have that, using Eq. 9.2.12

$$\lim_{k\to 0} D_k \to \frac{1}{k^2} \ \ \Rightarrow \ \ \Pi(0) = 0$$

showing that the theory is a massless. Hence we conclude that, to all order in perturbation theory, there is no mass renormalization for the gauge field.

9.10 Summary

An explicit one-loop calculation shows that, for the lattice theory, there is no mass renormalization of the lattice gauge field. There are four terms that are quadratically divergent, namely a term from the invariant Haar measure for the non-Abelian group, a term arising from the Faddeev-Popov ghost action and two contributions coming from the cubic and quartic nonlinear terms of the gauge field action.

The lattice introduces infinitely many new terms in the action, all of which are required for exact gauge invariance. The one-loop computation shows the proliferation of terms in the path integral: from the Haar measure, from the ghost field and from the lattice gauge field action. Any higher loop calculations need to be done using computer codes as the number of terms required become prohibitively large.

If one uses dimensional regularization for gauge fields defined on continuous spacetime, each quadratically divergent term for the gauge field is individually (and consistently) set to zero – due to Veltman's identity [Baaquie (2018)]. In contrast, the lattice theory generates many terms that are quadratically divergent. Their cancellation is remarkable because each of the divergent term is a numerical constant of the lattice theory – and the sum of all these numerical constants for one-loop cancel exactly. As mentioned earlier, the one-loop calculation provides evidence that the crux of the renormalizability of the non-Abelian gauge fields is gauge invariance, and which holds for a lattice spacetime – and by extension for manifolds with arbitrary geometries.

From the point of view of the renormalization group, in the calculation performed, all the higher momentum modes of the gauge field degrees of freedom were integrated out (to one loop), and their cumulative effect was obtained on the mass term for the zero momentum mode. The fact that the mass term for the zero momentum mode is null implies that the long distance behavior of the theory, in perturbation theory, has no mass term.

Zero mass renormalization does not mean that the gauge field's spectrum of states has a zero mass. In fact, it is expected that the pure gauge field state space *does* have a state vector with a non-zero renormalized and physically observable mass. The non-Abelian gauge field's mass gap above the vacuum state is expected to arise from the collective excitations of the infra-red degrees of freedom that give rise to nonlinear and strong coupling effects – and not from the ultra-violet degrees of freedom that generate the quadratically divergent mass term.

Chapter 10

Gauge Field Block-Spin Renormalization

10.1 Introduction

Wilson's formulation of renormalization and the renormalization group of a quantum field theory reveals the essence of quantum fields that had long been buried in a thick-shell of mathematical formalism [Wilson and Kogut (1974)]. The divergences that appear in renormalizable quantum fields are a reflection of the infinitely many coupled degrees of freedom, with each degree of freedom describing the properties of the quantum field for a different scale. This visionary conception of 'what is a quantum field' has opened up many new and vast vistas of inquiry that continue to provide unexpected and novel results. This Chapter provides a concrete realization of Wilson's renormalization theory.

The procedure for carrying out the renormalization of a quantum field can be done for both the Hamiltonian and state space [Tirrito et al. (2019)] as well as for the path integral and Lagrangian formulation [Wilson and Kogut (1974)]. In the path integral formulation, renormalization is based on integrating out the high momentum degrees of freedom from the path integral – and in doing so generating the theory at low momentum [Wilson (1983); Wilson and Kogut (1974)]. The procedure for integrating out the higher momentum modes of the quantum field theory is termed as 'thinning out' of the degrees of freedom [Baaquie (2018)].

For the lattice theory, integrating out the high momentum modes can be realized by *combining* blocks of 'nearby' degrees of freedom – to form what is called a block degree of freedom – and in doing so, increase the effective spacing of the lattice on which the quantum field theory is defined. This formulation of renormalization is referred to as the Wilson-Kadanoff (1977) block-spin renormalization – and is discussed in Kardar (2007), with many references therein. A simple example of the block-spin renormalization, using the procedure of spin decimation, that has been discussed for the Ising spin in Kardar (2007) and Baaquie (2014).

The block-spin renormalization procedure is applied to study the renormalization of the lattice gauge field. The full four- dimensional theory is far too difficult to solve using this technique. Instead, an approximation is used, based on the work of Migdal (1975), that makes the procedure tractable. The discussion in this Chapter is largely based on Baaquie (1976) and Baaquie (1977c).

In the Migdal approach, the lattice spacing on which the gauge field degrees of freedom are defined is *doubled* by a recursion equation; blocks on gauge field degrees of freedom on the initial lattice are combined to generate the effective renormalized gauge field action on the larger lattice. The recursion equation for doubling the lattice spacing for the four-dimensional theory is realized by first carrying out the block-spin transformation for the two-dimensional theory; the recursion equation for four-dimensions is then obtained by generalizing the final step required for completing the recursion equation.

The weak coupling limit of the Migdal approximation generates a gauge theory having asymptotic freedom in $d = 4$, and the strong coupling limit exhibits the expected confining behavior. Up until this point, one need not specify the gauge group. However, the most interesting feature of the recursion formula is that one can study the *entire* renormalization group trajectory; this entails studying the recurion equation for all values of the coupling, including the cross-over from the weak to strong coupling regime – which is essentially nonperturbative – by studying the theory numerically. The numerical solution of the recursion formula $SU(2)$ gauge group is the main focus of this Chapter.

The numerical study of the Migdal recursion equation shows that the non-Abelian gauge field has a smooth *cross-over* from its weak coupling asymptotically free behavior to its strong coupling confining limit. The fact that there is no phase transition separating the behavior of the weak and strong coupling limits of the gauge field provides evidence that quarks coupled to a non-Abelian gauge field smoothly cross-over from a free-like behavior at short distances to being permanently confined at large distances.

10.2   Two-Dimensional Lattice Gauge Field

The gauge field action functional, as in Eq. 7.1.5, is defined by (prime denotes $\mu \neq \nu$)

$$\mathcal{S} = \frac{1}{2g_0^2} \sum_n \sideset{}{'}\sum_{\mu\nu} \mathrm{Tr}(W_{n\mu\nu}) \;\; ; \;\; W_{n\mu\nu} = U_{n\mu} U_{n+\hat{\mu}\nu} U_{n+\hat{\nu}\mu}^{\dagger} U_{n\nu}^{\dagger}$$

The theory is defined on a hyper-cubic lattice with spacing a_0 and with g_0 being the bare coupling constant. The theory is quantized by performing the path integral of $e^{\mathcal{S}}$ over all possible values of all the degrees of freedom. Let $dU_{n\mu}$ be the invariant group measure of the group element $U_{n\mu}$. Then the Feynman path integral of $e^{\mathcal{S}}$, as in Eq. 7.1.7, is defined by

$$Z = \prod_\mu \prod_\mu \int_G dU_{n\mu} e^{\mathcal{S}}$$

Note $\mathcal{S}$ is invariant under local lattice gauge transformations defined by

$$U_{n\mu} \to V_n U_{n\mu} V_{n+\hat{\mu}}^{\dagger}$$

where V_n is also an element of the gauge group G.

The simplicity of Migdal's formula lies in that it is defined on a two-dimensional lattice. Going to higher dimension involves some approximations. Consider lattice gauge field defined on an infinite size two-dimensional lattice. The path integral of the action functional is defined by

$$Z = \prod_{i=1,2} \prod_{n} \int dU_{ni} e^{\mathcal{S}[U]}$$

$$\mathcal{S} = \frac{1}{2g_0^2} \sum_{n} \mathrm{Tr}(W_{n12} + W_{n21}) = \frac{1}{2g_0^2} \sum_{n} \mathrm{Tr}(W_n + W_n^\dagger)$$

and

$$W_n = W_{n12} = W_{n21}^\dagger = U_{n1} U_{n+\hat{1},2} U_{n+\hat{2},1}^\dagger U_{n2}^\dagger \qquad (10.2.1)$$

The matrix W_n is the *plaquette* of the gauge field.

For clarity define the renormalization group transformation in terms of $\{W_n\}$, as they all are independent variables. (This is not true for $d > 2$.) To do this, consider the change of variables from

$$W_n \to U_{n1}$$

and which yields, from Eq. 10.2.1

$$W_n = U_{n1}\left(U_{n+\hat{1},2} U_{n+\hat{2},1}^\dagger U_{n2}^\dagger\right) \quad \Rightarrow \quad dW_n = dU_{n1}$$

Then

$$Z = \left\{\prod_n \int dU_{n2}\right\}\left[\prod_n \int dW_n \exp\left\{\frac{1}{2g_0^2}\,\mathrm{Tr}(W_n + W_n^\dagger)\right\}\right]$$

$$= \prod_n \int dW_n \exp\left\{\frac{1}{2g_0^2}\,\mathrm{Tr}(W_n + W_n^\dagger)\right\}$$

We see that the path integral above is totally decoupled, and each integration over variable $\{W_n\}$ can be done independently, with action given by

$$\mathcal{S} = \frac{1}{2g_0^2} \sum_{n} \mathrm{Tr}(W_n + W_n^\dagger) \qquad (10.2.2)$$

The action given in Eq. 10.2.2 is the $SU(\mathcal{N})$ lattice gauge field theory in $d = 2$ spacetime dimensions. The action has been studied by Gross and Witten (1980), who have shown that the system has a third-order phase transition in the large $\mathcal{N}$ limit.

10.3 The Renormalization Group Transformation

The initial lattice spacing is a_0. The renormalization group transformation defines, iteratively, a sequence of actions $\{\mathcal{S}_\ell\}_{\ell=0}^\infty$ that describe the behavior of the gauge field on lattice of spacings $\{2^\ell a_0\}$; all degrees of freedom at distances less than $2^\ell a_0$ have been integrated out for the action $\mathcal{S}_\ell$. In particular, only correlation functions

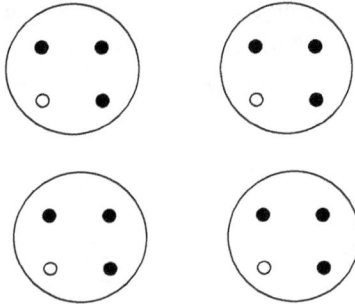

Fig. 10.1   Block-spin construction for Migdal's model. Four degrees of freedom at the old lattice points are combined to form the block degree of freedom at the new lattice point; circled sites form the new effective lattice.

having distances greater than $2^\ell a_0$ can be computed from $\mathcal{S}_\ell$. (The 2 in $2^\ell a_0$ is specific to Migdal's transformation.)

The renormalization group transformation is defined by obtaining $\mathcal{S}_1$ from $\mathcal{S}_0$; the procedure will then be generalized for obtaining $\mathcal{S}_{\ell+1}$ from $\mathcal{S}_\ell$.

Consider the original lattice: blocks of four variables at four adjacent lattice sites are combined to obtain a new variable at one new lattice point, which is marked by an empty circle (see Figure 10.1). There is a block gauge field variable defined on the circled site, and the new action $\mathcal{S}_1$ is a function of these new block-spin variables. Since the new action $\mathcal{S}_1$ is also completely local, one needs to consider only one set of four old lattice sites and how the transformation acts on them.

Label the lattice sites as 1, 2, 3 and 4 (see Figure 10.2). There is a variable W_n, which is a matrix belonging to the gauge group, at lattice site n. The piece of $\mathcal{S}_0$ belonging to the four lattice sites is given by

$$\mathcal{S}_{cell} = \frac{1}{2g_0^2} \sum_{i=1}^{4} \mathrm{Tr}(W_i + W_i^\dagger) = \sum_i A(W_i) \qquad (10.3.1)$$

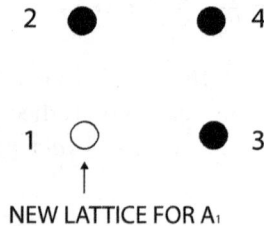

Fig. 10.2   Block-spin construction for a single new block-spin variable at the new lattice site (which is circled).

The action $\mathcal{S}_1$ of the circled site is defined by integrating $e^{\mathcal{S}_{cell}}$ over all the W_i's with the *constraint* that the product of all the W_i's – that is, $W_1 W_2 W_3 W_4$ – be kept fixed and is equal the new block-spin variable W. Note W is also an element of the gauge group. The (nonlinear) transformation that maps $\mathcal{S}_0$ into $\mathcal{S}_1$ is the renormalization group transformation. In symbols, the block-spin action is given by

$$e^{\mathcal{S}_{block}[W]} = \prod_{i=1}^{4} \int dW_i \delta[W - W_1 W_2 W_3 W_4] e^{\mathcal{S}_{cell}} \qquad (10.3.2)$$

The δ-function, from Eq. 2.5.6, is defined as follows

$$\int dU \delta[U - V] = 1 \quad ; \quad \int dU f(U) \delta[U - V] = f(V) \qquad (10.3.3)$$

where for U, V are elements of the Lie group.

From Eq. 2.5.5, $\delta[U - V]$ has the important property that, for group element σ

$$\delta[\sigma(U - V)\sigma^+] = \delta[U - V] \qquad (10.3.4)$$

Eqs. 10.3.3 and 10.3.4 will be required for later discussions.

Eq. 10.3.2 defines a renormalization group transformation that leaves Z invariant; this follows from Eqs. 10.3.2 and 10.3.3 since

$$\int dW e^{\mathcal{S}_{block}[W]} = \prod_{i=1}^{4} \int dW_i e^{\mathcal{S}_{cell}}$$

using

$$\int dW \delta[W - W_1 W_2 W_3 W_4] = 1$$

The reason four lattice sites and not two lattice points are combined to define the new effective lattice is to preserve the symmetry of parity.

Note that one could have defined the rather trivial renormalization group by integrating out variables at lattice sites 2, 3 and 4, and identifying W as W_4. However, this W in no sense represents the combined average value of the fields at the four sites, and leads to trivial results. Also, this trivial transformation has no generalization to $d > 2$.

Noteworthy 10.1: Block gauge field

The definition of block gauge field given by

$$W = W_1 W_2 W_3 W_4$$

crucially hinges on the fact that all the W_i's are elements of a compact group. Suppose for simplicity that the gauge group is $U(1)$ and that $W_i = \exp\{i\theta_i\}$; then the block field is given by

$$\theta = \left(\sum_{i=1}^{4} \theta_i \right) \bmod 2\pi$$

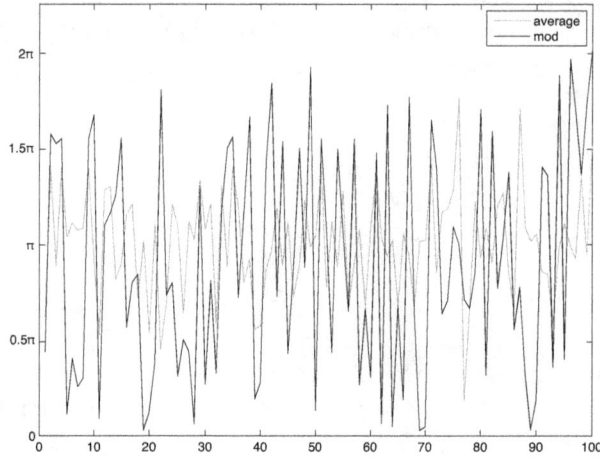

Fig. 10.3　Average $\theta_{\text{Avg}} = (\sum_i \theta_i)/4$ and the modulus $\theta = (\sum_i \theta_i) \text{mod} 2\pi$ for random samples of the θ_i's.

The necessity of taking the sum mod 2π is due to the compactness of the degree of freedom. Consider instead taking the average of the four θ_i's

$$\theta_{\text{Avg}} = \frac{1}{4} \sum_{i=1}^{4} \theta_i \;\; ; \;\; \theta_i \in [0, 2\pi]$$

Figure 10.3 shows the graph for θ and θ_{Avg}.

Taking the average of the θ_i's is another definition of the block variable. Both these definitions are comparable since both block variables vary between the expected limits of $[0, 2\pi]$. However, the method of defining the block variable using θ_{Avg} does not generalize to non-Abelian gauge fields. Hence the definition of block spin in Eq. 10.3.2 is appropriate and in fact reflects the average of the four ingredient degrees of freedom given by W_i's.

Since $\mathcal{S}_{cell}$ is a trace function of the W_i's, from Eqs. 2.5.5 and 10.3.4, and for σ in the gauge group

$$\mathcal{S}_{block}[\sigma W \sigma^+] = \mathcal{S}_{block}[W] : \text{gauge invariant} \tag{10.3.5}$$

Hence $\mathcal{S}_{block}$ is a trace function of the W, and one can conclude by induction that the new action is always a trace function of the new block-spin variables, given that the original action is a trace function. Let the new (circled) lattice sites be labeled by m; then, as shown in Figure 10.4, action $\mathcal{S}_1$ is defined on a lattice with double the spacing given by $2a_0$, and is given by

$$\mathcal{S}_1 = \sum_m \mathcal{S}_{block}[W_m] \tag{10.3.6}$$

This completes the definition of the renormalization group transformation.

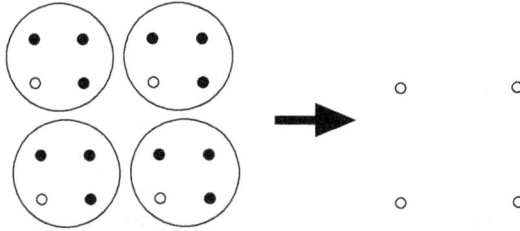

Fig. 10.4 The lattice spacing is *doubled* in going from the original degrees of freedom to the block degrees of freedom.

We now study Eq. 10.3.2 in more detail. First note that any product of all the W_i's gives the same $\mathcal{S}_{block}[W]$ since $\mathcal{S}_{cell}[W]$ is symmetric under any permutation of the four lattice sites. Performing the W_4 integration gives

$$e^{\mathcal{S}_{block}[W]} = \int dW_1 dW_2 dW_3 e \exp\left\{ \sum_{i=1}^{3} A(W_i) + A(W_3^\dagger W_2^\dagger W_1^\dagger W) \right\}$$

Consider the change of variable: $W_2 W_3 = V; dW_2 = dV$. Then

$$e^{\mathcal{S}_{block}[W]} = \int dW_1 dV dW_3 e^{A(W_1)+A(VW_3^\dagger)+A(W_3)} e^{A(V^\dagger W_1^\dagger W)} \tag{10.3.7}$$

Define convolution for functions $f(U)$ and $g(U)$ of the group elements U by

$$(f * g)(\sigma) = \int dU f(U) g(U^\dagger \sigma) \tag{10.3.8}$$

Then, from Eq. 10.3.7, and using that $A(W)$ is a trace function of W yields

$$e^{\mathcal{S}_{block}[W]} = \int d\sigma (e^A * e^A)(\sigma)(e^A * e^A)(W\sigma^\dagger) \tag{10.3.9}$$

$$= (e^A * e^A * e^A * e^A)(W) \tag{10.3.10}$$

In other words, $e^{\mathcal{S}_{block}}$ is given by a four-fold convolution of e^A (which is the action for a single site on the original lattice as in Eq. 10.3.1). Eq. 10.3.9 shows that the four-fold convolution can itself be performed in two steps; the first step consists of performing the two-fold convolution (which corresponds to combining two lattice sites), and then convolving this function with itself to obtain $e^{\mathcal{S}_{block}}$. This interpretation breaks the four-fold convolution into two identical steps of performing two-fold convolutions, and is more appropriate for numerical calculations.

Note in obtaining Eq. 10.3.10 no special assumptions were made about the initial action except that it be completely local and a trace function of the local variable. This property is true for each iteration, and hence Eq. 10.3.10 defines the general renormalization group transformation.

Eq. 10.3.10 is generalized to arbitrary dimension $d > 2$ by the following procedure. To do so, consider the recursion formula for d dimension. The renormalization group transformation for $d > 2$ is assumed to produce a sequence of effective actions $\mathcal{S}_\ell$ that are identical to the original action in all respects except that the action at any given site n can be an arbitrary function of the fundamental plaquettes on the new lattice.

More precisely, it is assumed that for each effective lattice with spacing $2^\ell a_0$ there is a set of gauge field variables that are defined on the effective lattice correspond uniquely to links connecting the effective lattice points. These variables range through the gauge group just like the original variables. The fundamental plaquette is now defined to be the trace of the product of these matrices around a square, as shown in Figure 10.5.

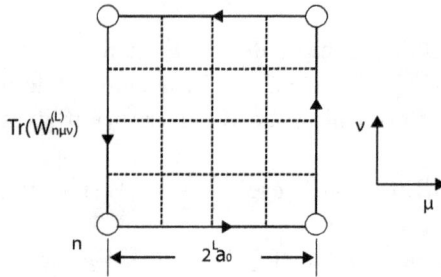

Fig. 10.5   The effective fundamental plaquette for the gauge field after ℓ iterations; the plaquette has side $2^\ell a_0$, where a_0 is the original lattice spacings.

$\mathcal{S}_\ell$ is assumed to be of the form

$$\mathcal{S}_\ell = \sum_{n\in\text{effective lattice}} \sideset{}{'}\sum_{\mu\nu} \mathcal{S}_{n\mu\nu}^{(\ell)} \tag{10.3.11}$$

where $\mathcal{S}_{n\mu\nu}^{(\ell)}$ is an arbitrary function of the fundamental plaquette $\text{Tr}(W_{n\mu\nu}^{(\ell)})$. A lot of drastic approximations are involved in the ansatz of Eq. 10.3.11, and are discussed in Section 10.4.

The recursion equation is generalized for obtaining $\mathcal{S}_\ell$ when $d > 2$. Note for $d = 2$, the recursion equation combines the plaquettes on four lattice sites to produce the effective plaquette W, with $\mathcal{S}_{block}$ being an arbitrary trace function of the block plaquette variable. Let $\mathcal{S}_{block}^{(\ell)}$ be the effective action for a given lattice point, after ℓ iterations, for the two-dimensional case. Then the recursion equation is defined, for all $n, \mu, \nu(\mu \neq \nu)$, by the following

$$e^{\mathcal{S}_{n\mu\nu}^{(\ell)}} = [e^{\mathcal{S}_{block}^{(\ell)}}]^{2^{d-2}} : \text{ Migdal's recursion formula} \tag{10.3.12}$$

Note Eq. 10.3.12 guarantees translational invariance and symmetry under exchange of axes for $\mathcal{S}_\ell$. The effective action $\mathcal{S}_\ell$ is also gauge invariant with respect to gauge transformations defined on the effective lattice, since under a gauge transformation given by $W_{n\mu\nu}^{(\ell)} \to \Phi_n W_{n\mu\nu}^{(\ell)} \Phi_n^\dagger$, the effective action $\text{Tr}(W_{n\mu\nu}^{(\ell)})$ is invariant.

In summary, the recursion equation assumes – for any effective lattice of spacing $2^\ell a_0$ – that $\mathcal{S}_\ell$ is described purely by plaquettes defined on the effective lattice. Eq. 10.3.12 gives a prescription (which is nonlinear and nontrivial) of how to obtain these plaquettes for $d > 2$ dimensions from a two-dimensional lattice. This prescription defines a model for the renormalization group of the lattice gauge field.

10.3.1 *Renormalization group and fixed points*

The mapping from $\mathcal{S}_0$ to $\mathcal{S}_1$ is the renormalization group transformation, denoted by $\mathcal{R}$ and written as

$$e^{\mathcal{S}_1} = \mathcal{R}\big[e^{\mathcal{S}_0}\big]$$

Since in defining $\mathcal{R}$ no specific property of the action was invoked, one has in general

$$e^{\mathcal{S}_\ell} = \mathcal{R}\big[e^{\mathcal{S}_{\ell-1}}\big] = \mathcal{R}^\ell\big[e^{\mathcal{S}_0}\big]$$

The collection of $\mathcal{R}$'s form a semi-group (no inverse) and is called the **renormalization group**.

A fixed point of the renormalization group, denoted by $\mathcal{S}_*$ is invariant under the renormalization transformation and given by

$$e^{\mathcal{S}_*} = \mathcal{R}\big[e^{\mathcal{S}_*}\big]$$

The recursion equation has two fixed points, namely the weak coupling and strong coupling fixed points, denoted by $\mathcal{S}_*^w$ and $\mathcal{S}_*^s$ respectively. From Eq. 10.3.2, by inspection, the weak coupling fixed point is given by

$$e^{\mathcal{S}_*^w} = \delta[W - \mathbb{I}]$$

and the strong coupling fixed point is given by

$$e^{\mathcal{S}_*^s} = 1$$

The weak coupling fixed point is a completely ordered action with all degrees of freedom having the same value; in contrast, the strong coupling fixed point is completely disordered with all possible values of the degrees of freedom being equally likely.

A fixed point is called *unstable* if under renormalization the series of actions $\{\mathcal{S}_\ell\}$ flow away from the fixed point and *stable* if the sequence of actions flows into the fixed point [Baaquie (2018)]. For the case of gauge fields in four dimensions, the weak coupling fixed point is unstable and the strong coupling fixed point is stable.

10.4 The Recursion Equation for d Dimensions

The recursion formula for the exact renormalization group transformation for the d-dimensional lattice gauge theory is discussed. In particular, it will be shown how the prescription for introducing dimension given in Eq. 10.3.12 arises from a crude

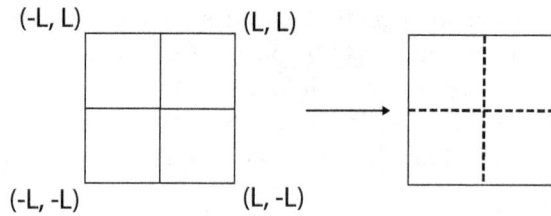

Fig. 10.6   The link picture for forming a 2L-plaquette from four L-plaquettes.

approximation of the exact theory. The discussion essentially repeats the arguments given by Migdal (1975).

For simplicity, start with the original action; let $U_{n\mu}$ be represented by the link $n \to n + \hat{\mu}$. A gauge field loop is defined by the trace of the product of $U_{n\mu}$'s taken around a closed path. The renormalization group transformation defined in terms of the W-variables in Eq. 10.3.2 for $d = 2$ can also be defined directly in terms of link integrations. For $d = 2$, this means combining four L-plaquettes to form the new effective 2L-plaquette (L = length of a side of the plaquette) by integrating out from the action all the links which are interior for the 2L-plaquette, as shown in Figure 10.6. This procedure gives the 2L-plaquette as a four-fold convolution of the L-plaquette, which is the result obtained for $d = 2$.

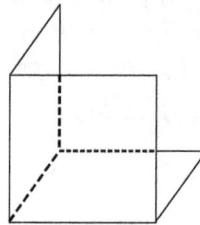

Fig. 10.7   Lattice gauge field links in 3 spacetime dimensions.

To repeat this procedure for higher dimensions, first consider the case of $d = 3$. The lattice theory provides in the original action, for *each* lattice site, 3 L-plaquettes with 2 links common between any 2 L-plaquettes, as in Figure 10.7. In all, 9 links are needed to define these L-plaquettes. To apply the above procedure, consider all the links in a $2L$ cube, figuratively shown in Figure 10.8.

The renormalization group transformation is defined by combining $2^3 = 8$ L sided cubes to form a single $2L$ sided cube by integrating out all the links in the *interior* of surface of the $2L$ cube – holding the links on the edges and surface of the $2L$ cube fixed – as shown in Figure 10.8(b). The resulting action for the $2L$-cube depends, in general, on all possible gauge field loops made out of the links on the *edges* and *faces* of the $2L$-cube. An example of such a non-planar loop is given in Figure 10.9.

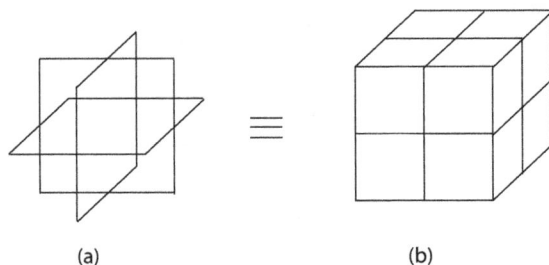

Fig. 10.8   Combining links of 8 cubes having a side of L to form a single $2L$-sided cube.

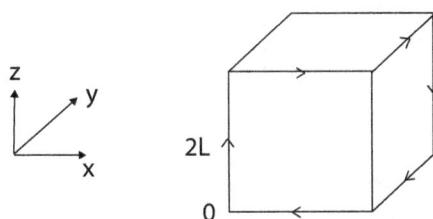

Fig. 10.9   Non-planar gauge field loop.

Migdal makes the approximation that only the *planar gauge field loops* need to be retained, and the rest of the loops are removed from the action. Consider only the planar gauge field loops that lie in xy-plane. As shown in Figure 10.10(a), there are two such gauge field loops, with their z-axis being $0, 2L$. The xy planar gauge field loops are also a function of the axis perpendicular to the plane xy. Note each planar gauge field loop is similar to the two-dimensional action $\mathcal{S}_{block}$; hence, the effective action for the xy-gauge field loops, as shown in Figure 10.10(a), is

$$e^{\mathcal{S}_{xy}^{(1)}} \approx e^{\mathcal{S}_{block}(0)} e^{\mathcal{S}_{block}(2L)}$$

To recover the original form of the action, there can be at most a single gauge field loop in the xy-plane; hence, a further approximation is made that the separation of the planar loops can be neglected, as shown in Figure 10.10(b) and the splitting of the gauge field loops in the perpendicular direction is ignored. This yields the approximate result that

$$e^{\mathcal{S}_{xy}^{(1)}} \approx \{e^{2\mathcal{S}_{block}(0)}\} = \{e^{\mathcal{S}_{block}(0)}\}^2$$

The approximations made to obtain Eq. 10.4.1 reproduces, for $d = 3$, the recursion equation given in Eq. 10.3.12; the transformation can be iterated.

For the case of arbitrary d-dimensions, one needs to evaluate the number of planar gauge field loops that result from combining 2^d L hyper-cubes to form a single $2L$ hyper-cube. Consider the origin of the coordinate to be the unique point

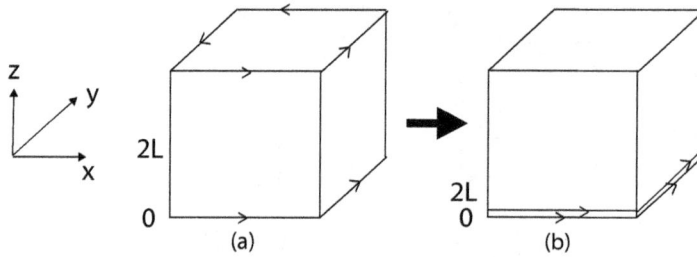

Fig. 10.10   Two planar loops in 3 dimensions, which are approximated to be identical.

that is common to all the $2^d L$ hyper-cubes. Consider the plane defined by the $\mu\nu$ axes. The transverse coordinates of the xy-gauge field planar loops are specified by a $(d-2)$-dimensional vector, which is of the form

$$x_\perp = (\underbrace{\pm, \pm, \cdots \pm}_{d-2}) \cdot L$$

For notational convenience the planar graphs are placed at $\pm L$. In other words, each dimension that is orthogonal to a plane contributes 2 planar graphs. The total combination is 2^{d-2}.

As in the three-dimensional case, all non-planar gauge field loops are ignored as well as the separation of the planar gauge field loops in the directions perpendicular to the plane. In summary, the number of planar-gauge field loops is 2^{d-2} (since this is the number of $x_\perp$'s). The new effective action, for any site n, is given by

$$e^{\mathcal{S}_{n\mu\nu}^{(1)}} = \{e^{\mathcal{S}_{block}}\}^{2^{d-2}}$$

The result obtained is the recursion equation, for d spacetime dimensions, given in Eq. 10.3.12. Since this approximation is repeated for each iteration, the result is valid for arbitrary number of iterations.

To recapitulate, the approximations of the exact theory made by Migdal are, for each iteration, the following.

- All *non-planar gauge field loops* are removed from the effective action.
- Modification of the lattice gauge field action.
- All the planar loops are taken to be identical by ignoring the splitting of the *original planar gauge field loops* into 2^{d-2} planar gauge field loops. Hence, 2^{d-2} is a pre-factor multiplying the single planar loop.
- Considering all planar gauge field loops to be identical, due to lattice translational invariance, reduces the theory to a two-dimensional lattice gauge field. Hence, the plaquette can be considered as the independent degree of freedom and the computation for d dimensions can be carried out in $d = 2$.

Migdal (1975) states that these approximations are reasonable when L is less than the typical Compton wavelength of bound states, and that these approximations lead to a description of the critical behavior of the field (for $d = 4$) that is good to $\sim 30\%$.

10.5 Strong Coupling Approximation for $SU(2)$

Consider the strong coupling limit for the recursion equation, as this is simpler than the weak coupling limit. In fact, for $d = 4$, the lowest order results can be obtained directly using the strong coupling expansion.

Let $G_1 = g_0^2/2$; the strong coupling approximation means studying Eq. 10.3.9 for $G_1 \gg 1$. The initial action is given by

$$S_1 = \frac{1}{4G_1} \sum_n \mathrm{Tr}(W_n + W_n^\dagger) \tag{10.5.1}$$

The recursion equation generates the sequence of actions $\{S_\ell\}_{\ell=1}^\infty$.

For $SU(2)$, $\mathrm{Tr}(U) = \mathrm{Tr}(U^\dagger)$ and yields

$$V(1, W) = \exp\left\{\frac{1}{4G_1} \mathrm{Tr}(W_n + W_n^\dagger)\right\} = \exp\left\{\frac{1}{2G_1} \mathrm{Tr}(W_n)\right\} \tag{10.5.2}$$

and to leading order in G_1

$$V(2, W) = \exp\left\{\frac{1}{4G_2} \mathrm{Tr}(W_n + W_n^\dagger)\right\} = \exp\left\{\frac{1}{2G_2} \mathrm{Tr}(W_n)\right\} \tag{10.5.3}$$

The recursion equation, from Eq. 10.3.12, yields

$$V(2, W) = \left[V(1, \cdot) * V(1, \cdot) * V(1, \cdot) * V(1, \cdot)(W)\right]^{2^{d-2}} \tag{10.5.4}$$

$V(2, W)$ is evaluated in two steps; let

$$\tilde{V}(1, \tau) = (V(1, \cdot) * V(1, \cdot))(\tau) = \int d\sigma V(1, \sigma) V(1, \sigma^\dagger \tau) \tag{10.5.5}$$

Consider the evaluation of $\tilde{V}(1, \tau)$. Let $G_1 \gg 1$; Eq. 10.5.2 yields[1]

$$
\begin{aligned}
\tilde{V}(1, \tau) &= \int dW \exp\left\{\frac{1}{2G_1} \mathrm{Tr}(W) + \frac{1}{2G_1} \mathrm{Tr}(W^\dagger \tau)\right\} \\
&= \int dW \left(1 + \frac{C}{G_1^2} + \left(\frac{1}{2G_1}\right)^2 \mathrm{Tr}(W) \mathrm{Tr}(W^\dagger \tau)\right) + O(G_1^{-3}))
\end{aligned}
$$

From the orthogonality given in Eq. 2.8.8

$$\int dW\, D^\ell(W)_{\alpha\beta} D^{\ell'}(W^\dagger)_{\gamma\delta} = \frac{\delta^{\ell\ell'} \delta_{\alpha\delta} \delta_{\beta\gamma}}{d_\ell} \tag{10.5.6}$$

where $d_\ell =$ dimension of the ℓth representation. For the fundamental representation $d_\ell = 2$ and yields

$$\int dW\, \mathrm{Tr}(W) \mathrm{Tr}(W^\dagger \tau) = \frac{1}{2} \mathrm{Tr}(\tau)$$

Therefore

$$\tilde{V}(l1, \tau) \cong (\text{const.}) \exp\left\{\left(\frac{1}{2G_1}\right)^2 \cdot \frac{1}{2} \mathrm{Tr}(\tau)\right\} + O(G_1^{-3}) \tag{10.5.7}$$

---

[1] $C \propto \int dW\, \mathrm{Tr}^2(W)$.

Repeating the convolution with $\tilde{V}(l, \tau)$ as the input function, yields from Eq. 10.5.3 (up to irrelevant constants)

$$V(2, W) = \left[\exp\left\{ \left[\frac{1}{2}\left(\frac{1}{2G_1}\right)^2 \right]^2 \frac{1}{2} \mathrm{Tr}(W) \right\} \right]^{2^{(d-2)}} = \exp\left\{ \frac{1}{2G_2} \mathrm{Tr}(W) \right\}$$

Hence

$$G_2 = 2^{8-d} G_1^4 : \text{strong coupling approximation} \qquad (10.5.8)$$

The generalization of Eq. 10.5.8 is

$$G_{I+1} = 2^{8-d} G_I^4 \qquad (10.5.9)$$

Suppose one starts with a $G_1 \gg 1$. The sequence of actions, labeled by the coupling constant G_I and having the functional form $V(I, W) = \exp\{\mathrm{Tr}(W)/2G_I\}$ goes, in the strong coupling limit, to the strong coupling fixed point for which $G* = \infty$; hence

$$V(\infty, W) = V_*(W) = e^{\mathcal{S}_*^s} = 1$$

This fixed point is stable for all $d \leq 8$, and hence there is no need go to higher order to determine the stability/instability of the fixed point for $d = 4$.

10.6   Weak Coupling Expansion for $SU(2)$

The weak coupling sector needs a second order calculation for the coupling constant renormalization.[2] To lowest order in $d = 4$ dimensions, it will be seen that in the weak coupling expansion $G_{I+1} = G_I$: hence the stability/instability of the weak coupling fixed point cannot be determined. This feature of the Migdal recursion equation is the one that occurs for $d = 4$ Yang-Mills gauge fields and shows that the procedure of putting in dimension $d > 2$ by the factor of 2^{d-2} correctly captures the qualitative features of the $d = 4$ theory.

The weak coupling fixed point is given by

$$V_*(W) = e^{\mathcal{S}_*^w} = \delta[W - \mathbb{I}]$$

which is equivalent to

$$V_*(W) = \lim_{G_I \to 0} \exp\left\{ \frac{1}{2G_I} \mathrm{Tr}(W - \mathbb{I}) \right\}$$

The second order calculation requires one to start with $G_I \simeq 0$; the action is written as

$$V(1, W) = \exp\left\{ \frac{1}{2G_I} \mathrm{Tr}(W - \mathbb{I}) + \frac{H_I}{2G_I} \mathrm{Tr}^2(W - \mathbb{I}) \right\} \qquad (10.6.1)$$

and the computation yields

$$V(I + 1, W) = \exp\left\{ \frac{1}{2G_{I+1}} \mathrm{Tr}(W - \mathbb{I}) + \frac{H_{I+1}}{2G_{I+1}} \mathrm{Tr}^2(W - \mathbb{I}) \right\} \qquad (10.6.2)$$

---

[2]This is because the action is a marginal operator in the classification of operators in the framework of the renormalization group [Baaquie (2018)].

The weak coupling approximation is valid for $G_I, G_{I+1} \ll 1$, which in turn implies $\text{Tr}(W-1) \sim O(G_I)$; take $H_I = O(1)$. For notational simplicity, let $G = G_I$ and $H = H_I$. From Eq. 10.5.5

$$\tilde{V}(I,T) = \int dU V(I,U) V(I,U^\dagger T) = \int dU e^{E(U,T)} \qquad (10.6.3)$$

where

$$E(U,T) = \frac{1}{2G} \text{Tr}(U - \mathbb{I}) + \frac{H}{2G} \text{Tr}^2(U - \mathbb{I}) + \frac{1}{2G} \text{Tr}(UT^\dagger - \mathbb{I}) + \frac{H}{2G} \text{Tr}^2(UT^\dagger - \mathbb{I}) \qquad (10.6.4)$$

Make the change of variables

$$U \rightarrow U' = UT^{-1/2} \quad ; \quad dU' = dU$$

and yields

$$\tilde{V}(1,T) = \int dU' e^{E(U',T)} \qquad (10.6.5)$$

Dropping the prime on U' gives

$$E(U,T) = \frac{1}{2G} \text{Tr}(UT^{1/2} + UT^{-1/2} - 2\mathbb{I}) + \frac{H}{2G}[\text{Tr}^2(UT^{1/2} - \mathbb{I}) + \text{Tr}^2(UT^{-1/2} - \mathbb{I})] \qquad (10.6.6)$$

Pauli matrices $\vec{\sigma}$ yield

$$U = e^{i\vec{u}\cdot\vec{\sigma}} = \cos u + i\hat{u} \cdot \hat{\sigma} \sin u$$

and

$$T = e^{i\vec{\tau}\cdot\vec{\sigma}} \quad : \quad T^{1/2} + T^{-1/2} = 2\cos(\tau/2) \qquad (10.6.7)$$

Hence

$$\text{Tr}(UT^{\pm 1/2}) = 2\{\cos(u)\cos(\tau/2) \mp \hat{u} \cdot \hat{\tau} \sin(u)\sin(\tau/2)\} \qquad (10.6.8)$$

Consider spherical coordinates for $\vec{u}$. Choose the $\vec{u}$ coordinate axes for $\vec{u}$ and $\vec{\tau}$ such that $\hat{\tau}$ points in the 3-direction, and let

$$x = \hat{u} \cdot \hat{\tau} = \cos(\theta) \quad \Rightarrow \quad \int_0^\pi \sin(\theta)d\theta = \int_{-1}^{+1} dx \qquad (10.6.9)$$

Since $E(U,\tau)$ is not a function of ϕ, the azimuthal angle in the measure on $SU(2)$ decouples from the integration and yields

$$\int dU = \frac{1}{2\pi^2} \int_0^\pi \sin^2 u\, du \int_0^\pi \sin(\theta)d\theta \int_0^{2\pi} d\phi = \frac{1}{\pi} \int_0^\pi \sin^2 u\, du \int_{-1}^{+1} dx \quad (10.6.10)$$

Let $\sigma \equiv \tau/2$; then, Eqs. 10.6.6 and 10.6.8 yield

$$E(U,\tau) = \frac{2}{G}(\cos u \cos \sigma - 1)$$

$$+ \frac{H}{2G} 4[(\cos\sigma\cos u + x\sin\sigma\sin u - 1)^2 + (\cos\sigma\cos u - x\sin\sigma\sin u - 1)^2]$$

$$= \frac{2}{G}(\cos u \cos \sigma - 1) + \frac{4H}{G}[(\cos\sigma\cos u - 1)^2 + (x\sin\sigma\sin u)^2] \qquad (10.6.11)$$

Note that $\text{Tr}(W - \mathbb{I}) = O(G)$ implies $u, \sigma = O(\sqrt{G})$; hence the second term in Eq. 10.6.11 is $O(G)$ and its exponential can be expanded in a power series, giving

$$
\tilde{V}(I, \tau) = \int dU\, e^{E(U, \tau)} = \frac{1}{\pi} \int_0^\pi du \sin^2 u\, e^{2(\cos u \cos \sigma - 1)/G}
$$
$$
\times \int_{-1}^{+1} dx \left[1 + \frac{4H}{G}(\cos \sigma \cos u - 1)^2 + x^2 \frac{4H}{G}(\sin \sigma \sin u)^2 + O(G^2) \right]
$$
$$
= \frac{2}{\pi} \int_0^\pi du \sin^2 u\, e^{2\cos u \cos \sigma / G} \left[1 + \frac{4H}{G}(\cos \sigma \cos u - 1)^2 \right.
$$
$$
\left. + \frac{4H}{3G}(\sin \sigma \sin u)^2 + O(G^2) \right] \tag{10.6.12}
$$

Let

$$
\tilde{V}(I, \tau) = \exp\{C(\tau)\} = \exp\left\{ \frac{1}{G'}(\cos \tau - 1) + \frac{2H'}{G'}(\cos \tau - 1)^2 \right\} \tag{10.6.13}
$$

Then, as shown in Appendix Eqs. 10.12.7 and 10.12.8

$$
\frac{1}{G'} = \frac{1}{2G} - \frac{3}{2} - \frac{5}{2}H + O(G) \; ; \quad \frac{2H'}{G'} = \frac{1}{4G}\left(H - \frac{1}{4}\right) + O(G) \tag{10.6.14}
$$

Rewriting the above equations yields

$$
G' = 2G\left[1 + \left(\frac{3}{4} + 5H\right)G\right] + O(G^3) \; ; \quad H' = \frac{1}{4G}\left(H - \frac{1}{4}\right) + O(G) \tag{10.6.15}
$$

The convolution is performed again using $\tilde{V}(I, \tau)$ as the input function; let

$$
V_A(I, W) = (\tilde{V}(I, \cdot)) * (\tilde{V}(I, \cdot))(W)
$$
$$
\equiv \exp\left\{ \frac{1}{G''}(\cos W - 1) + \frac{2H''}{G''}(\cos W - 1)^2 \right\} \tag{10.6.16}
$$

Eq. 10.6.15 yields

$$
G'' = 2G'\left[1 + \left(\frac{3}{4} + 5H'\right)G'\right] = 4G\left[1 + \left(\frac{13}{8} + \frac{15}{2}H\right)G\right] + O(G^3) \tag{10.6.17}
$$
$$
H'' = \frac{1}{4}\left(H' - \frac{1}{4}\right) = \frac{1}{16}\left(H - \frac{5}{4}\right) + O(G)
$$

The recursion equation given in Eq. 10.3.12 and Eq. 10.6.16 yield

$$
V(I + l, W) = [V_A(I, W)]^{2^{d-2}} \tag{10.6.18}
$$
$$
= \exp\left\{ \frac{1}{G_{I+1}}(\cos W - 1) + \frac{2H_{I+1}}{G_{I+1}}(\cos W - 1)^2 \right\} \tag{10.6.19}
$$

and hence

$$
G_{I+1} = 2^{2-d}G'' \tag{10.6.20}
$$
$$
\frac{2H_{I+1}}{G_{I+1}} = 2^{2-d}\frac{2H''}{G''} \tag{10.6.21}
$$

Using the fact that $G = G_I, H = H_I$ and simplifying gives the final result for the weak coupling recursion equation

$$
G_{I+1} = 2^{4-d}G_I\left[1 + \left(\frac{13}{8} + \frac{15}{2}H_I\right)G_I\right] + O(G_I^3) \tag{10.6.22}
$$
$$
H_{I+1} = \frac{1}{16}\left(H_I - \frac{5}{4}\right) + O(G_I) \tag{10.6.23}
$$

For $d < 4$, the coupling constant G_I is increasing to lowest order, and the second order result is unnecessary. For $d = 4$, the recursion equation yields[3]

$$G_{I+1} = G_I + \left(\frac{13}{8} + \frac{15}{2}H_I\right)G_I^2 + O(G_I^3) \qquad (10.6.24)$$

$$H_{I+1} = \frac{1}{16}\left(H_I - \frac{5}{4}\right) + O(G_I)$$

Consider the effect the H_I term has on G_{I+I}. From Eq. 10.6.23

$$H_{I+1} = \frac{1}{4^{2I}}\left(H_1 + \frac{1}{12}\right) - \frac{1}{12} + O(G_I) \qquad (10.6.25)$$

Since $H_1 = O(1)$, H_I very rapidly goes to the fixed point $H* = -\frac{1}{12}$. Let $H_I \simeq H*$; then Eq. 10.6.17 yields

$$G_{I+1} = G_I + G_I^2 > G_I \qquad (10.6.26)$$

Eq. 10.6.24 shows that the instability of the weak coupling fixed point $G_* = 0$ is not affected by the H_I term. Since $G_1 > 0$, which implies $G_I > 0$, Eq. 10.6.24 shows that $G_* = 0$ is the only fixed point; the fixed point is independent of H_I, given that $H_I = O(1)$.

The fixed point function for $G_* = 0$ is $V_*(W) = \delta[W - 1]$. Hence, starting from the input initial action $S_0(W) = (\cos W - l)/G_1, G_1 \ll 1$ leads to the result that $G_{I+I} > G_I$; in other words, the $G* = 0$ fixed point is once unstable, and one leaves the fixed point with each iteration. The region for G_I that lies in between the weak coupling region and the strong coupling region is the intermediate coupling *cross-over region*, and is essentially nonperturbative.

Noteworthy 10.2: Recursion equation and non-Abelian group

The instability of the $G_* = 0$ fixed point crucially hinges on the non-Abelian nature of the gauge group. This is clearly seen in this model by noting that the term $\frac{13}{8}G_I^2$ in the weak coupling recursion given in Eq. 10.6.24 arises from the invariant measure of the $SU(2)$ gauge group. If this term had been absent in Eq. 10.6.24, the coefficient of G_I^2 in Eq. 10.6.26 would become negative and the $G_* = O$ fixed point would no longer be unstable; in fact, if one does the second order calculation for the Abelian case, one finds that G_I is still marginal, and no conclusions of the stability/instability of the $G_* = O$ fixed point for the $U(1)$ Abelian gauge field can be drawn.

Examining Eq. 10.6.26, one could naively expect that the sequence of coupling constants converge to the strong coupling fixed point (for which $G* = \infty$). The numerical solution shows that, in fact, this is exactly what happens showing that the gauge field goes over smoothly from its weak coupling asymptotic free behavior to its confining strong coupling behavior. For $d \leq 4$, there are only two fixed points for Migdal's recursion equation – given by the unstable weak coupling fixed point and the stable strong coupling fixed point.

---

[3]Note that our result does not agree with the result given in Migdal (1975).

10.7 Weak Coupling Approximation for $d = 4 + \epsilon$

Consider the recursion equation for $d = 4 + \epsilon > 4(\epsilon \simeq 0)$. Eq. 10.6.22, using $H_I = H* = -\frac{1}{12}$, yields

$$G_{I+1} = 2^{4-d}G_I(1 + G_I) = 2^{-\epsilon}G_I(1 + G_I)$$

The fixed point equation is

$$G_* = 2^{-\epsilon}G_*(1 + G_*)$$

and yields the following fixed points

$$G_* : 0, 2^\epsilon - 1, \infty$$

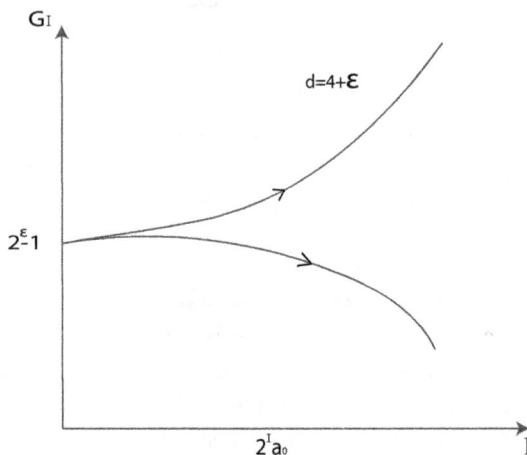

Fig. 10.11   The ultra-violet fixed points is twice unstable. The infra-red fixed point leads to either a strongly coupled theory or a free massless theory. The two phases are separated by a critical phase with $G* = 2^\epsilon - 1$.

The fixed point $G_* = 2^\epsilon - 1$ is twice unstable. For $G_0 < G_*$, $G_I \to 0$ as $I \to \infty$; for $G_0 > G_*, G_I \to \infty$ as $I \to \infty$. See Figure 10.11. Both the fixed points $G_* = 0$ or $G_* = \infty$ are now stable, with the fixed point at $G_* = 2^\epsilon - 1$ being twice unstable. The system undergoes a phase transition at $G_* = 2^\epsilon - 1$, and the behavior of the gauge field at the $G_* = 2^\epsilon - 1$ fixed point is described by a scale invariant theory.

For $d = 4 + \epsilon > 4$, the asymptotically free domain of $G \simeq 0$ is separated by a phase transition from the strongly coupled domain of $G \simeq \infty$, and the gauge field does not continuously go from the asymptotically free theory to the strongly coupled theory; that is, for $d > 4$, the gauge-field cannot simultaneously exhibit free-like behavior at short distances, and strong coupling behavior at large distance. The UV fixed point with $G* = 2^\epsilon - 1$ is a strongly coupled theory. The renormalization group trajectories are shown in Figure 10.11.

Furthermore, as $\epsilon \to 0$, the $G* = 2^\epsilon - 1$ fixed point coalesces with $G* = 0$ fixed point – making it once unstable – and leading to the gauge field having no phase transition separating the weak from the strong coupling sector.

10.8   Weak Coupling $SU(2)$ Gauge Field: β-Function

From above, the weak coupling recursion equation for $d = 4 + \epsilon > 4$ is given by

$$G_{I+1} = 2^{4-d}G_I(1 + G_I) = 2^{-\epsilon}G_I(1 + G_I) \tag{10.8.1}$$

To convert to the original coupling constant recall that

$$G_I = \frac{1}{2}g_I^2 \; ; \quad a \equiv 2^I a_0$$

Using

$$a_{I+1} = 2a_I \;\; \Rightarrow \;\; a_{I+1} = a + a \;\; \Rightarrow \;\; G_{I+1} = \frac{1}{2}g_{I+1}^2(a_I + a_I)$$

Eq. 10.8.1 yields

$$\frac{1}{2}g^2(a + a) = 2^{-\epsilon}\frac{1}{2}g^2(a)\left(1 + \frac{1}{2}g^2(a)\right)$$

In terms of the coupling constant $g(a)$, Eq. 10.8.1 yields

$$g^2(a + a) = g^2(a) + a\frac{\partial g^2(a)}{\partial a} + \cdots = 2^{-\epsilon}g^2(a)\left(1 + \frac{1}{2}g^2(a)\right) \tag{10.8.2}$$

Let a momentum scale μ be defined by

$$\mu = \frac{1}{a} \;\; \Rightarrow \;\; a\frac{\partial g^2(a)}{\partial a} = -2g\mu\frac{\partial g}{\partial \mu} \equiv -2g\beta$$

where the β-function is defined by

$$\beta = \mu\frac{\partial g}{\partial \mu}$$

Hence, from Eq. 10.8.2, the Migdal β-function is given by

$$\beta_M = (1 - 2^{-\epsilon})g - \frac{1}{2}2^{-\epsilon}g^3 \tag{10.8.3}$$

Taking the limit of $\epsilon \to 0$ yields, from Eq. 10.8.3, the following

$$\beta_M = -0.5g^3$$

The result we have obtained for the gauge field β-function is a special case of the $SU(\mathcal{N}_c)$ gauge field. The following exact result, for $SU(2)(\mathcal{N}_c = 2)$ and $g \approx 0$ is given, from Peskin and Schroeder (1995)

$$\beta_{\text{Exact}} = -\frac{11\mathcal{N}_c}{48\pi^2}g^3 = -\frac{11}{24\pi^2}g^3 \simeq -0.05g^3$$

The Migdal approximation yields a β-function that is qualitatively similar to the exact result. It is not clear how one can improve the Migdal approximation or whether it can be used to make any empirical predictions.

10.9   Numerical Solution of the Recursion Equation

Recall from Eq. 10.5.5 we have

$$\tilde{V}(I,\tau) = \int d\sigma V(I,\mathrm{Tr}(\sigma))V(I,\mathrm{Tr}(\sigma^\dagger\tau))$$

$$= \int d\sigma V(I,\mathrm{Tr}(\sigma))V(I,\mathrm{Tr}(\sigma\tau)) \qquad (10.9.1)$$

In obtaining Eq. 10.9.1, the following was used: (a) that $V(I,\tau)$ and $\tilde{V}(I,\tau)$ are trace functions of τ and (b) for $SU(2)$, $\mathrm{Tr}(\sigma^\dagger) = \mathrm{Tr}(\sigma)$ and $d(\sigma^\dagger) = d(\sigma^{-1}) = d\sigma$. From Eq. 10.5.4

$$V(I+1,\tau) = \left\{ \int dW \tilde{V}(I,\mathrm{Tr}(W))\tilde{V}(I,\mathrm{Tr}(W\tau)) \right\}^{2^{d-2}} \qquad (10.9.2)$$

Since the invariant measure dW is always ≥ 0, it follows that $V(I+l,\tau) \geq 0$ if $V(I,\tau) \geq 0$. The positivity of $V(I,\tau)$ ensures that the effective action is always real, given that $V(I,\tau) \geq 0$ – and yields results for numerical calculations that are rapidly convergent.

Eqs. 10.9.1 and 10.9.2 are studied numerically. Since both equations are essentially the same, except for the additional step of introducing dimension, for now let us focus on Eq. 10.9.1. For notational convenience, $V(I,\tau)$ is considered to be a function of $\frac{1}{2}\mathrm{Tr}(\tau)$. From Eq. 10.6.10

$$\int d\sigma = \frac{1}{\pi} \int_0^\pi \sin^2 \sigma d\sigma \int_{-1}^{+1} dx$$

where $x = \cos\theta$; choosing the spherical coordinate system for σ such that $\hat{\tau}$ is in the 3-direction, yields

$$\frac{1}{2}\mathrm{Tr}(\sigma) = \cos\sigma \quad ; \quad \frac{1}{2}\mathrm{Tr}(\sigma\tau) = \cos\sigma\cos\tau - x\sin\sigma\sin\tau$$

Hence

$$\tilde{V}(I,\tau) = \frac{1}{\pi} \int_0^\pi d\sigma \sin^2 \sigma V(I,\cos\sigma) \int_{-1}^{+1} dx V(I,\cos\sigma\cos\tau - x\sin\sigma\sin\tau) \quad (10.9.3)$$

10.9.1   *Change of integration variable*

Recall, from Eq. 10.6.9, $x = \cos(\theta)$. Make a change of variable from x to ξ; let

$$\cos\xi = \cos\sigma\cos\tau - x\sin\sigma\sin\tau \Rightarrow -\sin\xi d\xi = -\sin\sigma\sin\tau dx$$

$$dx = \frac{\sin\xi}{\sin\sigma\sin\tau}d\xi \qquad (10.9.4)$$

The upper and lower limits of integration for variable ξ are given as follows[4]

---

[4]A careless treatment of the limits of ξ results in the wrong limits and the recursion equation fails to give the correct answer.

- Upper limit

$$x = +1 : \cos \xi = \cos(\sigma + \tau)$$

$$\Rightarrow \quad \xi = (\sigma + \tau) \bmod 2\pi \equiv \min\{\sigma + \tau, 2\pi - \sigma - \tau\}$$

- Lower limit

$$x = -1 : \cos \xi = \cos(\sigma - \tau)$$

$$\Rightarrow \quad \xi = |\sigma - \tau|$$

Note

$$\xi \in [0, \pi] \quad \Rightarrow \quad \sigma, \tau \in [0, \pi]$$

From Eq. 10.9.3

$$\sin \tau \tilde{V}(I, \tau) = \frac{1}{\pi} \int_0^\pi d\sigma \sin \sigma V(I, \cos \sigma) \int_{|\sigma - \tau|}^{(\sigma+\tau)\bmod 2\pi} d\xi \sin \xi V(I, \cos \xi) \quad (10.9.5)$$

Let

$$W(I; \sigma, \tau) \equiv \int_{|\sigma - \tau|}^{(\sigma+\tau)\bmod 2\pi} d\xi \sin \xi V(I, \cos \xi) \quad (10.9.6)$$

Then

$$\sin \tau \tilde{V}(I, \tau) = \frac{1}{\pi} \int_0^\pi d\sigma \sin \sigma V(I, \cos \sigma) W(I; \sigma, \tau) \quad (10.9.7)$$

The two integrations to be performed numerically are Eqs. 10.9.5 and 10.9.7. The change of variable made in Eq. 10.9.4 is important for the numerical calculation. As things stand in Eq. 10.9.3, for performing the integration on a grid of points for the x integration the function $V(I, \sigma)$ is required at points that are not the same as the points necessary to perform the σ-integration. And if some interpolated values of $V(I, \tau)$ are used for performing the x-integration, the errors introduced are large.

Hence, if we are to use values of the function $V(I, \sigma)$ or $\tilde{V}(I, \sigma)$ at grid points which are computed by the recursion formula, then Eq. 10.9.3 cannot be used.

What Eq. 10.9.4 does is that it produces the function $\tilde{V}(I, \sigma)$ at the *same* grid points that are to be used for performing the σ- and ξ-integrations. In other words, the change of variable from x to ξ results in Eq. 10.9.5 that uses the value of $V(I, \tau)$ only at the grid points that are fixed and which are the same as the ones used for performing both the ξ, σ-integration. Hence, no interpolation for values of $V(I, \tau)$ are necessary.

Also, in computing $W(I; \sigma, \tau)$, the limits on the integral always fall on the fixed grid points, with no interpolation being made for the values of $V(I, \tau)$. To summarize, the change of variable from x to ξ allows us to perform, for each iteration, both integrations in Eqs. 10.9.5 and 10.9.7 on the same fixed grid points for which the recursion formula gives values of the computed function.

Note after the change of variable in Eq. 10.9.4, the function that is required is generically

$$f(I,\theta) = \sin\theta \tilde{V}(I,\cos\theta)$$

This appears in the integrand and is also the results of the numerical integration, as can be seen from Eqs. 10.9.5 and 10.9.7. In fact, Eq. 10.9.5 can be written as

$$\tilde{f}(I,\tau) = \frac{1}{\pi}\int_0^\pi d\sigma f(I,\sigma)\int_{|\sigma-\tau|}^{(\sigma+\tau)\mathrm{mod}2\pi} d\xi f(I,\xi)$$

The functions $\tilde{f}(I,\tau)$, $f(I,\theta)$ are defined on the same grid points, with the variation in the limits of integration on the given grid, as seen in Eq. 10.9.5, generating the result $\sin\tau V(I,\tau)$.

10.9.2   *Numerical algorithm*

Eqs. 10.9.5 and 10.9.7 are ideally suited for a numerical solution. Each step in the recursion is identical as far as the structure of the integrals goes – with the input function changing from $V(I,\sigma)$ to $V(I+1,\sigma)$. The function $V(I+1,\sigma)$ is computed from $V(I,\sigma)$, and in turn is the input for calculating $V(I+2,\sigma)$, and so on. The initial function $V(1,\sigma)$ for the recursion is the exponential of the bare action given by

$$V(1,\sigma) = \exp\left\{\frac{1}{G_1}(\cos\sigma - 1)\right\}$$

The input coupling constant G_1 is a parameter that can be varied.

For the purpose of performing the integrations numerically, the variables σ and ξ are discretized. The range of integration, for the first step in the recursion, is divided into N number of grid points. The total number of grid points may vary with each step in the recursion. The number N is an input, whereas $N(I)$ is fixed by the computer program; the total number of grid points for the Ith step in the recursion is denoted by $N(I)$, with *maximum value* of $N(I)$ being $2N$ number of grid points.

The only other input for the computer program is the maximum number of iterations to be performed, denoted by M. In summary, a computer program based on the numerical algorithm discussed above has the following four parameters as input

- The initial plaquette action with coupling constant G_1
- The dimension of spacetime d
- The maximum number of grid points for the integration $2N$
- The maximum number of allowed iterations M.

10.9.3 *Total grid size $S(I)$*

Consider the case of $G_1 \simeq 10^{-3}$; then

$$V(1,\sigma) = \exp\left\{\frac{1}{G_1}(\cos\sigma - 1)\right\} \sim \exp\left\{-\frac{1}{2G_1}\sigma^2\right\}$$

From above, it can be seen that $V(1,\sigma)$ rapidly goes to zero when $\sigma \gg 1/\sqrt{G_1}$. The size of the grid $S(1)$ is chosen such that $V(1,S(1)) \geq G_1^2$ on all points inside this space; for G_1 small, $S(1) \sim \sqrt{G_1}$. The choice of $V(1,N) \sim G_1^2$ is necessary to do the second order calculation of G_I for $d = 4$.[5] With this in mind, for each iteration I, the range of integration of variable σ is given by the total grid size $S(I)$ such that

$$V(I, S(I)) > G_I^2$$

For reasons to be explained later, the total grid size $S(I)$ must always be of the form $\pi/2^{n-m}$ (n, m = integers); n is chosen from the equation

$$S(1) = \frac{\pi}{2^n} \sim \sqrt{G_1}$$

Given $S(1)$ and N, the initial grid spacing $\epsilon(1)$ is initially chosen so that

$$\epsilon(1) = \frac{S(1)}{N}$$

In general[6]

$$S(I) = \frac{\pi}{2^{n-m}} \quad ; \quad \epsilon(I) = \frac{S(I)}{N(I)}$$

Qualitatively, $S(I) < \pi/2$ for the weak coupling region and $\pi/2 < S(1) < \pi$ for the cross-over region. As discussed above, the initial total grid size is fixed such that $V(1,N) = G_1^2$. The computed function's normalization is chosen such that

$$V(I,1) = \tilde{V}(I,1) = 1 \tag{10.9.8}$$

As the iteration is performed to compute $V(2,\sigma), V(3,\sigma)$, and so on, the interval of σ on which these functions are $> G_2^2, G_3^2$ and so on also increases – since for $d = 4$ the coupling constants G_I are increasing. Figures 10.12(a) and (b) show the action and total grid size as the one iterates and the running (effective) coupling constant increases.

Eq. 10.9.7 is used to compute the function $\tilde{V}(I,\tau)$ for values of τ until the point when $\tilde{V}(I, S(I)) = G_I^2$; $N(I)$ is defined using the equation

$$\epsilon(I)N(I) = S(I) \quad ; \quad N \leq N(I) \leq 2N$$

This procedure of increasing the total number of lattice points is continued until $N(I) > 2N$; when this happens, all the odd lattice points are dropped from the calculation, and the lattice spacing is consequently *doubled*. Figure 10.12(a) shows

---

[5] For $d < 4$, G_1^2 can be replaced by say a fixed value equal to 0.001 without any large errors.
[6] The computer program is slightly more complicated and $S(I)$ may not be of the form $\pi/2^{n-m}$ for the first few steps.

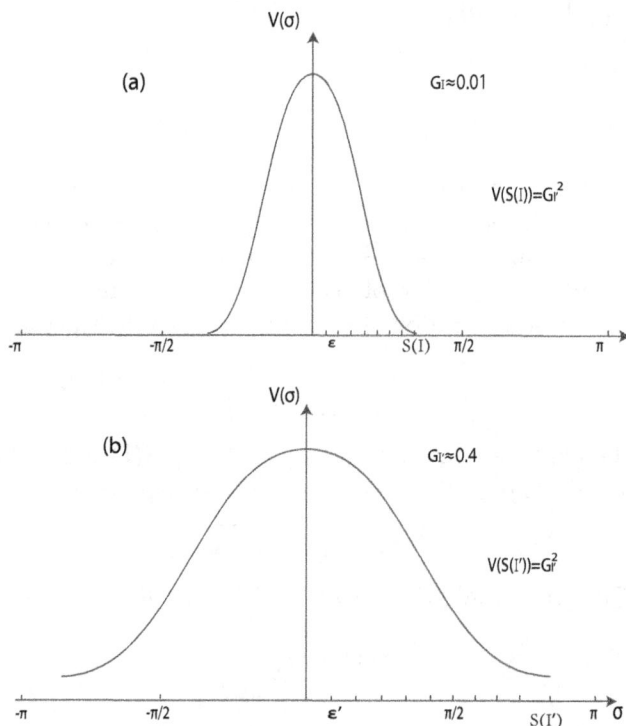

Fig. 10.12   (a) *Weak coupling regime:* $0 < S(I) < \pi/2$. The action and total grid size for $G_I \simeq 0.01$ and with $\epsilon = S(I)/N(I)$. (b) *Cross-over regime:* $\pi/2 < S(I) < \pi$. The action and total grid size for $I' > I$ and $G'_I \simeq 0.1$ and with $2\epsilon = S(I')/N(I')$.

grid spacing ϵ and, for example, Figure 10.12(b) in which the grid spacing ϵ' has doubled compared to previous recursions. Furthermore, the doubling of the total number of lattice points m times implies that

$$ S(I) = \frac{\pi}{2^{n-m}} \;\; ; \;\; m = 0, 1, \cdots, n $$

The procedure of increasing the total number of lattice points is continued until $S(I) = \pi$; when $S(I) = \pi$, the number of lattice points is permanently fixed, and the $S(I)$ no longer is allowed to increase. This is the reason why $S(1)$ had to have the form $\pi/2^n$; because after doubling the lattice spacing n times, we end up with $S(I) = \pi$.

The value of $S(I) = \pi$ is reached when the gauge field becomes strongly coupled. The strong coupling regime is shown in Figure 10.13. This method of doubling the lattice spacing is crucial in allowing one to go from the weak to the strong coupling domain. The reason being that if the lattice spacing was kept fixed at its initial value, and the starting is from the weak coupling sector, then one would eventually need about 2^{10} grid points to cover the interval $[0, \pi]$. This would make the calculation infeasible.

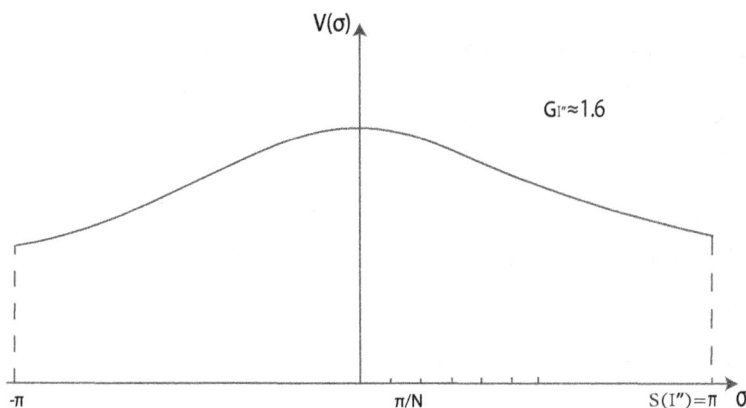

Fig. 10.13 *Strong coupling limit*: $S(I) = \pi$. The total space is fixed to be $S(I) = \pi$ with fixed grid spacing $\epsilon = \pi/N$.

The function $W(I; \sigma, \tau)$ is computed using Simpson's rule. Let $f(\xi) = \sin \xi V(I, \cos \xi)$; then from Eq. 10.9.6

$$W(I; \sigma, \tau) \equiv \int_{|\sigma-\tau|}^{(\sigma+\tau)\text{mod}2\pi} f(\xi)d\xi = W(I; \tau, \sigma) \qquad (10.9.9)$$

Since $W(I; \sigma, \tau)$ is *symmetric* under $\sigma \leftrightarrow \tau$, only the case of $\sigma > \tau$ needs to be considered. Hence, the numerical integration that needs to be performed is only the following

$$W(I; \sigma, \tau) \equiv \int_{\sigma-\tau}^{(\sigma+\tau)\text{mod}2\pi} f(\xi)d\xi$$

and yields the the following interval for ξ-integration:

- For $\sigma + \tau < 2\pi$; $\sigma + \tau - (\sigma - \tau) = 2\tau$
- For $\sigma + \tau > 2\pi$; $2\pi - (\sigma + \tau) - (\sigma - \tau) = 2(\pi - \sigma)$

In other words, the interval for the ξ-integration is always even. The total number of grid points $N(I)$ is always arranged to give an even number of intervals so that Simpson's rule, given below, can be used for the ξ-integration[7]

$$\int_1^3 f(\xi)d\xi = \text{const}[f_1 + 4f_2 + f_3] \qquad (10.9.10)$$

A straightforward extension can be made for arbitrary number of even intervals. The trapezoidal rule is used for evaluating

$$\int_0^\pi d\sigma \sin \sigma V(I, \cos \sigma) W(I; \sigma, \tau)$$

---

[7]The overall constant is removed by the normalization given in Eq. 10.9.8.

since the trapezoidal rule is more accurate for evaluating the integral of a periodic function over its period than is Simpson's rule. However, the gains of using the trapezoidal rule are minor, and it is preferred mostly due to its simplicity. The reason for using Simpson's rule for $W(1; \sigma, \tau)$ is that if the trapezoidal rule is used, then there are large systematic errors in going from the weak to the strong coupling domain.

The definition of the running coupling constant for the entire renormalization trajectory is made by assuming that the functional form of the effective action is given by

$$V_E(I, \cos \alpha) = \exp \left\{ \frac{1}{G_I}(\cos \alpha - 1) + \frac{2H_I}{G_I}(\cos \alpha - 1)^2 \right\} \tag{10.9.11}$$

The coupling constant G_I is determined by using the above ansatz and can be evaluated by comparing the numerical value of the computed function $V(I, \cos \alpha)$ at three lattice points. The ansatz is exact for the weak and strong coupling limits, and is a natural interpolation for the intermediate cross-over region, for which G_I is evaluated for a number of adjacent points. It is found that the variation in using different definitions for the coupling constants in the cross-over region is about 10%. The coupling constant G_I is taken to be the one evaluated using $V(I, 2), V(I, 3)$ and $V(I, 4)$.

The numerical accuracy of the computer program is checked with the weak and strong coupling analytic results. The program is accurate, for each iteration, to 1% when calculations are performed with ~ 100 grid points. The computer program produces the sequence of functions $V(I, \sigma)$ from which G_I can be computed. In Figure 10.14 the results are plotted.

10.10 Numerical Results

The sequence of coupling constants for different initial values of G_1 lie on the same trajectory; changing the value of G_1 simply shifts the sequence of coupling constants along the trajectory on which the other sequences lie. Taking the limit of $G_1 \rightarrow 0$ yields the renormalized coupling constant as a function of the effective lattice spacing.

The renormalization group trajectory for the sequence of coupling constants $\{G_I\}_{I=0}^{\infty}$ has three distinct regions, namely the weak coupling regime, the strong coupling regime and the cross-over region. These three are smoothly connected and can be identified only qualitatively.

By convention, the intermediate region is identified to be such that the deviation from the weak and strong approximations is $> 10\%$. The numerical result, for $d = 4$ (the numbers depend on d) is shown in Figure 10.14, yields the following.

- Weak coupling regime is for $0 \leq G_I \leq 0.3$.
  The coupling constant increases gradually.

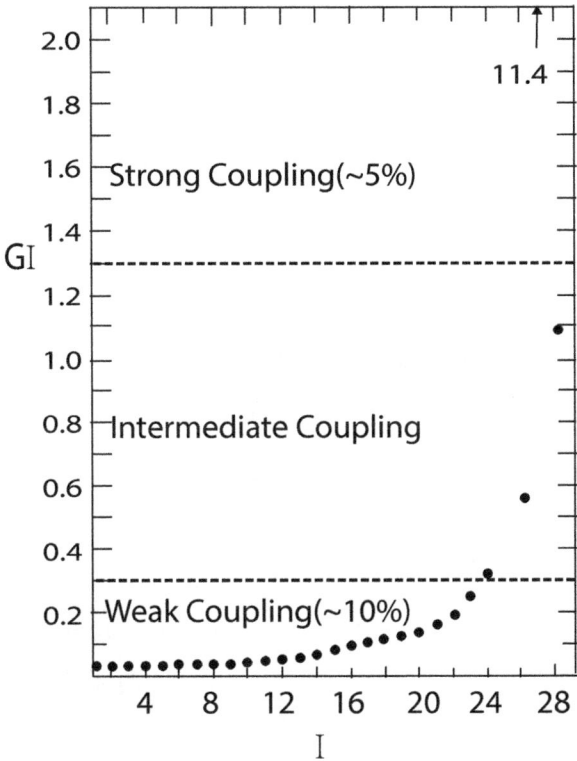

Fig. 10.14 Graph of coupling constant renormalization trajectory, with G_I vs. I, where $I =$ number of iterations, and G_I is the effective coupling constant for the gauge field.

- Intermediate regime is for $0.3 \leq G_I \leq 1.5$.
  The coupling constant abruptly increases over a small range.
- Strong coupling regime is for $1.5 \leq G_I \leq \infty$.
  The coupling constant increases rapidly.

The sequence of effective actions are monotonically increasing functions; recall the effective actions are normalized such that $V(I,1) \equiv 1$; that is $V(I+1,\alpha) \geq V(I,\alpha)$, $\alpha > 1$. Since $V(I = \infty, \alpha) = V_*(\alpha) = 1$, we see that the sequence of effective actions converge uniformly to the stable strong coupling fixed point action. In fact, it is found numerically that, to $\sim 15\%$ accuracy, the form of the initial action is valid for all I.

$$V(I,\alpha) = \exp\left\{\frac{1}{G_I}(\cos\alpha - 1)\right\}$$

The strong coupling fixed point reached by the numerical calculation depends on the total range of integration used for the intermediate and strong coupling regimes.

The exact theory gives $0 \leq \sigma \leq \pi$. In the numerical calculation, if the range

$$0 \leq \sigma \leq a \quad : \quad a > \pi \ \text{ or } \ a < \pi$$

is used, then a fixed point was reached which was different from $V_*(\sigma) = 1$. This is the reason why $S(1)$ needs to have the form $\pi/2^n$; because after doubling the lattice spacing n times, the total grid space ends up with $S(I) = \pi$ – and which results in the correct strong coupling fixed point.

The range $0 \leq \sigma \leq \pi$ means that each group element of $SU(2)$ is covered once and only once in the numerical integration: the strong coupling fixed point is very sensitive to the entire structure of the non-Abelian gauge group. This fact is also apparent in the exact lattice gauge theory, where the strong coupling expansion involves the entire gauge group.

10.11   Summary: Confinement and Asymptotic Freedom

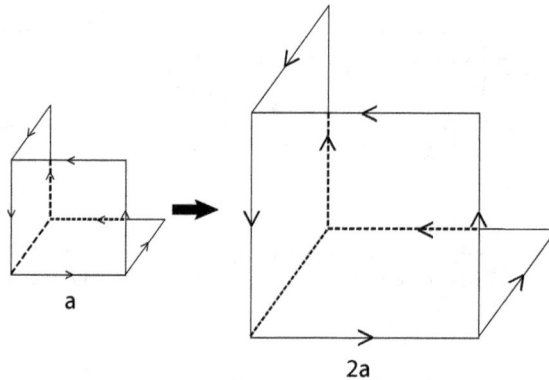

Fig. 10.15   Gauge field renormalization: the link representing the gauge field changes with the increase in the size of the lattice.

Migdal's recursion formula is, in essence, a scheme for gauge field renormalization, with the recursion equation being an expression of the renormalization group transformation. The procedure for renormalization encodes the intuitive formulation of Wilson (1983) that a quantum field theory consists of infinitely many coupled length scales. Each lattice size has its own length scale, which is its lattice spacing, and there is a separate effective coupling constant G_I for the effective action and gauge field degrees of freedom for *that length scale*.

For example, Figure 10.15 shows that under renormalization, the link variable representation of the lattice gauge field becomes longer and longer for increasing lattice spacing, providing an intuitive realization of gauge field renormalization. Figure 10.15 shows that the renormalization of the gauge field leads to the intuitive result that classical gauge fields are represented by very long link variables and hence their variation over space is smooth. It is for this reason that classical gauge fields can be approximated by differentiable functions – as is the case for the Maxwell Abelian gauge field.

The physical interpretation of G_I is that it is the strength of the gauge field felt by particles which couple to it, say quarks, when these quarks are separated by a distance of $2^I a_0$ (a_0 = original spacetime lattice spacing). Asymptotic freedom implies that $G_1 \to 0$ as $a_0 \to 0$; for the sake of discussion, let $G_1 \simeq 10^{-3}$ for $a_0 \simeq 10^{-16}$cm. The recursion equation shows that as the quarks are separated to larger distances, the strength of gauge field increases continuously.

The fact that, for $d = 4$, the model has an unstable weak coupling fixed point and a stable strong coupling fixed point implies that the quarks behave almost like free particles at short distances, and become strongly coupled at large distances.

The absence of any other fixed points shows that the weak and strong coupling behavior of the gauge field is not separated by any phase transition, and the quarks go continuously from their short distance weak coupling behavior to their strongly coupled behavior, and which gives rise to the bound states of the quarks.

Since the coupling constant becomes arbitrarily large for an arbitrarily large distance, the quark-antiquark separate to a definite distance, after which the quark-gauge field system produces quark-antiquark pairs – since this becomes energetically more favorable than any further separation of the quark-antiquark in question. This explains why quarks in a bound state cannot be separated to macroscopic distances – and results in the permanent confinement of quarks.

Of course, in the above discussion, it has been assumed that the qualitative behavior of the pure gauge field is not destroyed when the quark field is coupled to it. Given this assumption, Migdal's model provides a theoretical example of the physical idea of quark confinement, and gives a tractable example of an asymptotically free field theory going over to a strongly coupled system. As is well known, this rather remarkable behavior is unique to non-Abelian gauge fields and is shown by no other known quantum field.

10.12 Appendix

Eq. 10.6.12 yields

$$\tilde{V}(I,\tau) = \text{const} \frac{e^{\frac{2}{G}}}{\pi} \int_0^\pi du \sin^2 u \exp\left\{ \frac{2(\cos u \cos\sigma - 1)}{G} \right\}$$
$$\times \left[1 + \frac{4H}{G}(\cos\sigma\cos u - 1)^2 + \frac{4H}{3G}(\sin\sigma\sin u)^2 + O(G^2) \right] \quad (10.12.1)$$

The computation of the Appendix will give the result stated in Eq. 10.6.14.

Recall $\sigma = \tau/2$. Let

$$a = \frac{\cos\sigma}{G} : 0(1/G) \quad (10.12.2)$$

Since $\sigma, u = 0(\sqrt{G})$ the following approximations are made:

-

$$\frac{2}{G}(\cos\sigma\cos u - 1) = \frac{2}{G}(\cos\sigma - 1) - au^2 + O(G)$$

-

$$\sin^2 u = u^2 + O(G) = u^2[1 + O(G^2)] \tag{10.12.3}$$

-

$$(\cos u \cos\sigma - 1)^2 = (\cos\sigma - 1)^2 - (\cos\sigma - 1)u^2 + O(G^2)$$

-

$$\sin^2\sigma\sin^2 u = (\sin^2\sigma)u^2 + O(G^2)$$

Let

$$P = 1 + \frac{4H}{G}(\cos\sigma - 1)^2 \;\; ; \;\; Q = \frac{4H}{G}\left\{\frac{1}{3}\sin^2\sigma - (\cos\sigma - 1)\right\}$$

Then above equations and Eq. 10.12.1 yield (up to constants)

$$\tilde{V}(I,\tau) = e^{2(\cos\sigma - 1)/G}\int_0^\pi du u^2 e^{-au^2}[P + Qu^2] \tag{10.12.4}$$

$$\simeq e^{2(\cos\sigma - 1)/G}\frac{1}{2}\int_{-\infty}^{\infty} du u^2 e^{-au^2}[P + Qu^2]$$

$$= \frac{e^{2(\cos\sigma - 1)/G}}{\cos^{3/2}\sigma}\left[1 + \frac{4H}{G}(\cos\sigma - 1)^2 + 6H\left\{\frac{1}{3}\sin^2\sigma - (\cos\sigma - 1)\right\} + O(G^2)\right]$$

From Eq. 10.6.13

$$\tilde{V}(I,\tau) = \exp\{C(\tau)\}$$

Therefore, up to a constant, from Eq. 10.12.4

$$C(\tau) = \frac{2}{G}(\cos\sigma - 1) - \frac{3}{2}\ln\cos\sigma + \frac{4H}{G}(\cos\sigma - 1)^2$$

$$+ 6H\left\{\frac{1}{3}\sin^2\sigma - (\cos\sigma - 1)\right\} + O(G^2) \tag{10.12.5}$$

Since $\sigma = \tau/2$, we have

$$\sin^2\sigma = -\frac{1}{2}(\cos\tau - 1) \;\; ; \;\; \cos\sigma = \sqrt{1 + \frac{1}{2}(\cos\tau - 1)}$$

$$\ln\cos\sigma = \frac{1}{4}(\cos\tau - 1) + O(G^2)$$

Therefore

$$C(\tau) = \left[\frac{1}{2G} - \frac{3}{8}\frac{5H}{2}\right](\cos\tau - 1) + \frac{1}{4G}(\cos\tau - 1)^2\left(H - \frac{1}{4}\right) \tag{10.12.6}$$

But

$$C(\tau) = \frac{1}{G'}(\cos\tau - 1) + \frac{2H'}{G'}(\cos\tau - 1)^2$$

Hence, the final result is given by

$$\frac{1}{G'} = \frac{1}{2G} - \frac{3}{8}\frac{5H}{2} \tag{10.12.7}$$

$$\frac{2H'}{G'} = \frac{1}{4G}\left(H - \frac{1}{4}\right) \tag{10.12.8}$$

Chapter 11

Lattice Gauge Field Hamiltonian

The Hamiltonian for QCD (quantum chromodynamics) has been widely studied using the lattice and continuum formulations. The mathematical treatment of gauge fixing the Yang-Mills Hamiltonian goes back to Schwinger (1962). The paper by Christ and Lee (1980) gives a clear and complete treatment of gauge-fixing the continuum non-Abelian gauge field Hamiltonian.

The lattice gauge field Hamiltonian has been derived by Kogut and Susskind (1975). The lattice gauge field Hamiltonian can be regulated to all orders, and can be used for calculations involving two loops or higher. To use the lattice Hamiltonian for the weak-coupling approximation, it is necessary to choose a gauge, for example, the Coulomb gauge. Gauge-fixing the lattice gauge theory Hamiltonian essentially *involves only the lattice gauge field* and the quarks (Dirac field) enter only through the quark-color-charge operator. For this reason, gauge-fixing is analyzed before the complete derivation of the color-charge-operator, discussed later in Chapter 14.

Gauge-fixing the lattice action is similar to the continuum case, as can be seen from the path integral formulation discussed in Section 7.5. Similarly, gauge-fixing the lattice Hamiltonian is very similar in spirit to gauge-fixing the non-Abelian continuum Hamiltonian. The discussion on gauge-fixing the lattice gauge-field Hamiltonian is based on the treatment of Baaquie (1985), and is similar to the continuum treatment of Christ and Lee (1980). The Coulomb gauge for the continuum Abelian gauge field Hamiltonian is discussed at length in Baaquie (2018) and it is recommended that the Abelian case be reviewed to facilitate navigating the more complex non-Abelian case.

The lattice gauge field is defined using finite-group elements of $SU(\mathcal{N})$ as the fundamental degrees of freedom, whereas the continuum uses only the Lie algebra of $SU(\mathcal{N})$. This difference introduces a lot of extra complications leading to significant differences between the lattice and continuum gauge field Hamiltonians – both for the kinetic operator and the potential term. In particular, the lattice gauge field has an additional complication that is absent in the continuum case, which is the distinction between the left- and right-group multiplication required for defining lattice gauge transformation.

Given an appropriate generalized interpretation of the basic symbols, it will turn out that the form of the gauge-fixed continuum and lattice Hamiltonians are very similar.

11.1   Lattice Gauge Field Hamiltonian

Consider a three-dimensional Euclidean lattice with *spatial* lattice spacing given by a; $U_{ni}, i = 1, 2, 3$ is the $SU(\mathcal{N})$ link degree of freedom from lattice site n to $n + \hat{i}$ ($\hat{i}$ is the unit lattice vector in the ith direction) and ψ_n, $\bar{\psi}_n$, is the lattice Dirac field.

A detailed derivation of the full lattice gauge theory Hamiltonian H is given in Chapter 14. In Section 14.3, the Hamiltonian H for $SU(\mathcal{N})$ is obtained in the axial gauge, specified by

$$U_{n0} = \mathbb{I} \; ; \quad n : \text{four-dimensional lattice}$$

From Eq. 14.5.5, Hamiltonian H is given by

$$H = H_G[U] + H_D[\bar{\psi}, \psi, U]$$

where H_D is the quark-gauge-field part.

The derivation of the lattice gauge field Hamiltonian is given based on a summary of Section 14.3; it is obtained from the asymmetric lattice gauge-field Lagrangian by taking the limit of $\epsilon \to 0$. The asymmetric Lagrangian, from Eq. 7.1.3, is

$$\mathcal{L}(n; \epsilon) = \frac{1}{4g^2} \frac{a}{\epsilon} \sum_{i=1}^{3} \text{Tr}(W_{n,0i} + W_{n,0i}^\dagger) + \frac{1}{4g^2} \frac{\epsilon}{a} \sum_{i \neq j = 1}^{3} \text{Tr}(W_{n,ij}) \; ; \quad i, j = 1, 2, 3$$

The asymmetric lattice Lagrangian in the axial gauge, given in Eq. 7.1.3, yields from the Dirac-Feynman formula, the gauge field Hamiltonian

$$\lim_{\epsilon \to 0} \langle U' | e^{-\epsilon H_G} | U \rangle = e^{-\epsilon H_G(U')} \prod_{ni} \delta(U_{ni} - U'_{ni}) = \mathcal{N} \exp\{\mathcal{L}(n; \epsilon)\} \qquad (11.1.1)$$

$$= \lim_{\epsilon \to 0} \exp \left\{ \frac{1}{4g^2} \frac{a}{\epsilon} \sum_{i=1}^{3} \text{Tr}(U'_{n,i} U_{n,i}^\dagger + h.c.) + \frac{1}{4g^2} \frac{\epsilon}{a} \sum_{i \neq j = 1}^{3} \text{Tr}(W'_{n,ij} + W_{n,ij}) \right\}$$

For $\epsilon \to 0$ Eq. 4.3.5 yields

$$\exp \left\{ \frac{a}{\epsilon} \frac{1}{4g^2} \sum_{ni} \text{Tr} \left(U'_{ni} U_{ni}^\dagger + U_{ni}(U')_{ni}^\dagger \right) \right\}$$

$$= \exp \left\{ + \frac{g^2 \epsilon}{a} \sum_{n,i} \nabla^2 (U'_{ni}) \right\} \prod_{ni} \delta(U_{ni} - U'_{ni})$$

where ∇^2 is the Laplace-Beltrami operator for $SU(\mathcal{N})$ discussed in Section 2.9. The gauge-field Hamiltonian in the axial gauge, from Eqs. 11.1.1 and 14.5.4, is given by

$$H_G = -\frac{g^2}{a} \sum_{n,i} \nabla^2 (U_{ni}) - \frac{1}{2ag^2} \sum_{n,ij} \text{Tr}(U_{ni} U_{n+\hat{i},j} U_{n+\hat{j},i}^\dagger U_{nj}^\dagger)$$

$$= \frac{g^2}{a} \sum_{n,i} \sum_a E_a^2(U_{ni}) - \frac{1}{2ag^2} \sum_{n,ij} \text{Tr}(U_{ni} U_{n+\hat{i},j} U_{n+\hat{j},i}^\dagger U_{nj}^\dagger)$$

where $E_a(U_{ni})$ is the chromoelectric field operator (see Eqs. 11.1.4 and 11.1.13).

Gauge-transformations are given by

$$U_{ni} \rightarrow U_{ni}(\Phi) \equiv \Phi_n U_{ni} \Phi^\dagger_{n+\widehat{i}} \qquad (11.1.2)$$

The Hamiltonian acts only on gauge-invariant wave functionals Ψ. As given in Eq. 14.3.5, the wave functional Ψ is invariant under time-independent gauge-transformations given in (11.1.2); that is,

$$\Psi[U] = \Psi[U(\Phi)] \qquad (11.1.3)$$

Choose canonical coordinates B^a_{ni}; suppressing the lattice and vector indices, the link variable is given by

$$U_{ni} = \exp(iB^a X_a)$$

The differential operator $E_a(U)$ is discussed in Section 2.9. From Eqs. 2.11.4 and 2.11.7 (summing on repeated non-Abelian indices) yields[1]

$$E^L_a(U) = f_{ab}(U)\frac{\partial}{i\partial B^b} \equiv \frac{\delta}{i\delta_L B^a} \quad ; \quad E^R_a(U) = f^T_{ab}(U)\frac{\partial}{i\partial B^b} \equiv \frac{\delta}{i\delta_R B^a} \quad (11.1.4)$$

As discussed in Eq. 2.11.4, $E^L_a(U)$ is *left-invariant* in the sense that

$$E^L_a(VU) = E^L_a(U) \quad ; \quad V \in SU(\mathcal{N})$$

and $E^L_a(U)$ is *right-invariant* in the sense that

$$E^L_a(UV) = E^L_a(U) \quad ; \quad V \in SU(\mathcal{N})$$

From Eq. 2.11.8, the operators E^R_a and E^L_a are the lattice *chromoelectric* field operators, which are first-order Hermitian differential operators with the commutation equation, from Eq. 2.11.8, given by

$$[E^L_a, E^L_b] = iC^{abc}E^L_c \quad ; \quad [E^R_a, E^R_b] = -iC^{abc}E^R_c \quad ; \quad [E^R_a, E^L_b] = 0$$
$$E^R_a(U) = R_{ab}(U)E^L_b(U), \quad R_{ab}(U) = \text{Tr}(X_a U X_b U^\dagger)$$

where R_{ab} is the adjoint representation given by[2]

$$R_{ab}(U) = \text{Tr}(X_a U X_b U^\dagger) \qquad (11.1.5)$$

Recall X_a are the generators and C^{abc} the structure constants of $SU(\mathcal{N})$. The choromelectric operators are the non-Abelian quantum field generalization of the electric field of the Abelian gauge field.

In canonical coordinates the gauge transformation is given by

$$\Phi_n = \exp(i\phi^a_n X_a)$$

and from Eq. 2.11.11

$$e^{i\phi^a_n E^R_a(U)}U = e^{i\phi^a_n X_a}U = \Phi_n U \quad ; \quad e^{i\phi^a_n E^L_a(U)}U = Ue^{i\phi^a_n X_a} = U\Phi_n \qquad (11.1.6)$$

---

[1]Explicit expression for f_{ab} is given in Eqs. 2.9.4, but will not be required for the derivations of this Chapter.

[2]R_{ab} is the notation for the adjoint representation, which is different from the one used in Chapter 2, is chosen so as to not clash with the definition of the quark charge operator.

The gauge transformation, from Eq. 11.1.6, is hence given by

$$\exp\left(i\sum_{ni}\left\{\phi_n^a\left(E^R(U_{ni})-E_a^L(U_{n-\hat{i},i})\right)\right\}\right)\Psi[U]=\Psi\left[U(\Phi)\right] \tag{11.1.7}$$

Since, from Eq. 11.1.3

$$\Psi[U]=\Psi[U(\Phi)]$$

Eq. 11.1.7 yields Gauss's law

$$\left[\sum_i\left\{E_a^R(U_{ni})-E_a^L(U_{n-\hat{i},i})\right\}\right]|\Psi\rangle=0 \tag{11.1.8}$$

Gauge invariance of the Hamiltonian under time independent gauge transformations, as expressed in Eq. 14.2.14, yields the following

$$\left[H_G,\sum_i\left\{E_a^R(U_{ni})-E_a^L(U_{n-\hat{i},i})\right\}\right]=0$$

Hence, the constraint equation given in Eq. 11.1.8 holds for all values of time.

Gauss's law confirms the identification $E_a^L(U_{ni})$ as the chromoelectric operator of the gauge field corresponding to the link variable U_{ni}.

The lattice *quark-color-charge operator* $\rho_{na}(\bar{\psi},\psi,U)$ is discussed in Section 14.7 and is defined in Eq. 14.7.14; it satisfies

$$[\rho_{na},\rho_{mb}]=iC_{abc}\rho_{nc}\delta_{nm}$$

Hence, in the presence of the chromoelectric charge operator ρ_{na}, from (14.7.9) Gauss's law yields

$$0=\left[\sum_i[R_{ab}(U_{ni})E_b^L(U_{ni})-E_a^L(U_{n-\hat{i},i})]-\rho_{na}\right]|\Psi\rangle$$

$$\equiv\left[\sum_{m,i}D_{nmi}^{ab}E_b^L(U_{mi})-\rho_{na}]\right]|\Psi\rangle \tag{11.1.9}$$

where D_{nmi}^{ab} is the lattice covariant *backward* derivative.

Let $|n,a\rangle$ be a ket vector of lattice site n and non-Abelian index a; then, from Eq. 11.1.9, the real matrix D_i given by

$$D_{nmi}^{ab}=\langle n,a|D_i|m,b\rangle=R_{ab}(U_{ni})\delta_{nm}-\delta_{ab}\delta_{n-\hat{i}m} \tag{11.1.10}$$

It is seen from above that D_i performs a finite rotation R_{ab} on the ket vector and then displaces it in the backward direction.

The full Hamiltonian of the gauge field coupled to fermions is a sum of the kinetic and potential energy and is given by

$$H=K(U)+P[\bar{\psi},\psi,U] \tag{11.1.11}$$

where, from Eq. 11.4.5

$$K=-\frac{g^2}{2a}\sum_{n,i}\nabla^2(U_{ni}) \tag{11.1.12}$$

and from Eq. 2.11.13[3]

$$-\nabla^2(U)=\sum_a E_a^L(U)E_a^L(U) \tag{11.1.13}$$

---

[3] As shown in Eq. 2.11.13, the Laplacian can also be written as

$$-\nabla^2(U)=\sum_a E_a^R(U)E_a^R(U).$$

11.2 Gauge-Fixed Chromoelectric Operator

We can see from Gauss's law that all the U_{ni}'s are not required to describe the gauge-invariant wave functional Ψ. We gauge transform U_{ni} to a new set of variables V_{ni} which are constrained; the constrained variables V_{ni} will decouple from Gauss's law.

Consider the *change of variables* from $\{U_{ni}\}$ to $\{\Phi, V_{ni}\}$ given by

$$U_{ni} = \Phi_n V_{ni} \Phi^\dagger_{n+\widehat{i}} \qquad (11.2.1)$$

The new gauge degrees of freedom $\{V_{ni}\}$ are *constrained*, with $\{V_{ni}\}$ having one *constraint* for each n; the Coulomb gauge for the lattice is given by

$$\mathcal{X}^a_n(V) \equiv \operatorname{Im} \sum_i \operatorname{Tr} X_a(V_{ni} - V_{n-\widehat{i},i}) = 0 \qquad (11.2.2)$$

The explicit form of the gauge constraint will not be required for the derivations below.

In canonical coordinates we have

$$V_{ni} = \exp(iA^a_{ni}X_a), \quad \Phi_n = \exp(i\phi^a_n X_a), \quad U_{ni} = \exp(iB^a_{ni}X_a)$$

A small variation $A^a + dA^a$ yields

$$V(A + dA) = V(A)\left[1 + V^\dagger(A)\frac{\partial V(A)}{\partial A^a}dA^a\right]$$
$$= V(A)[1 + iX_a e^R_{ab}(A)dA^b] \qquad (11.2.3)$$

where, as in Eq. 2.3.1

$$e_{ab}(A) \equiv e^R_{ab}(A) = -i\operatorname{Tr}\left[V^\dagger \frac{\partial V}{\partial A^a}X_b\right] = e^L_{ba}(A)$$

Define

$$\delta_L A^a = e^L_{ab}(A)dA^b \ ; \ \delta_R A^a = e^R_{ab}(A)dA^b \qquad (11.2.4)$$

From Eq. 2.9.4

$$f_{ab} = f^R_{ab} \ ; \ f^L_{ab} = f^T_{ab} = f_{ba}$$

Then

$$V(A + dA) = V(A)(1 + iX_a\delta_R A^a) = (1 + iX_a\delta_L A^a)V(A)$$

As shown in Eq. 2.9.4

$$e^L_{a\alpha}f^L_{ab} = \delta_{ab} \ ; \ e^R_{a\alpha}f^R_{ab} = \delta_{ab} \qquad (11.2.5)$$

and hence matrix $e_{a\alpha}$ is the inverse of $f_{a\alpha}$.

Under the change of variables from U_{ni} to V_{ni}, given in Eqs. 11.2.1 and 11.2.2, the potential energy $P(\bar{\psi}, \psi, U)$ in Eq. (11.1.11) can be expressed as a function of only V_{ni}. For the kinetic energy K the expression for E^L_a is required.

Noteworthy 11.1: Non-Abelian change of variables

Recall from Eq. 2.9.5, the left-invariant chromoelectric operator is given by

$$E_a = f^L_{ab}(B)\frac{\partial}{i\partial B^b} \quad \Rightarrow \quad E^L_a U = UX_a$$

Consider a change of variables from B^a to A^a; the chain rule yields

$$\frac{\partial}{\partial B^b} = \sum_a \frac{\partial A^a}{\partial B^b}\frac{\partial}{\partial A^a} = \sum_{a,\alpha,\beta} \frac{\partial A^a}{\partial B^b}e^L_{a\beta}(A)f^L_{\beta a}(A)\frac{\partial}{\partial A^\alpha}$$

$$= \sum_{a,\alpha,\beta} \frac{e^R_{\beta a}(A)\partial A^a}{\partial B^b}f^L_{\beta\alpha}(A)\frac{\partial}{\partial A^\alpha} = \sum_\beta \frac{\delta_R A^\beta}{\partial B^b}\frac{\delta}{\delta_L A^\beta}$$

Hence

$$E_a = f^L_{ab}(B)\frac{\partial}{i\partial B^b} = \sum_\beta \frac{\delta_R A^\beta}{\delta_L B^b}\frac{\delta}{i\delta_L A^\beta} \tag{11.2.6}$$

Consider $h(B)$, a function of the B^a variables; then, similar to the derivation above (summing over repeated non-Abelian index)

$$dh(B) = \frac{\partial h}{\partial B^a}dB^a = \frac{\delta h}{\delta_L B^a}\delta_R B^a \tag{11.2.7}$$

Consider from Eq. 11.2.1

$$V_{ni} = \Phi^\dagger_n U_{ni}\Phi_{n+\widehat{i}} \tag{11.2.8}$$

The left hand side of Eq. 11.2.8, from Eq. 11.2.7, yields

$$dV_{ni} = \frac{\delta V_{ni}}{\delta_L A^a}\delta_R B A^a = iV_{ni}X_a\delta_R A^a$$

and the right hand side of Eq. 11.2.8 yields

$$d\left(\Phi^\dagger_n U_{ni}\Phi_{n+\widehat{i}}\right) = -i(X_a\Phi^\dagger_n U_{ni}\Phi_{n+\widehat{i}})\delta_R\phi^a_n + i(\Phi^\dagger_n U_{ni}X_a\Phi_{n+\widehat{i}})\delta_R B^a_{ni}$$

$$+i(\Phi^\dagger_n U_{ni}\Phi_{n+\widehat{i}}X_a)\delta^a_R\phi_{n+\widehat{i}}$$

Recall from Eq. 11.1.5

$$R_{ab}(U) = \text{Tr}(X_a U X_b U^\dagger)$$

From above, after some simplifications

$$\delta_R A^a_{ni} = \delta_R\Phi^a_{n+\widehat{i}} - R_{ab}(V^\dagger_{ni})\delta_R\phi^b_n + R_{ab}(\Phi^\dagger_{n+\widehat{i}})\delta_R B^b_{ni} \tag{11.2.9}$$

Using the chain rule and Eq. 11.2.6 yields

$$\frac{\partial}{\partial B^b_{mj}} = \sum_{n,i,a}\frac{\partial A^a_{ni}}{\partial B^b_{mj}}\frac{\partial}{\partial A^a_{ni}} + \sum_{n,a}\frac{\partial\phi^a_n}{\partial B^b_{mj}}\frac{\partial}{\partial\phi^a_n}$$

$$= \sum_{n,i,a,\alpha,\beta} e^R_{a\alpha}(A_{ni})\frac{\partial A^\alpha_{ni}}{\partial B^b_{mj}}f^L_{\alpha\beta}(A_{ni})\frac{\partial}{\partial A^\beta_{ni}} + \cdots \tag{11.2.10}$$

Therefore, from Eqs. (11.1.4) and (11.2.10)

$$E^L_b(U_{mj}) = \frac{\delta}{i\delta_L B^b_{mj}} = \sum_{n,i}\frac{\delta_R A^a_{ni}}{\delta_L B^b_{mj}}\frac{\delta}{i\delta_L A^a_{ni}} + \sum_n\frac{\delta_R\phi^a_n}{\delta_L B^b_{mj}}\frac{\delta}{i\delta_L\phi^a_n}$$

The coefficient functions $\delta_R A_{ni}^a / \delta_L B_{mj}^b$ of the above equation are evaluated. The constraint given in Eq. 11.2.2 is valid under variations of A_{ni}^a to $A_{ni}^a + dA_{ni}^a$ and yields

$$0 = \mathcal{X}_n^a(A) = \mathcal{X}_n^a(A + dA) \tag{11.2.11}$$

Hence, from Eqs. 11.2.7 and 11.2.11

$$\sum_{m,i} \Gamma_{nmi}^{ab}(A) \delta_R A_{mi}^b = 0 \tag{11.2.12}$$

For constraint given in Eqs. 11.2.2 and 11.2.12 yields

$$\Gamma_{nmi}^{ab} = \langle n, a | \Gamma_i | m, b \rangle = \delta \mathcal{X}_n^a / \delta_L A_{mi}^b = \omega_{ni}^{ab} \delta_{nm} - \omega_{n-\hat{i},i}^{ab} \delta_{n-\hat{i},m}$$

where, from Eqs. 11.2.2 and 11.2.12

$$\omega_{ni}^{ab} = \text{Tr}(X_a V_{ni} X_b + X_b V_{ni}^\dagger X_a)$$

The constraint Eq. 11.2.12 on δA_{ni}^a determines $\delta\phi/\delta B$. Consider from Eq. 11.2.2, the following variation:

$$V_{ni}(A + dA) = \Phi_n^\dagger(\phi + d\phi) U_{ni}(B + dB) \Phi_{n+\hat{i}}(\phi + d\phi)$$

and which yields, from Eq. 11.2.9

$$\delta_R A_{ni}^a = \delta_R \phi_{n+\hat{i}}^a - R_{ab}(V_{ni}^\dagger) \delta_R \phi_n^b + R_{ab}(\Phi_{n+\hat{i}}^\dagger) \delta_R B_{ni}^b \tag{11.2.13}$$

$$\equiv \sum_m \mathcal{D}_{nmi}^{ab} \delta_R \phi_m^b + R_{ab}(\Phi_{n+\hat{i}}^\dagger) \delta_R B_{ni}^b \tag{11.2.14}$$

From Eqs. 11.2.13 and 11.2.14, the lattice covariant *forward* derivative operator $\mathcal{D}_i$ is given by

$$\mathcal{D}_{nmi}^{ab} = \langle n, a | \mathcal{D}_i | m, b \rangle = \delta_{ab} \delta_{n+\hat{i},m} - R_{ab}(V_{ni}^\dagger) \delta_{nm}$$

Eqs. 11.2.12 and 11.2.14 yield

$$\sum_{m,i,b} \langle n, a | \Gamma_i \mathcal{D}_i | m, b \rangle \delta_R \phi_m^b + \sum_{m,i,b} \langle n, a | \Gamma_i R_i^T | m, b \rangle \delta_R B_{mi}^b = 0 \tag{11.2.15}$$

where T stands for transpose and

$$\langle n, a | R_i | m, b \rangle = \delta_{nm} R_{ab}(\Phi_{n+\hat{i}}) \tag{11.2.16}$$

Hence, from (11.2.15) we have

$$\frac{\delta_R \phi_n^a}{\delta_L B_{mj}^b} = -\left\langle n, a \left| \frac{1}{\Gamma \cdot \mathcal{D}} \Gamma_j R_j^T \right| m, b \right\rangle \tag{11.2.17}$$

where $(\Gamma \cdot \mathcal{D})^{-1}$ is the inverse of operator $\sum_i \Gamma_i \mathcal{D}_i$. From Eqs. 11.2.14 and 11.2.17

$$\frac{\delta_R A_{ni}^a}{\delta_L B_{mj}^b} = -\left\langle n, a \left| \left[\mathcal{D}_i \frac{1}{\Gamma \cdot \mathcal{D}} \Gamma_j R_j^T - R_j^T \delta_{ij} \right] \right| m, b \right\rangle \tag{11.2.18}$$

Hence, from Eqs. 11.2.11, 11.2.17, and 11.2.18

$$\frac{\delta}{\delta_L B_{mj}^b} = \sum_{n,i} \left\langle n, a \left| \left[\delta_{ij} - \mathscr{D}_i \frac{1}{\Gamma \cdot \mathscr{D}} \Gamma_j \right] R_j^T \right| m, b \right\rangle \frac{\delta}{\delta_L A_{ni}^a}$$

$$- \sum_n \left\langle n, a \left| \frac{1}{\Gamma \cdot \mathscr{D}} \Gamma_j R_j^T \right| m, b \right\rangle \frac{\delta}{\delta_L \phi_n^a} \qquad (11.2.19)$$

Eq. (11.2.19) provides the solution for expressing the unconstrained chromo-electric operator $\delta/\delta_L B$ in terms of the new *constrained* gauge-fixed chromoelectric operator $\delta/\delta_L A$ and the gauge transformation $\delta/\delta_L \phi$. In essence, this solves the problem of gauge fixing the lattice Hamiltonian. Note Eq. 11.2.18 yields the following identity

$$\sum_{n,i} \langle l, c | \Gamma_i | n, a \rangle \frac{\delta_R A_{ni}^a}{\delta_L B_{mj}^b} = 0$$

as expected from Eq. 11.2.12. Also, from Eq. 11.2.12 the gauge-fixed chromoelectric operator is *transverse* since

$$\sum_{m,i} \langle n, a | \Gamma_i | m, b \rangle \frac{\delta}{\delta_L A_{mi}^b} = 0 \qquad (11.2.20)$$

Hence, from Eqs. 11.1.4 and 11.2.20

$$\left[\frac{\delta}{\delta_L A_{ni}^b}, A_{mj}^b \right] = \left[\delta_{mm} \delta_{ij} \delta_{ac} - \left\langle n, a \left| \Gamma_i^T \frac{1}{\Gamma \cdot \Gamma^T} \Gamma_j \right| m, c \right\rangle \right] e_{cb}^L (A_{mj})$$

11.3 Gauss's Law

To check that constrained variables V_{ni} decouple from Gauss's law, recall from Eqs. 11.1.10 and 11.2.19, we have

$$\sum_{m,j} \langle l, c | D_j | m, b \rangle \frac{\delta}{\delta_L B_{mj}^b} = \sum_{m,j} \langle m, b | D_j^T | l, c \rangle \frac{\delta}{\delta_L B_{mj}^b}$$

$$= \sum_{n,ij} \left\langle n, a \left| \left[\delta_{ij} - \mathscr{D}_i \frac{1}{\Gamma \cdot \mathscr{D}} \Gamma_j \right] R_j^T D_j^T \right| l, c \right\rangle \frac{\delta}{\delta_L A_{ni}^a}$$

$$- \sum_{n,ij} \left\langle n, a \left| \frac{1}{\Gamma \cdot \mathscr{D}} \Gamma_j R_j^T D_j^T \right| l, c \right\rangle \frac{\delta}{\delta_L \phi_n^a} \qquad (11.3.1)$$

From the definitions of D_i and $\mathscr{D}_i$ given in Eqs. 11.1.10 and 11.2.15, respectively, one has the crucial operator identity

$$R_j^T D_j^T = -\mathscr{D}_j R^T \qquad (11.3.2)$$

where

$$\langle n, a | R | m, b \rangle = \delta_{nm} R_{ab}(\Phi_n) \qquad (11.3.3)$$

Noteworthy 11.2: Proof of $R_j^T D_j^T = -\mathscr{D}_j R^T$

From Eq. 11.1.10, the real matrix D_i is given by

$$D^{ab}_{nmi} = \langle n, a | D_i | m, b \rangle = R_{ab}(U_{ni})\delta_{nm} - \delta_{ab}\delta_{n-\hat{i}m}$$

and from Eq. 11.2.15 $\mathscr{D}_i$ is given by

$$\mathscr{D}^{ab}_{nmi} = \langle n, a | \mathscr{D}_i | m, b \rangle = \delta_{ab}\delta_{n+\hat{i},m} - R_{ab}(V^\dagger_{ni})\delta_{nm}$$

Furthermore, from Eqs. 11.2.16 and 11.3.3

$$\langle n, a | R_i | m, b \rangle = \delta_{nm} R_{ab}(\Phi_{n+\hat{i}}) \; ; \quad \langle n, a | R | m, b \rangle = \delta_{nm} R_{ab}(\Phi_n)$$

Eq. 2.3.7 gives the following multiplication law for the adjoint representation

$$R_{ac}(U) R_{cb}(V) = R_{ab}(UV)$$

Note that

$$R_{ab}(U^\dagger) = R_{ab}(U)$$

To prove the identity, the left hand side yields, using the definition of D_i given in Eq. 11.1.10

$$\langle n, a \mid R_j^T D_j^T \mid m, b \rangle = \sum_{l,c} \langle n, a \mid R_j^T \mid l, c \rangle \langle l, c \mid D_j^T \mid m, b \rangle$$

$$= \sum_{l,c} \delta_{nl} R_{ca}(\Phi_{l+j}) \big\{ R_{bc}(U_{nj})\delta_{ml} - \delta_{cb}\delta_{m-j,l} \big\}$$

$$= \delta_{nm} R_{ba}(U_{nj}\Phi_{n+j}) - \delta_{m-j,n} R_{ba}(\Phi_{n+j})$$

$$= \delta_{nm} R_{ba}(\Phi_n V_{nj}) - \delta_{m-j,n} R_{ba}(\Phi_{n+j}) \tag{11.3.4}$$

The right hand side, using the definition of $\mathscr{D}_i$ given in Eq. 11.2.15, yields

$$\langle n, a \mid \mathscr{D}_i R^T \mid m, b \rangle = \sum_{l,c} \langle n, a | \mathscr{D}_i | l, c \rangle \langle l, c \mid R^T \mid m, b \rangle$$

$$= \sum_{l,c} \big\{ \delta_{ac}\delta_{n+j,l} - R_{ac}(V^\dagger_{nj})\delta_{nl} \big\} \delta_{ml} R_{bc}(\Phi_m)$$

$$= \delta_{n+j,m} R_{ba}(\Phi_m) - \delta_{nm} R_{ca}(V_{nj}) R_{bc}(\Phi_m)$$

$$= \delta_{n+j,m} R_{ba}(\Phi_{n+j}) - \delta_{m-j,n} R_{ba}(\Phi_m V_{nj}) \tag{11.3.5}$$

Hence, from Eqs. 11.3.4 and 11.3.5, the following identity has been proven

$$R_j^T D_j^T = -\mathscr{D}_j R^T$$

Hence, from Eqs. 11.3.1 and 11.3.2 the first term in Eq. 11.3.1 is zero and

$$\sum_{m,j} D^{cb}_{lmj} \frac{\delta}{\delta_L B^b_{mj}} = R_{cb}(\Phi_l) \frac{\delta}{\delta_L \phi^b_l} = \frac{\delta}{\delta_R \phi^c_l} \tag{11.3.6}$$

Recall from Eq. 11.2.1 that
$$\psi_n = \Phi_n \zeta_n \ ; \qquad \bar{\psi}_n = \bar{\zeta}_n \ ; \quad \Phi_n^\dagger = U_{ni} = \Phi_n V_{ni} \Phi_{n+\hat{i}}^\dagger$$
From Eq. 11.1.9, Gauss's constraint is given by
$$\left[\sum_{m,i} D_{nmi}^{ab} E_b^L (U_{mi}) - \rho_{na} \right] |\Psi(\bar{\psi}, \psi, U)\rangle = 0$$
and from Eq. 11.3.6
$$\sum_{m,i} D_{nmi}^{ab} E_b^L (U_{mi}) = \frac{\delta}{i\delta_R \phi_n^a}$$

Hence, V_{ni} has decoupled from Gauss's constraint, and from Eqs. 11.1.9 and 11.3.6
$$\left[\frac{\delta}{i\delta_R \phi_n^a} - \rho_{na} \right] |\Psi(\bar{\zeta}, \zeta, V)\rangle = 0 \tag{11.3.7}$$
Using Eq. 11.1.9
$$\frac{\delta}{i\delta_R \phi^a} \exp(i\phi^\alpha \rho_\alpha) = \rho_a \exp(i\phi^\alpha \rho_\alpha) \tag{11.3.8}$$
and solving Eq. 11.3.7 using that $\Psi(\bar{\psi}, \psi, U)$ is gauge invariant yields
$$\Psi(\bar{\psi}, \psi, U) = \exp\left(i \sum_n \rho_{na} \phi_n^a \right) \Psi(\bar{\zeta}, \zeta, V) \tag{11.3.9}$$
Eq. 11.3.9 shows that the state space of the lattice gauge theory is gauge-invariant up to a phase. To verify $\Psi(\bar{\psi}, \psi, U)$ given in Eq. 11.3.9 obeys Gauss's law, note from Eqs. 11.3.6 and 11.3.8
$$\left[\sum_{m,i} D_{nmi}^{ab} E_b^L (U_{mi}) - \rho_{na} \right] |\Psi(\bar{\psi}, \psi, U)\rangle$$
$$= \left[\frac{\delta}{i\delta_R \phi_n^a} - \rho_{na} \right] \left(\exp[i \sum_n \rho_{na} \phi_n^a] \right) |\Psi(\bar{\zeta}, \zeta, V)\rangle = 0$$
The change of variables from $\{U_{ni}\}$ to $\{V_{ni}, \Phi_n\}$ has a Jacobian given by the Faddeev-Popov determinant. Similar to the evaluation of the ghost Faddeev-Popov in gauge-fixing path integral, it can be shown that the Jacobian is equal to
$$\mathscr{J}^{-1}[V] = \prod_n \int d\Phi_n \prod_{n,a} \delta[\mathcal{X}_n^a (\Phi_n V_{ni} \Phi_{n+\hat{i}}^\dagger)] \tag{11.3.10}$$

For weak coupling, $\mathscr{J}[V]$ has been evaluated to $O(A^2)$ in Section 7.8. Hence, suppressing the fermion field variables, for a gauge-invariant operator G and gauge-invariant state $|\Psi\rangle$, from Eqs. 11.2.1 and 11.3.9
$$\langle \Psi |G| \Psi \rangle = \prod_{n,i} \int dU_{ni} \Psi^*[U] G[U, \delta/\delta U] \Psi[U]$$
$$= \prod_{n,i} \int dV_{ni} \prod_{n,a} \delta[\mathcal{X}_n^a (V)] \left[\Psi^*[V] \mathscr{J}^{1/2}[V] \exp\left[i \sum_n \phi_n^a \rho_{na} \right] \right]$$
$$\times (\mathscr{J}^{1/2}[V] \widehat{G}[V, \delta/\delta V] \mathscr{J}^{-1/2}[V])$$
$$\left[\exp\left[i \sum_n \phi_n^a \rho_{na} \right] \mathscr{J}^{1/2}[V] \Psi[V] \right]$$

Hence, the effective wave functional defined by absorbing the Jacobian is given by

$$\widetilde{\Psi}[V] = \mathscr{J}^{1/2}[V]\Psi[V]$$

and the effective operator is

$$\widetilde{G} = \mathscr{J}^{1/2}[V] \exp\left[i\sum_n \phi_n^a \rho_{na}\right] G \exp\left[-i\sum_n \phi_n^a \rho_{na}\right] \mathscr{J}^{-1/2}[V] \qquad (11.3.11)$$

such that

$$\langle \Psi | G | \Psi \rangle = \langle \widetilde{\Psi} | \widetilde{G} | \widetilde{\Psi} \rangle \qquad (11.3.12)$$

11.4 Gauge-Fixed Lattice Gauge Field Hamiltonian

The kinetic operator is given from Eqs. 11.1.4, 11.1.12 and 11.1.13 by the following:

$$K = \frac{\delta}{i\delta_L B_{ni}^a} \frac{\delta}{i\delta_L B_{ni}^a} \qquad (11.4.1)$$

where all repeated indices are summed over. Symbolically write the transformation Eq. 11.2.19 as

$$\frac{\delta}{\delta_L B_p} = L_{pq} \frac{\delta}{\delta_L C_q} \qquad (11.4.2)$$

Then from Eqs. 11.4.1 and 11.4.2

$$K = L_{pq} \frac{\delta}{i\delta_L C_q}\left[L_{pq'} \frac{\delta}{i\delta_L C_{q'}}\right] = \frac{1}{L}\frac{\delta}{i\delta_L C_q}\left[LL_{qp}^T L_{pq'} \frac{\delta}{i\delta_L C_{q'}}\right] \qquad (11.4.3)$$

where

$$L = \det\|L_{ab}\| \qquad (11.4.4)$$

As for any curved space, the Laplacian for functions defined on the group space is given by

$$-\nabla^2 = \frac{1}{g}\frac{\partial}{\partial B^a}\left(\sqrt{g}\,g^{ab}\frac{\partial}{\partial B^b}\right) \qquad (11.4.5)$$

The transformation given by Eq. 11.2.19 yields

$$L = \mathscr{J}[V] \qquad (11.4.6)$$

and the Jacobian $\mathscr{J}$ is given by Eq. 11.3.10. The choice of operator ordering given by Eq. 11.4.3 allows for further simplifications. Recall that from Eq. 11.2.20 $\delta/\delta_L A_{ni}^a$ is 'transverse'; using this equation and Eq. 11.4.3 yields

$$K = \frac{1}{\mathscr{J}}\frac{\delta}{i\delta_L A_{ni}^a}\left[\mathscr{J}\frac{\delta}{i\delta_L A_{ni}^a}\right] + \frac{1}{\mathscr{J}}\left[\frac{\delta}{i\delta_L \phi_n^a} + \frac{\delta}{i\delta_L A_{n'i}^{a'}}\mathscr{D}_{n'ni}^{a'a}\right]$$
$$\times \mathscr{J}\left\langle n,a\left|\frac{1}{\Gamma\cdot\mathscr{D}}\Gamma_j\Gamma_j^T\frac{1}{\mathscr{D}^T\cdot\Gamma^T}\right|m,b\right\rangle\left[\frac{\delta}{i\delta_L \phi_m^b} + \mathscr{D}_{mm'k}^{Tbb'}\frac{\delta}{i\delta_L A_{m'k}^{b'}}\right] \qquad (11.4.7)$$

The effective Hamiltonian, using (11.3.11), is given by

$$\widetilde{H} = \mathscr{J}^{1/2} e^{i\phi_n^a \rho_{na}} H e^{-i\phi_n^a \rho_{na}} \mathscr{J}^{-1/2}$$

Note that

$$e^{-i\phi_n^a \rho_{na}} \frac{\delta}{i\delta_L \phi_m^b} e^{i\phi_n^a \rho_{na}} = \rho_{mb}$$

Hence, the final expression for the gauge-fixed lattice Hamiltonian is given by

$$
\begin{aligned}
\widetilde{H} = &+\frac{g^2}{2a} \left[\mathscr{J}^{-1/2} \frac{\delta}{i\delta_L A_{ni}^a} \left[\mathscr{J} \frac{\delta}{i\delta_L A_{ni}^a} \mathscr{J}^{-1/2} \right] \right. \\
&+ \mathscr{J}^{-1/2} \left[\frac{\delta}{i\delta_L A_{n'i}^{a'}} \mathscr{D}_{n'ni}^{a'a} - \rho_{na} \right] \mathscr{J} \left\langle n,a \left| \frac{1}{\Gamma \cdot \mathscr{D}} \Gamma \cdot \Gamma^T \frac{1}{\mathscr{D}^T \cdot \Gamma^T} \right| m,b \right\rangle \\
&\left. \left[\mathscr{D}_{mm'k}^{Tbb'} \frac{\delta}{i\delta_L A_{m'k}^{b'}} - \rho_{mb} \right] \mathscr{J}^{-1/2} \right] + P(\bar{\zeta},\zeta,V) \quad (11.4.8)
\end{aligned}
$$

Note that the redundant gauge degree of freedom Φ_n has completely decoupled from the Hamiltionian and state space.

The quark-color charge ρ_{na} has the instantaneous non-local and non-Abelian lattice Coulomb potential given by

$$(\Gamma \cdot \mathscr{D})^{-1} \Gamma \cdot \Gamma^T (\mathscr{D}^T \cdot \Gamma^T)^{-1}$$

The state functionals depend on only the constrained variables V_{ni}, that is

$$\widetilde{\Psi} = \widetilde{\Psi}(\bar{\zeta},\zeta,V) \qquad (11.4.9)$$

Recall from Eq. 11.2.21 the commutation equation is given by

$$\left[\frac{\delta}{\delta_L A_{ni}^a}, A_{mj}^b \right] = \left[\delta_{nm}\delta_{ij}\delta_{ac} - \left\langle n,a \left| \Gamma_i^T \frac{1}{\Gamma \cdot \Gamma^T} \Gamma_j \right| m,c \right\rangle \right] e_{cb}^L(A_{mj}) \qquad (11.4.10)$$

Eqs. 11.4.8, 11.4.9 and 11.4.10 completely define the gauge-fixed Hamiltonian for the $SU(\mathcal{N})$ lattice gauge field. The redundant gauge degrees of freedom $\{\Phi_n\}$ have been completely decoupled from the system, as expected.

The expression for $\widetilde{H}$ in Eq. 11.4.8 is exact, and is equally valid for strong and weak couplings. Comparing Eqs. 11.4.1 and 11.4.7, shows that the coordinates $\{U_{ni}\}$ are analogous to Cartesian coordinates for the gauge field whereas coordinates $\{V_{ni}\}$ are analogous to curvilinear coordinates.

11.5   Hamiltonian and Covariant Gauge: Faddeev-Popov Quantization

The Hamiltonian for the lattice gauge field was discussed in Section 11.2 using the Coulomb gauge – which is manifestly non-covariant. The Hamiltonian can be chosen for a *covariant gauge*, but this can lead to non-physical state vectors that have zero and even negative norm, and which is not allowed by the Born interpretation of the state vector representing probabilities.

The essential ideas of the Faddeev-Popov formalism are discussed for the simpler case of the photon field. For the Abelian gauge field, the Gupta-Bleuler procedure allows for a covariant gauge-fixing of the Hamiltonian, and is based on constraining state space so as to avoid the non-physical states. The Gupta-Bleuler procedure cannot be applied to non-Abelian gauge fields. However, with the introduction of the Faddeev-Popov ghost fields, a covariant formulation of the Hamiltonian and state space can be defined using the Faddeev-Popov quantization that generalizes to the non-Abelian case. The formalism is based on BRST cohomology and discussed below.

All calculations are carried out for the Abelian gauge field defined on Euclidean spacetime. Only the Abelian gauge field is discussed as it illustrates the key features of the covariant state space. The continuum formulation is used due to its notational simplicity, and because the lattice generalization can be read off from the continuum derivation. The discussion follows the derivations given in Baaquie (2018).

The full power of the Faddeev-Popov formulation comes to the fore in the study of Yang-Mills non-Abelian gauge fields.

The gauge-fixed action, with a covariant gauge-fixing term $-\alpha(\partial_\mu A_\mu)^2/2$, from Eq. 7.7.7, is given by

$$
\begin{aligned}
\mathcal{S}_{GF} &= -\frac{1}{4}\int d^4x F_{\mu\nu}^2 - \frac{\alpha}{2}\int d^4x(\partial_\mu A_\mu)^2 + \int d^4x \partial_\mu \bar{c}\partial_\mu c \\
&= \mathcal{S} + \mathcal{S}_\alpha + \mathcal{S}_{FP}
\end{aligned}
\tag{11.5.1}
$$

The path integral for the gauge-fixed Abelian theory is given by[4]

$$
Z = \int DA_\mu D\bar{c}Dc \exp\{\mathcal{S}_{GF}\}
$$

11.6 Ghost State Space and Hamiltonian

The state space of the quantum field is determined by the time derivative terms in the action; the reason being that the time derivative couples the gauge field at two different instants and at each instant, the gauge field is a coordinate of the underlying state space. The ghost action is similar to the action for the complex scalar field [Baaquie (2018)]; the state space has the fermion coordinate eigenstates and is given by

$$
|\bar{c},c\rangle = |\bar{c}\rangle \otimes |c\rangle \equiv \prod_{\vec{x}} |\bar{c}(\vec{x})\rangle \otimes |c(\vec{x})\rangle
$$

The completeness equation is given by

$$
\int D\bar{c}Dc |\bar{c},c\rangle\langle\bar{c},c| = \mathbb{I}
$$

The inner product is given by

$$
\langle\bar{c},c|\bar{c}',c'\rangle = \delta(\bar{c}-\bar{c}')\delta(c-c')
$$

[4]Fermion integration is reviewed in Chapter 6.

The rules of fermion calculus yield

$$\delta(\bar{c} - \bar{c}')\delta(c - c') = -(\bar{c} - \bar{c}')(c - c') \qquad (11.6.1)$$

To prove above statement, consider an arbitrary function $f(\bar{c}, c)$ of $\bar{c}, c$ with the following Taylor expansion

$$f(\bar{c}, c) = \alpha + \beta\bar{c} + \gamma c + \omega\bar{c}c$$

Using the rules of fermion integration given in Section 6.2 yields

$$\int d\bar{c}dc f(\bar{c}, c)(\bar{c} - \bar{c}')(c - c') = \int d\bar{c}dc\,(\alpha\bar{c}c - \beta\bar{c}\bar{c}'c - \gamma c\bar{c}' + \omega\bar{c}c\bar{c}'c' + \cdots)$$

$$= \int d\bar{c}dc c\bar{c}\,(\alpha + \beta\bar{c}' + \gamma c' + \omega\bar{c}'c' + \cdots) = -f(\bar{c}', c')$$

where $\cdots$ refers to terms that go to zero. Hence Eq. 11.6.1 has been verified. In general, for N complex fermions, one has

$$\prod_{n=1}^{N} \delta(\bar{c}_n - \bar{c}'_n)\delta(c_n - c'_n) = (-1)^N \prod_{n=1}^{N} (\bar{c}_n - \bar{c}'_n)(c_n - c'_n)$$

The gauge-fixed action in the Feynman gauge, with $\alpha = 1$, from Eq. 11.5.1 is given by

$$\mathcal{S}_{GF} = -\frac{1}{2}\int d^4x(\partial_\mu A_\nu)^2 + \int d^4x\partial_\mu\bar{c}\partial_\mu c \qquad (11.6.2)$$

In the covariant gauge, the gauge field state space requires all four components of the gauge field. The completeness equation is given by

$$\mathbb{I} = \prod_{\mu=0}^{3}\prod_{\vec{x}} \int dA_\mu(\vec{x})|A_\mu(\vec{x})\rangle\langle A_\mu(\vec{x})| \equiv \prod_{\mu=0}^{3} \int DA_\mu|A_\mu\rangle\langle A_\mu|$$

The completeness equation for the gauge plus ghost field is given by

$$\mathbb{I} = \int DA_\mu D\bar{c}Dc \prod_\mu |\bar{c}, c; A_\mu\rangle\langle\bar{c}, c; A_\mu| \qquad (11.6.3)$$

The state space of the gauge and ghost field system is given by a tensor product of the two state spaces

$$\mathcal{V} = \mathcal{V}_A \otimes \mathcal{V}_{\bar{c},c}$$

The connection of the state space and action to the Hamiltonian is similar to the case of the gauge-field given in Eq. 11.1.1. The Dirac-Feynman formula for Euclidean time is given by

$$\langle\bar{c}, c; A_\mu|e^{-\epsilon H}|\bar{c}', c'; A'_\mu\rangle = \mathcal{N}(\epsilon)e^{\epsilon\mathcal{L}(\bar{c},\bar{c}',c,c';A'_\mu,\epsilon)} \qquad (11.6.4)$$

where $\mathcal{N}(\epsilon)$ is a normalization.

The action consists of two decoupled free fields; the Hamiltonian is the sum of the gauge field and ghost Hamiltonians and given by

$$H = H_A \oplus H_G \qquad (11.6.5)$$

The Hamiltonian of the continuum gauge field is obtained from Eq. 11.1.1 and, for $\alpha = 1$, is given by (index i summed over)

$$H_A = -\frac{1}{2} \int d^3x \frac{\delta^2}{\delta A_\mu^2(\vec{x})} + \frac{1}{2} \int d^3x (\partial_i A_\nu(\vec{x}))^2 \quad ; \quad i = 1, 2, 3$$

$$= \frac{1}{2} \int d^3x [\pi_\mu^2 + (\partial_i A_\nu)^2] \quad ; \quad \pi_\mu(\vec{x}) = -i\frac{\delta}{\delta A_\mu(\vec{x})} \tag{11.6.6}$$

To obtain the ghost field Hamiltonian, note that the ghost Lagrangian density, from Eq. 11.6.2, is given by

$$\epsilon \mathcal{L}_G = \frac{1}{\epsilon}(\bar{c} - \bar{c}')(c - c') + \epsilon \partial_i \bar{c} \partial_i c \tag{11.6.7}$$

Fermion calculus yields the following identity

$$\langle \bar{c}, c| \exp\left\{ -\epsilon \frac{\delta^2}{\delta\bar{c}\delta c} \right\} |\bar{c}', c'\rangle = \exp\left\{ -\epsilon \frac{\delta^2}{\delta\bar{c}\delta c} \right\} \delta(\bar{c} - \bar{c}')\delta(c - c')$$

$$= -\left(1 - \epsilon \frac{\delta^2}{\delta\bar{c}\delta c}\right)(\bar{c} - \bar{c}')(c - c') = -\left[(\bar{c} - \bar{c}')(c - c') + \epsilon\right]$$

$$= -\epsilon \exp\left\{ \frac{1}{\epsilon}(\bar{c} - \bar{c}')(c - c') \right\} \tag{11.6.8}$$

Note the rather unfamiliar result that the factor of ϵ is inverted in the exponent since the ϵ term appears in the numerator on the right hand side, unlike the case for real variables where Gaussian integration results in the ϵ factor being in the denominator.

Similar to Eq. 11.1.1, the ghost field Hamiltonian is given, up to irrelevant constants, by the Dirac-Feynman formula

$$\exp\{-\epsilon H_G\}\langle \bar{c}, c|\bar{c}', c'\rangle = \langle \bar{c}, c| \exp\{-\epsilon H_G\}|\bar{c}', c'\rangle = \exp\{\epsilon \mathcal{L}_G\} \tag{11.6.9}$$

Hence, from Eqs. 11.6.7, 11.6.8 and 11.6.9, the ghost field Hamiltonian is given by

$$H_G = \int d^3x \left[\frac{\delta^2}{\delta\bar{c}(\vec{x})\delta c(\vec{x})} - \partial_i \bar{c}(\vec{x})\partial_i c(\vec{x}) \right] \tag{11.6.10}$$

11.6.1 *BRST cohomology: state space*

The Coulomb gauge, discussed in Section 11.2, is manifestly noncovariant and explicitly breaks Lorentz invariance. The gauge condition is $A_0 = 0$ supplemented by imposing the transversality condition on the state space given by $\partial_i A_i = 0$. The normal mode method of gauge-fixing also breaks Lorentz invariance since only the space components of the gauge field appear in the normal mode expansion, with the time component of the gauge field removed by gauge-fixing.

One can also obtain a state space description of the photon field using a *covariant gauge* that respects Lorentz symmetry. For example, consider the Lorentz gauge defined by $\partial_\mu A_\mu = 0$; one can consistently quantize the photon field using the Gupta-Bleuler formalism. However, as mentioned earlier the Gupta-Bleuler approach cannot be generalized to Yang-Mills non-Abelian gauge fields.

The BRST method of quantization is a modern formulation of the Gupta-Bleuler approach that is equally valid for the Yang-Mills fields. The ghost fields are used to remove the extra degrees of freedom for the case of a covariant (Lorentz invariant) gauge, and this formalism can be generalized to the lattice gauge field, as in Chapter 7.

The BRST method has the following ingredients. The gauge-fixed action has a BRST symmetry with a fermionic BRST charge operator Q_B. The BRST conserved charge obeys $[H, Q_B] = 0$, where H is the Hamiltonian of the gauge fixed theory given in Eq. 11.6.5. The state space is enlarged to include the ghost field $\mathcal{V}_A \otimes \mathcal{V}_{\bar{c},c}$ with completeness on this state space given by Eq. 11.6.3.

The physical state space is a subspace of $\mathcal{V}_A \otimes \mathcal{V}_{\bar{c},c}$ and is defined for $|\Phi\rangle \in \mathcal{V}_A \otimes \mathcal{V}_{\bar{c},c}$. The physical gauge invariant states $|\Phi\rangle$ obey further conditions of BRST cohomology, given below, that defines the physical gauge invariant state subspace.

- The operator Q_B is nilpotent since $Q_B^2 = 0$.
- The exact states are of the form $Q_B|\chi\rangle$ that are automatically annihilated by Q_B. These exact states are states that correspond to pure gauge transformations such that $A_\mu = \partial_\mu \phi$.
- Physical states $|\Phi\rangle$ are *constrained* to be annihilated by Q_B and are said to closed under Q_B and obey $Q_B|\Phi\rangle = 0$. Since charge Q_B is conserved, the constraint is conserved over time.
- Physical states $|\Phi\rangle$ are not exact, that is $|\Phi\rangle \neq Q_B|\chi\rangle$.
- States that can be written as $|\Phi\rangle + Q_B|\chi\rangle$ are all equivalent to $|\Phi\rangle$ and can be shown to differ by only a gauge transformation.
- Physical states $|\Phi\rangle$ are precisely the gauge invariant states of the photon field.

The BRST formalism is shown in Section 11.8.1 to reduce to the Gupta-Bleuler scheme for the case of the Abelian gauge field.

11.7   BRST Charge Q_B

The gauge fixed action given in Eq. 11.6.2 is BRST invariant, which is the result of choosing a gauge and its compensating Faddeev-Popov counter term. The Noether current due to BRST invariance yields the BRST charge Q_B [Das (2006), Zinn-Justin (1993)].

It is convenient to write the gauge-fixed action in terms of an auxiliary field as this simplifies the derivation. From Eq. 11.6.2, in the Feynman gauge with $\alpha = 1$, the gauge fixed action

$$\mathcal{S}_{GF} = -\frac{1}{4}\int d^4x F_{\mu\nu}^2 - \frac{1}{2}\int d^4x (\partial_\mu A_\mu)^2 + \int d^4x \partial_\mu \bar{c} \partial_\mu c$$

is re-written using an auxiliary field G as follows

$$e^{S_{GF}} = \int DG e^{\tilde{S}} \quad ; \quad \tilde{S} = \int d^4x \tilde{\mathcal{L}} \tag{11.7.1}$$

$$\tilde{\mathcal{L}} = -\frac{1}{4}F_{\mu\nu}^2 - \frac{1}{2}G^2 + i\partial_\mu G A_\mu + \partial_\mu \bar{c}\partial_\mu c \tag{11.7.2}$$

Consider a fermionic parameter λ such that

$$\lambda^2 = 0 \quad ; \quad \{\lambda, \bar{c}\} = 0 = \{\lambda, c\}$$

The BRST transformation consists of an infinitesimal gauge transformation for A_μ together with the following transformation of the ghost and auxiliary fields

$$\delta A_\mu = \lambda \partial_\mu c \quad ; \quad \delta \bar{c} = -i\lambda G \quad ; \quad \delta c = 0 = \delta G \tag{11.7.3}$$

Note one can define a BRST transformation with $\delta \bar{c} = 0$ and $\delta c \neq 0$; this gives rise to the same BRST charge.

The BRST variation of Eq. 11.7.1 is given by

$$\delta \tilde{S} = i\lambda \int \partial_\mu G \partial_\mu c + i\lambda \int G \partial^2 c = 0$$

BRST invariance yields a conserved fermionic charge Q_B using the Noether current. The BRST current is given by the varying both A_μ and $\bar{c}, c$ in the Lagrangian. Using

$$\frac{\partial}{\partial(\partial_\mu A_\nu)}\left(-\frac{1}{4}F_{\mu\nu}^2\right) = F_{\nu\mu} = -F_{\mu\nu}$$

the variation of the Lagrangian in Eq. 11.7.1 under the BRST symmetry given in Eq. 11.7.3 yields

$$j_\mu = \delta A_\nu \frac{\partial \mathcal{L}}{\partial(\partial_\mu A_\nu)} + \delta \bar{c}\frac{\partial \mathcal{L}}{\partial(\partial_\mu \bar{c})} = \lambda \partial_\nu c F_{\nu\mu} - i\lambda G \partial_\mu c \tag{11.7.4}$$

Since the BRST current is conserved $\partial_\mu j_\mu = 0$, dropping the parameter λ yields the conserved BRST charge

$$Q_B = \int d^3x j_0 = \int d^3x \left(\partial_i c F_{i0} - iG\partial_0 c\right)$$

$$= \int d^3x \left(-c\partial_i F_{i0} - iG\partial_0 c\right)$$

The Euler-Lagrange equation for A_μ

$$\partial_\mu \left(\frac{\partial \mathcal{L}}{\partial(\partial_\mu A_\nu)}\right) = \frac{\partial \mathcal{L}}{\partial A_\nu} \quad \Rightarrow \quad \partial_\mu F_{\nu\mu} = i\partial_\nu G$$

leads to the following

$$Q_B = \int d^3x \left(ic\partial_0 G - iG\partial_0 c\right)$$

The Euler-Lagrange equation for G

$$\partial_\mu \left(\frac{\partial \mathcal{L}}{\partial(\partial_\mu G)}\right) = \frac{\partial \mathcal{L}}{\partial G} \quad \Rightarrow \quad i\partial_\mu A_\mu = -G$$

yields the final expression for BRST charge

$$Q_B = \int d^3x \left(c(\vec{x})\partial_0(\partial_\mu A_\mu(\vec{x})) - (\partial_\mu A_\mu(\vec{x}))\partial_0 c(\vec{x})\right) \tag{11.7.5}$$

Note that all operators are considered to be time dependent Heisenberg operators and the time derivatives are taken on the Heisenberg operators.

11.8 Q_B and State Space

The conserved charge is an operator Q_B that acts on the state space of the gauge fixed action. To obtain an explicit representation of the operator Q_B the operator representation of the quantum fields A_μ and $\bar{c}, c$ are written in terms of creation and destruction operators.

Since the auxiliary field has been removed in Eq. 11.7.5, Q_B follows from the gauge fixed action given by Eq. 11.5.1; fixing $\alpha = 1$ yields

$$S_{GF} = -\frac{1}{2}\sum_{\mu,\nu}\int d^4x(\partial_\mu A_\nu)^2 + \int d^4x\,\partial_\mu\bar{c}\partial_\mu c \tag{11.8.1}$$

The gauge field consists of four free scalar fields and the ghost field is the fermionic version of the free complex scalar field. Hence we can use the results of free fields with some minor modifications.

The normal mode expansion in Euclidean time $t = -i\tau$ of the gauge field, similar to the case for the Euclidean free field, is the following [Baaquie (2018)]

$$A_\mu(\tau,\vec{x}) = \int_{\vec{p}}\frac{1}{\sqrt{2E_{\vec{p}}}}(e^{-E_{\vec{p}}\tau+i\vec{p}\cdot\vec{x}}a_{\vec{p}\mu} + e^{E_{\vec{p}}\tau-i\vec{p}\cdot\vec{x}}a_{\vec{p}\mu}^\dagger)\;;\quad \int_{\vec{p}} = \int\frac{d^3p}{(2\pi)^3} \tag{11.8.2}$$

and where

$$E_{\vec{p}} = |\vec{p}\,| \tag{11.8.3}$$

The equal time canonical commutation equations yields

$$[a_{\vec{p}\mu}, a_{\vec{k}\nu}^\dagger] = (2\pi)^3\delta^3(\vec{p}-\vec{k})\delta_{\mu-\nu}$$

with the rest of the commutators being zero.

The ghost field are similar to the complex scalar fields and one has the expansion for Euclidean spacetime given by

$$c(\tau,\vec{x}) = \int_{\vec{p}}\frac{1}{\sqrt{2E_{\vec{p}}}}(e^{-E_{\vec{p}}\tau+i\vec{p}\cdot\vec{x}}A_{\vec{p}} + e^{E_{\vec{p}}\tau-i\vec{p}\cdot\vec{x}}B_{\vec{p}}^\dagger)$$

$$\bar{c}(\tau,\vec{x}) = \int_{\vec{p}}\frac{1}{\sqrt{2E_{\vec{p}}}}(e^{-E_{\vec{p}}\tau+i\vec{p}\cdot\vec{x}}B_{\vec{p}} + e^{E_{\vec{p}}\tau-i\vec{p}\cdot\vec{x}}A_{\vec{p}}^\dagger) \tag{11.8.4}$$

The Fourier expansion of $c, \bar{c}$ given in Eq. 11.8.4 above is for Euclidean spacetime. Under charge conjugation

$$A_{\vec{p}} \leftrightarrow B_{\vec{p}} \quad \Rightarrow \quad c \leftrightarrow \bar{c}$$

The equal time canonical anticommutation equations

$$\left\{\frac{\partial c(\tau,\vec{x})}{\partial\tau}, c(\tau,\vec{x}')\right\} = \delta^3(\vec{x}-\vec{x}') = \left\{\frac{\partial\bar{c}(\tau,\vec{x})}{\partial\tau}, \bar{c}(\tau,\vec{x}')\right\}$$

yield

$$\{A_{\vec{p}}, A_{\vec{p}'}^\dagger\} = \{B_{\vec{p}}, B_{\vec{p}'}^\dagger\} = \delta^3(\vec{p}-\vec{p}')\;;\quad \{A_{\vec{p}}, A_{\vec{p}'}\} = 0 = \{A_{\vec{p}}, B_{\vec{p}'}^\dagger\} = 0 = \{B_{\vec{p}}, B_{\vec{p}'}\}$$

and with all the other anticommutators being zero.

In the Feynman gauge, the action as given in Eq. 11.8.1 for the gauge field is equivalent to four decoupled scalar fields. Hence, the gauge field Hamiltonian, in Euclidean time $(t = -i\tau)$, as given in Eq. 11.6.6, yields

$$H_A = \frac{1}{2}\sum_{\mu=0}^{3}\int_{\vec{x}}\left[-\left(\frac{\partial A_\mu(\tau,\vec{x})}{\partial\tau}\right)^2 + (\vec{\nabla}A_\mu(\tau,\vec{x}))^2\right] \quad ; \quad \int_{\vec{x}} = \int d^3x \quad (11.8.5)$$

The ghost Hamiltonian is similar to the complex scalar field and given by

$$H_G = \int_{\vec{x}}\left[\frac{\delta^2}{\delta\bar{c}(\vec{x})\delta c(\vec{x})} - \partial_i\bar{c}(\vec{x})\partial_ic(\vec{x})\right]$$
$$= \int_{\vec{x}}\left[\partial_0\bar{c}(\tau,\vec{x})\partial_0c(\tau,\vec{x}) - \partial_i\bar{c}(\vec{x})\partial_ic(\vec{x})\right] \quad (11.8.6)$$

The signs of the kinetic and potential term of H_G are fixed by the rules of fermion integration. Note the differing signs for H_A, H_G for the kinetic and potential terms; this is the reason that the zero point energy of the bosonic and fermionic fields have opposite signs.[5]

Substituting the Fourier expansion for $A_\mu, \bar{c}, c$ given in Eqs. 11.8.2 and 11.8.4 yields the following

$$H = H_A + H_G$$
$$= \int_{\vec{p}}E_{\vec{p}}\left[a^\dagger_{\vec{p}\mu}a_{\vec{p}\mu} + A^\dagger_{\vec{p}}A_{\vec{p}} + B^\dagger_{\vec{p}}B_{\vec{p}}\right] + E_0 \quad (11.8.7)$$

Substituting Eqs. 11.8.2 and 11.8.4 into the expression for Q_B, given in Eq. 11.7.5, and after some algebra, yields the conserved (time independent) BRST charge operator

$$Q_B = -\int_{\vec{p}}|\vec{p}\,|(a^\dagger_{\vec{p}0}A_{\vec{p}} + a_{\vec{p}0}B^\dagger_{\vec{p}}) + i\sum_{i=1}^{3}\int_{\vec{p}}p_i(a^\dagger_{\vec{p}i}A_{\vec{p}} + a_{\vec{p}i}B^\dagger_{\vec{p}}) \quad (11.8.8)$$

To simplify the notation define

$$p_0 \equiv i|\vec{p}\,| \quad (11.8.9)$$

Note Eq. 11.8.9 is what one expects from the continuation of Minkowski to Euclidean time. The BRST charge can be written in the following invariant form

$$Q_B = i\sum_{\mu=0}^{3}\int_{\vec{p}}p_\mu\left(a^\dagger_{\vec{p}\mu}A_{\vec{p}} + a_{\vec{p}\mu}B^\dagger_{\vec{p}}\right) \quad (11.8.10)$$

To show that Q_B is conserved, consider the following.

$$[Q_B, H_A] = i\int_{\vec{p}}(-p_\mu a^\dagger_{\vec{p}\mu}A_{\vec{p}\mu} + p_\mu a_{\vec{p}\mu}B^\dagger_{\vec{p}}) \quad (11.8.11)$$

$$[Q_B, H_G] = i\int_{\vec{p}}(p_\mu a^\dagger_{\vec{p}\mu}A_{\vec{p}\mu} - p_\mu a_{\vec{p}\mu}B^\dagger_{\vec{p}}) = -[Q_B, H_A] \quad (11.8.12)$$

---

[5] Hence $E_0 = (\frac{1}{2}\cdot 4 - 1)(2\pi)^3\delta^{(3)}(0)\int_{\vec{p}}E_{\vec{p}} = (2\pi)^3\delta^{(3)}(0)\int_{\vec{p}}E_{\vec{p}}$.

From above and Eq. 11.8.7

$$[Q_B, H] = [Q_B, H_A + H_G] = 0$$

and hence we have confirmed that for the quantized theory BRST charge is conserved.

BRST charge is not a Hermitian operator, with its Hermitian conjugate given by

$$Q_B^\dagger = -\int_{\vec{p}} |\vec{p}\,| (a_{\vec{p}0} A_{\vec{p}}^\dagger + a_{\vec{p}0}^\dagger B_{\vec{p}}) - i\sum_{i=1}^{3} \int_{\vec{p}} p_i (a_{\vec{p}i} A_{\vec{p}}^\dagger + a_{\vec{p}i}^\dagger B_{\vec{p}})$$

The significance of $Q_B^\dagger$ is discussed by Malik (2001).

To verify that $Q_B^2 = 0$, using the anticommuting property of $A_{\vec{p}}, B_{\vec{p}}^\dagger$ yields the following

$$Q_B^2 = i^2 \sum_{\mu,\nu=0} \int_{\vec{p},\vec{k}} p_\mu k_\nu [a_{\vec{p}\mu}, a_{\vec{k}\nu}^\dagger] B_{\vec{p}}^\dagger A_{\vec{k}} = i^2 \sum_{\mu=0} \int_{\vec{p}} p_\mu p_\mu B_{\vec{p}}^\dagger A_{\vec{p}} \qquad (11.8.13)$$

But from Eq. 11.8.9

$$p_\mu p_\mu = p_0^2 + \sum_{i=1}^{3} p_i p_i = -|\vec{p}\,|^2 + (\vec{p}\,)^2 = 0$$

Hence, from Eq. 11.8.13 and above

$$Q_B^2 = 0$$

We have the important result that as an *operator* Q_B is nilpotent, namely that $Q_B^2 = 0$ is an *operator equation* that is valid both on-shell and off-shell.

As discussed in Section 11.6.1, every vector $|\Phi\rangle$ in the physical gauge invariant state space is annihilated by Q_B, that is (the physical state is not exact)

$$Q_B|\Phi\rangle = 0 \qquad (11.8.14)$$

11.8.1   *Gupta-Bleuler condition*

It is shown how the definition of state space given by BRST quantization reduces to the Gupta-Bleurel constraint on state space when the gauge field state space is considered by itself, without the presence of the ghost field.

From Eqs. 11.8.13 and 11.8.14

$$0 = Q_B|\Phi\rangle = i\int_{\vec{p}} (p_\mu a_{\vec{p}\mu}^\dagger A_{\vec{p}} + p_\mu a_{\vec{p}\mu} B_{\vec{p}}^\dagger)|\Phi\rangle$$

Since the gauge and ghost field are decoupled, the physical state vector is taken to be a tensor product

$$|\Phi\rangle = |\Phi_A\rangle|\Phi_G\rangle$$

Ghost number is conserved and $|\Phi_G\rangle$ is taken to have zero ghost number. From the expression for Q_B given in Eq. 11.8.13, the ghost state is taken to be the ground state $|\Phi_G\rangle = |\Omega_G\rangle$ that is annihilated by $A_{\vec{p}}, B_{\vec{p}}$, defined by[6]

$$A_{\vec{p}}|\Omega_G\rangle = 0 = B_{\vec{p}}|\Omega_G\rangle$$

The BRST constraint reduces to

$$Q_B|\Phi\rangle = -i\int_{\vec{p}} p_\mu a_{\vec{p}\mu}|\Phi_A\rangle B_{\vec{p}}^\dagger|\Omega_G\rangle = 0$$

Since $B_{\vec{p}}^\dagger|\Omega_G\rangle \neq 0$, to achieve $Q_B|\Phi\rangle = 0$ the following constraint is imposed on the gauge field state vectors

$$Q_B|\Phi\rangle = 0 \quad \Rightarrow \quad p_\mu a_{\vec{p}\mu}|\Phi_A\rangle = 0$$

This constraint can be written more transparently in real space; from Eq. 11.8.2

$$\partial_\mu A_\mu(\vec{x}) = \partial_\mu A_\mu^{(-)}(\vec{x}) + \partial_\mu A_\mu^{(+)}(\vec{x})$$

where $A_\mu^{(-)}$ has all the annihilation operators and $A_\mu^{(+)}$ has all the creation operators. The BRST constraint yields the following

$$p_\mu a_{\vec{p}\mu}|\Phi_A\rangle = 0 \quad \Rightarrow \quad \partial_\mu A_\mu^{(-)}(\vec{x})|\Phi_A\rangle = 0 \;:\; \text{Gupta-Bleuler condition}$$

Note that, since

$$\langle\Phi_A|\partial_\mu A_\mu^{(+)} = 0$$

the Gupta-Bleuler condition implies that for any physical state $|\Phi_A\rangle$

$$\langle\Phi_A|\partial_\mu A_\mu|\Phi_A\rangle = \langle\Phi_A|\left(\partial_\mu A_\mu^{(+)} + \partial_\mu A_\mu^{(-)}\right)|\Phi_A\rangle = 0 \qquad (11.8.15)$$

In other words, instead of imposing the covariant gauge condition of the *operator equation* $\partial_\mu A_\mu = 0$, in the Gupta-Bleuler approach the weaker condition is imposed, as in Eq. 11.8.15, that the *expectation value* of the operator $\partial_\mu A_\mu$ for any gauge invariant (physical) state has to be zero.

In conclusion, the physical gauge invariant state space for both the Abelian and non-Abelian gauge fields, is given by the constraint

$$Q_B|\Phi\rangle = 0 \;;\; |\Phi\rangle = |\Phi_A\rangle|\Omega_G\rangle$$

The physics of the BRST constraint on state space for non-Abelian gauge fields has been discussed in Peskin and Schroeder (1995).

---

[6]It can be verified by using the ghost Hamiltonian H_G given in Eq. 11.8.6 that

$$H_G|\Omega_G\rangle = 0 \quad \Rightarrow \quad \langle\bar{c},c|\Omega_G\rangle = \mathcal{N}\exp\left\{-\int\frac{d^3p}{(2\pi)^3}|\vec{p}|\bar{c}(\vec{p})c(\vec{p})\right\}$$

where $\bar{c}(\vec{p}); c(\vec{p})$ are the Fourier transform of the coordinate basis $\bar{c}(\vec{x}); c(\vec{x})$.

11.9   Summary

The two underpinnings of the lattice gauge field – namely that it is defined directly using the finite group element of the gauge group and secondly that it has a lattice cutoff – introduce many new features that are absent in the continuum formulation.

For the case of the Hamiltonian, the finite group element leads to the chromoelectric field operators being differential operators on the Lie group manifold. Gauge-fixing necessitates the removal of the extra gauge degrees of freedom from the kinetic operator. The gauge-constraint (Coulomb gauge) requires constraining the differential operators and requires properties of the lattice chromo-electric field operator that do not arise in the continuum formulation. In particular, the change of variables eliminating the redundant gauge degree of freedom from the chromo-electric operator depends on the structure of the Lie group space – and which is unlike the continuum case where only the Lie algebra is required for gauge-fixing.

The BRST operator provides an exactly conserved charge that was used for choosing a covariant gauge for the gauge-field Hamiltonian. The closed but not exact states of BRST cohomology precisely pick out the all gauge-invariant states: this procedure works for both Abelian and non-Abelian gauge fields. The case of the Abelian gauge field was exactly solved, and the computation was carried out in continuum spacetime notation. The notation for the continuum case is more convenient than the lattice case, and the continuum formulation – for the Abelian gauge field – can be carried over to the lattice case by inspection.

PART 3

Lattice Dirac Field

Chapter 12

Dirac Lattice Path Integral

The Dirac field, which is a particular case of a field based on fermionic variables, was first proposed by P. A. M. Dirac in 1928 to describe the relativistic electron in four-dimensional spacetime [Dirac (1999)]. Every electron has spin 1/2 and hence needs two independent fermionic variables for its description; its antiparticle, namely the positron is also spin 1/2 and needs another two independent fermionic variables. Hence in total a Dirac fermion needs four fermionic degrees of freedom and is consequently described by a four-component fermionic variable, and which transforms as a spinor under Lorentz transformation.

The electron carries electric charge and hence can couple to the Abelian gauge field. The Dirac field has many realizations, with particular varieties of the Dirac field being realized in weak interactions as leptons, and in strong interactions by quarks – both of which are described by Dirac fields that carry non-Abelian charges [Baaquie and Willeboordse (2015)]. The Dirac field is sometimes referred to as a quark field when one is emphasizing its non-Abelian nature.

This Chapter is based on the results obtained in Baaquie (1983b). The Dirac field is studied using lattice fermions introduced by Wilson (1974) and further elaborated by Ginsparg and Wilson (1982). In particular, the lattice is used for calculating the evolution kernel of the Dirac field. The continuum limit is taken, and the complete energy eigenfunctionals of the free Dirac field – as well its propagator – are then evaluated using the kernel.

12.1 Introduction

The time evolution kernel, or the probability amplitude, is defined for quantum fields by the matrix elements of $\exp(-itH/\hbar)$ between arbitrary initial and final field configurations, where t is the physical time and H the Hamiltonian operator. In Euclidean time, the probability amplitude is given by $\exp(-\tau H)$. The kernel contains the complete content of the quantum field theory.

For the free boson field, the kernel has been evaluated by Feynman and Hibbs (1965). A similar calculation is carried out for the free Dirac field in Baaquie (1983b) There are a number of problems with the fermions which are not present in the boson case. Note that for the free boson field, the evolution kernel is given by the finite time classical action – with the appropriate boundary conditions on the classical field. However, in the case of the Dirac field the classical Dirac Lagrangian is zero, giving an incorrect result for the kernel. The result stems from the ambiguity in incorporating the boundary conditions for the finite time Dirac action and this short-coming is resolved using the lattice formulation that results in a *boundary term* for the finite time continuum action. Furthermore, the lattice formulation shows that the appropriate boundary conditions for the continuum Dirac field are the same as for the lattice theory. The ambiguities of the continuum fermion path integral for the Dirac field have been noted by many authors [Creutz (1983); Montvay and Munster (1994); Roth (1997)].

Wilson's formulation of the lattice fermions incorporates the boundary values in an unambiguous manner and specifies a particular way of defining the fermion initial and final field configurations. With the hindsight gained from the lattice, a derivation is given in Section 12.2 for the coordinates of the Dirac field directly from the continuum formulation.

12.2   Dirac Field Coordinates

What are the field coordinates, namely the degrees of freedom, for the Dirac field's state vector? For starters, consider the Dirac fermion field directly in the continuum formulation without recourse to the lattice.

The Dirac equation was obtained by Dirac by starting from the field equation for a scalar field given by the Klein-Gordon equation and reducing the second derivatives to first derivatives [Baaquie (2018)]. The resulting field equation in Minkowski time is given by

$$\left(i\gamma_0 \frac{\partial}{\partial t} + i\vec{p}\cdot\vec{r} + m\right)\psi = 0$$

and the conjugate Dirac equation is

$$\bar{\psi}\left(-i\gamma_0 \frac{\partial}{\partial t} - i\vec{p}\cdot\vec{r} + m\right) = 0$$

where $\bar{\psi}$, ψ are the Dirac four-component spinors, m is the mass term and γ_μ the gamma matrices. Boosting the spinors to the rest-frame results in setting the three-momentum $\vec{p} = 0$ in above equations, and yields the classical solution given by

$$\left(i\gamma_0 \frac{\partial}{\partial t} + m\right)\psi_c = 0 \quad ; \quad \bar{\psi}_c\left(-i\gamma_0 \frac{\partial}{\partial t} + m\right) = 0 \qquad (12.2.1)$$

Using the 2×2 block representation, and denoting the two component upper and lower doublet spinors by subscripts u and ℓ yields

$$\gamma_0 = \begin{pmatrix} 1 & 0 \\ 0 & -1 \end{pmatrix} \quad ; \quad \psi = \begin{pmatrix} \psi_u \\ \psi_\ell \end{pmatrix} \quad ; \quad \bar{\psi} = (\bar{\psi}_u, \bar{\psi}_\ell) \qquad (12.2.2)$$

Consider the following ansatz $\bar{\psi}_c$, ψ_c

$$\psi_c = \begin{pmatrix} e^{imt}\psi_u \\ e^{-imt}\psi_\ell \end{pmatrix} \qquad (12.2.3)$$

and

$$\bar{\psi}_c = \left(e^{-imt}\bar{\psi}_u,\ e^{imt}\bar{\psi}_\ell \right) \tag{12.2.4}$$

It can be readily seen that the solutions given in Eqs. 12.2.3 and 12.2.4 satisfy the Dirac field equation given in Eq. 12.2.1.

The time dependence of a state with energy E is given by e^{-iEt}; from Eqs. 12.2.3 and 12.2.4, the time evolution of the degrees of freedom is given by the following

$$\bar{\psi}_u\ ;\ \psi_\ell:\ \text{evolving forward in time}$$

and

$$\bar{\psi}_\ell\ ;\ \psi_u:\ \text{evolving backward in time}$$

Hence, we conclude that

- the field coordinates propagating *forward* in time are the conjugate coordinate eigenstates

$$\langle\bar{\psi}_u, \psi_\ell|$$

- and those propagating *backward* in time are the coordinate eigenstates

$$|\bar{\psi}_\ell, \psi_u\rangle$$

The same conclusion is reached in Section 12.3 using the Wilson lattice fermions.

Under spatial rotations, for the choice of the gamma matrices given in Eq. 12.3.2, the upper and lower components of $\bar{\psi}$ and ψ transform independently. Hence, each of the upper and lower components $\bar{\psi}_u, \psi_\ell$, provide a two-dimensional spin 1/2 spinor representation of the rotation group in three-dimensional space.

12.3 Dirac Lattice Lagrangian

The reasoning used by Dirac to discover his equation was discussed in Section 12.2. One can use a different line of reasoning developed in Chapter 6 to also arrive at the Dirac equation. Recall from Eq. 6.11.4, fermion calculus and the nature of the fermion and antifermion state space yields the Minkowski time Dirac Lagrangian in one spacetime dimension as given by

$$\mathcal{L}_M = -\bar{\psi}\left(i\gamma_0^M \frac{\partial}{\partial t_M} + m\right)\psi$$

Extending the time coordinate t to space and time $t, x_i\ ;\ i = 1, 2, 3$ requires constraints be put on the variation of the field variables in the space direction, which is realized by having derivative in the space direction, with their own Gamma matrices. Requiring a Lorentz invariant Lagrangian yields

$$\mathcal{L}_M(x) = -\bar{\psi}\left(i\gamma_0^M \frac{\partial}{\partial t_M} - i\gamma_i^M \frac{\partial}{\partial x_i} + m\right)\psi\ ;\ i = 1, 2, 3$$
$$= -\bar{\psi}(i\gamma_\mu^M \partial^\mu + m)\psi\ ;\ \mu = 0, 1, 2, 3 \tag{12.3.1}$$

where $\bar{\psi}(x)$ and $\psi(x)$ are *four-component* anticommuting spinors, m is the mass; γ_μ^M are the Minkowski space gamma matrices that satisfy the following anticommutation equation

$$\{\gamma_\mu^M, \gamma_\nu^M\} = \eta_{\mu\nu}\mathbb{I}$$

where $\mathbb{I}$ is a 4×4 unit matrix and $\eta_{\mu\nu}$ is the Lorentz metric given by

$$\eta_{\mu\nu} = \mathrm{diag}(1, -1, -1, -1)$$

Define for Euclidean time τ *Euclidean gamma matrices* γ_μ with the following representation (σ_i are the Pauli matrices)[1]

$$t = -i\tau, \qquad \gamma_0^M = \gamma_0 = \begin{bmatrix} 1 & 0 \\ 0 & -1 \end{bmatrix}, \qquad -i\gamma_i^M = \gamma_i = i\begin{bmatrix} 0 & -\sigma_i \\ \sigma_i & 0 \end{bmatrix} \qquad (12.3.2)$$

and the Hermitian gamma matrices have the following anticommutation equations

$$\{\gamma_\mu, \gamma_\nu\} = \delta_{\mu\nu}\mathbb{I} \quad ; \quad \gamma_\mu^\dagger = \gamma_\mu$$

Hence, from Eqs. 12.3.1 and 12.3.2 the Dirac Euclidean Lagrangian is given by

$$\mathcal{L}(x) = -\bar{\psi}(x)(\gamma_\mu\partial_\mu + m)\psi(x) = -\bar{\psi}\Big(\gamma_0\frac{\partial}{\partial\tau} + \gamma_i\frac{\partial}{\partial x_i} + m\Big)\psi \qquad (12.3.3)$$

Re-writing the Dirac Lagrangian in a symmetric form (more appropriate for the lattice theory) yields the four-dimensional continuum spacetime action given by

$$S = \int d^4x \mathcal{L}(x)$$

$$\mathcal{L}(x) = -\frac{1}{2}\left[\bar{\psi}(x)\gamma_\mu\partial_\mu\psi(x) - \partial_\mu\bar{\psi}(x)\gamma_\mu\psi(x)\right] - m\bar{\psi}(x)\psi(x) \qquad (12.3.4)$$

Discretize spacetime into a four-dimensional lattice, with $x = na$, where n is a four-dimensional lattice point and with lattice spacing a; let $\hat{\mu}$ be a unit vector in the μ-direction. More precisely

$$x = na \;\; ; \;\; n = (n_0, n_1, n_2, n_3) : \text{ lattice point}$$

$$\hat{\mu} : \text{ unit vector with } \mu = 0, 1, 2, 3$$

Using an approximation for the first derivative yields the *naive* lattice action of the form

$$S = \sum_n \tilde{\mathcal{L}}_n \qquad (12.3.5)$$

$$\tilde{\mathcal{L}}_n = ma^4\bar{\psi}(na)\psi(na)$$

$$-a^4\frac{1}{2a}\sum_\mu \Big\{\bar{\psi}(na)\gamma_\mu[\psi(na + \hat{\mu}a) - \psi(na)] - [\bar{\psi}(na + \hat{\mu}a) - \bar{\psi}(na)]\gamma_\mu\psi(na)\Big\}$$

$$= ma^4\bar{\psi}(na)\psi(na) - \frac{a^3}{2}\sum_\mu \Big\{\bar{\psi}(na)\gamma_\mu\psi(na + \hat{\mu}a) - \bar{\psi}(na + \hat{\mu}a)\gamma_\mu\psi(na)\Big\}$$

---

[1]Dirac matrices have infinitely many allowed representations.

Define dimensionless lattice fermion field variables $\bar\psi_n, \psi_n$ by the following (the constant K is defined for simplifying the final result)

$$2K = (4 + ma)^{-1}$$

$$\psi_n = \left(\frac{a^3}{2K}\right)^{1/2} \psi(na)$$

$$\bar\psi_n = \left(\frac{a^3}{2K}\right)^{1/2} \bar\psi(na)$$ (12.3.6)

From Eq. 12.3.5, the naive lattice Lagrangian is given by

$$\tilde{\mathcal{L}}_n = -2Kma\bar\psi_n\psi_n + K\sum_{\mu}\left\{\bar\psi_{n+\hat\mu}\gamma_\mu\psi_n - \bar\psi_n\gamma_\mu\psi_{n+\hat\mu}\right\}$$ (12.3.7)

For zero space and one discrete time dimension, the following Dirac lattice Lagrangian, given in Eq. 6.11.3, was obtained from the underlying fermionic Hilbert space

$$-\bar\psi_n\psi_n + \frac{1}{2(1+ma)}\left\{\bar\psi_{n+\hat0}(1+\gamma_0)\psi_n + \bar\psi_n(1-\gamma_0)\psi_{n+\hat0}\right\} \quad : \quad d = 1 \quad (12.3.8)$$

One expects to recover the expression given in Eq. 12.3.8 as the one discrete time dimension limit of the four-dimensional Dirac lattice Lagrangian. To have four-dimensional hyper-cubic symmetry for the lattice Euclidean Lagrangian consistent with Eq. 12.3.8 entails adding the following term to the naive lattice Lagrangian given in Eq. 12.3.7

$$K\sum_{\mu}\left\{\bar\psi_{n+\hat\mu}\psi_n + \bar\psi_n\psi_{n+\hat\mu}\right\}$$ (12.3.9)

Note that, in the $a \to 0$ limit, this term yields an extra contribution to the mass term of the continuum Lagrangian that, from Eq. 12.3.9 – to leading order in the lattice spacing a – is given by

$$K\sum_{\mu=0}^{3}\left\{\bar\psi_{n+\hat\mu}\psi_n + \bar\psi_n\psi_{n+\hat\mu}\right\} = 8K\bar\psi_n\psi_n + O(a^2)$$ (12.3.10)

Adding Eq. 12.3.10 to the naive Lagrangian given in Eq. 12.3.7 does not change the continuum limit. Hence, define the Dirac lattice Lagrangian by the following

$$\mathcal{L}_n = \tilde{\mathcal{L}}_n + K\sum_{\mu=0}^{3}\left\{\bar\psi_{n+\hat\mu}\psi_n + \bar\psi_n\psi_{n+\hat\mu}\right\} - 8K\bar\psi_n\psi_n$$ (12.3.11)

Using Eq. 12.3.7 for $\tilde{\mathcal{L}}_n$ and with

$$8K + 2Kma = 1 \quad \Rightarrow \quad 2K = \frac{1}{4+ma}$$

yields the Dirac lattice Lagrangian given by

$$\mathcal{L}_n = -\bar\psi_n\psi_n + K\sum_{\mu=0}^{3}\left\{\bar\psi_n(1-\gamma_\mu)\psi_{n+\hat\mu} + \bar\psi_{n+\hat\mu}(1+\gamma_\mu)\psi_n\right\}$$ (12.3.12)

The lattice Dirac field is written in terms of *Wilson fermions* that consist of introducing the projecting operators $1 \pm \gamma_\mu$ for defining the nearest neighbor coupling of the Dirac field. The four-dimensional Dirac lattice Lagrangian given in Eq. 12.3.12 has the requisite hyper-cubic symmetry that reduces to the one-dimensional time lattice result given in Eq. 12.3.8; in particular, the hyper-cubic symmetry ensures that the fermionic degrees of freedom that propagate are the ones required by the Lagrangian in Eq. 12.3.8.

Although the lattice Lagrangian $\mathcal{L}$ in Eq. 12.3.12 has the same $a \to 0$ limit as the earlier naive Lagrangian $\tilde{\mathcal{L}}$ given in Eq. 12.3.7, the full $a \neq 0$ lattice theories for the two Lagrangians are radically different. The lattice Lagrangian $\mathcal{L}$ allows the propagation of the correct four degrees of freedom, namely the two spin states each for the particle and the antiparticle as can be seen by the argument that led to the Lagrangian given in Eq. 12.3.8; in the continuum limit, Lagrangian $\mathcal{L}$ yields the correct state space for the spin $1/2$ Dirac particle.

In contrast, there are eight propagating degrees of freedom for the naive Lagrangian $\tilde{\mathcal{L}}$; the reason being that the projection operators $1 \pm \gamma_0$, which project out two degrees of freedom for every step in time, are absent in $\tilde{\mathcal{L}}$. If one takes continuum limit of the states space generated by $\tilde{\mathcal{L}}$, one obtains the incorrect limit of *two* distinct species of spin $1/2$ Dirac particles for the quantum spectrum of states.

Note that the time lattice spacing has been taken to be equal to the spatial lattice spacing, and the symmetric limit of $a \to 0$ is taken for obtaining the continuum limit. One could have equivalently started with continuous three-space and a time lattice, and which is required for obtaining the continuum Dirac Hamiltonian. The symmetric hyper-cubic symmetric spacetime lattice is more appropriate for the path integral formulation.

12.4   Lattice Fermions and Chiral Symmetry

Massless fermions have chiral symmetry. In the Standard Model of particle physics, the neutrino is massless and so it is necessary to have a discretization scheme for the fermions that respects chiral symmetry. An excellent summary of chiral symmetry for lattice fermions is given by Chandrasekharan and Wiese (2004). In this Section, the calculation is done in Euclidean spacetime since it provides for a well-defined path integral.

The inversion of spacetime, from $x \to -x$ is realized for fermions by the transformation

$$\psi(x) \to \psi(-x) = \gamma_5 \psi(x)$$

The gamma five matrix γ_5 in Euclidean space is given by

$$\gamma_5 = \gamma_0 \gamma_1 \gamma_2 \gamma_3 \quad \Rightarrow \quad \{\gamma_\mu, \gamma_5\} = 0$$

Chiral symmetry is defined by local rotations of the fermion fields and given by

$$\psi \to e^{i\alpha\gamma_5}\psi \quad ; \quad \bar{\psi} \to \bar{\psi}e^{i\alpha\gamma_5}$$

An action having chiral symmetry yields

$$S[\bar{\psi}e^{i\alpha\gamma_5}, e^{i\alpha\gamma_5}\psi] = S[\bar{\psi}, \psi]$$

A mass term in the lattice action, under a chiral transformation, yields the following

$$\bar{\psi}_n\psi_n \rightarrow \bar{\psi}_n e^{2i\alpha\gamma_5}\psi_n \neq \bar{\psi}_n\psi_n$$

and breaks chiral symmetry. By this criterion, Wilson fermions break chiral symmetry since the Lagrangian given in Eq. 12.3.12 has a mass-like terms.

To obtain chiral symmetry starting with Wilson fermions requires fine-tuning the mass-term in the lattice action, and is not very suitable for numerical simulations.

Since one is interested in lattice theories that yield chiral symmetry in the continuum limit, one can ask a separate question: which lattice theories have a (hidden) chiral symmetry coming from the continuum action? In a landmark paper, Ginsparg and Wilson (1982) provided a definition of chiral symmetry for lattice fermions that are obtained by a block-spin renormalization group transformation, starting from continuum fermions having an exact chiral symmetry.

It was later realized that the Ginsparg-Wilson lattice fermions define what is termed as *perfect lattice fermions*: fermions that arise from the block-spin of continuum chiral invariant fermions, and hence do not have any artifacts of the lattice – although the continuum results might not be apparent in the lattice.[2]

Consider d-dimensional Euclidean spacetime with chiral fermions $\chi, \bar{\chi}$ having an action $S_I(\chi, \bar{\chi})$ that is invariant under chiral rotations

$$S_I(e^{i\alpha\gamma_5}\chi, \bar{\chi}e^{i\alpha\gamma_5}) = S_I(\chi, \bar{\chi})$$

The continuum fermions $\chi, \bar{\chi}$ are combined into 'block-spin' fermions χ_n, which are defined on a discrete spacetime lattice, denoted by lattice points n. The block-spin fermions are obtained by an averaging procedure over a hypercube $E(n)$, which is centered on n, and weighted by a density function $\rho(x)$. The block-spin fermions are given by the following

$$\chi_n \equiv \int_{E(n)} d^dx \rho(x)\chi(x) \quad ; \quad \bar{\chi}_n \equiv \int_{E(n)} d^dx \rho^*(x)\bar{\chi}(x)$$

with

$$\sum_n \oplus E(n) = \mathcal{R}^d$$

Let the fermions on the lattice be given by $\psi_n, \bar{\psi}_n$ with action for lattice fermions be $S(\psi, \bar{\psi})$; the block-spin renormalization group transformation yields the following

$$e^{-S(\psi, \bar{\psi})} = \int D\chi D\bar{\chi} \exp\left[-(\bar{\psi}_m - \bar{\chi}_m)A_{mn}(\psi_n - \chi_n) - S_I(\chi, \bar{\chi}) \right]$$

with sum over all repeated lattice indices. Take $A_{mn} = a\delta_{m-n}$ for simplicity as it does not change the result. Hence

$$e^{-S(\psi, \bar{\psi})} = \int D\chi D\bar{\chi} \exp\left[-a(\bar{\psi}_n - \bar{\chi}_n)(\psi_n - \chi_n) - S_I(\chi, \bar{\chi}) \right]$$

---

[2]Block-spin renormalization for gauge fields is discussed in Chapter 10; a simple one-dimensional case is discussed in Baaquie (2014).

The kernel of the block-spin transformation

$$\exp\left[-a(\bar{\psi}_n - \bar{\chi}_n)(\psi_n - \chi_n)\right]$$

explicitly breaks chiral symmetry as this is required for the renormalization group transformation to have a nonsingular fixed point [Ginsparg and Wilson (1982)].

Consider a global chiral transformation on the lattice

$$\psi_n \to e^{i\epsilon\gamma_5}\psi_n \ ; \ \bar{\psi}_n \to \bar{\psi}_n e^{i\epsilon\gamma_5}$$

Since $S_I(\chi, \bar{\chi})$ is invariant under the chiral transformation, one can do the change of fermion integration variables

$$\chi_n \to e^{i\epsilon\gamma_5}\chi_n \ ; \ \bar{\chi}_n \to \bar{\chi}_n e^{i\epsilon\gamma_5}$$

and this yields

$$\exp\{-S(e^{i\epsilon\gamma_5}\psi, \bar{\psi}e^{i\epsilon\gamma_5})\}$$
$$= \int D\chi D\bar{\chi} \exp\left[-a(\bar{\psi}_n - \bar{\chi}_n)e^{2i\epsilon\gamma_5}(\psi_n - \chi_n) - S_I(\chi, \bar{\chi})\right] \quad (12.4.1)$$

Suppose $S_I(\chi, \bar{\chi})$ is a quadratic function of $\chi, \bar{\chi}$; this in turn implies that $S(\psi, \bar{\psi})$ is a quadratic function of $\psi, \bar{\psi}$ given by

$$S(\psi, \bar{\psi}) = \bar{\psi}\mathcal{D}^{-1}\psi \equiv \bar{\psi}_m \mathcal{D}_{mn}^{-1}\psi_n \ ; \ \mathcal{D}_{mn}^{-1} = -\mathcal{D}_{nm}^{-1} \quad (12.4.2)$$

and

$$S(e^{i\epsilon\gamma_5}\psi, \bar{\psi}e^{i\epsilon\gamma_5}) = S(\psi, \bar{\psi}) + i\epsilon\bar{\psi}\{\mathcal{D}^{-1}, \gamma_5\}\psi \quad (12.4.3)$$

Note from above that $\mathcal{D}^{-1}$ breaks chiral symmetry since $\{\mathcal{D}^{-1}, \gamma_5\} \neq 0$. The right hand side of Eq. 12.4.1, using fermion differentiation, yields

$$\int D\chi D\bar{\chi} \exp\left[-a(\bar{\psi}_n - \bar{\chi}_n)(1 + 2i\epsilon\gamma_5)(\psi_n - \chi_n) - S_I(\chi, \bar{\chi})\right]$$
$$= \int D\chi D\bar{\chi}\left(1 - 2ia\epsilon(\bar{\psi}_n - \bar{\chi}_n)\gamma_5(\psi_n - \chi_n)\right) \exp\left[-a(\bar{\psi}_n - \bar{\chi}_n)(\psi_n - \chi_n) - S_I(\chi, \bar{\chi})\right]$$
$$= \left(1 + 2\frac{i}{a}\epsilon\frac{\partial}{\partial\psi}\gamma_5\frac{\partial}{\partial\bar{\psi}}\right)e^{-S(\psi, \bar{\psi})} \quad (12.4.4)$$

where the final result has been obtained using $\text{Tr}(\gamma_5) = 0$. Furthermore, from Eq. 12.4.2

$$\frac{\partial}{\partial\psi}\gamma_5\frac{\partial}{\partial\bar{\psi}}e^{-S(\psi, \bar{\psi})} = -\bar{\psi}\mathcal{D}^{-1}\gamma_5\mathcal{D}^{-1}\psi e^{-S(\psi, \bar{\psi})} \quad (12.4.5)$$

The left hand side of Eq. 12.4.1 yields, using Eq. 12.4.3

$$e^{-S(e^{i\epsilon\gamma_5}\psi, \bar{\psi}e^{i\epsilon\gamma_5})} = \left(1 - i\epsilon\bar{\psi}\{\mathcal{D}^{-1}, \gamma_5\}\psi\right)e^{-S(\psi, \bar{\psi})} \quad (12.4.6)$$

Hence, from Eqs. 12.4.5 and 12.4.6

$$\frac{2}{a}\bar{\psi}\mathcal{D}^{-1}\gamma_5\mathcal{D}^{-1}\psi = \bar{\psi}\{\mathcal{D}^{-1}, \gamma_5\}\psi$$

and which yields the final result

$$\mathcal{D}\gamma_5 + \gamma_5 \mathcal{D} = \frac{2}{a}\gamma_5 \quad : \quad \text{Ginsparg-Wilson Relation}$$

The Ginsparg-Wilson result provides a criterion for chiral lattice fermions that apparently break chiral symmetry – but which are guaranteed to yield the correct chiral symmetric continuum limit without the need for any fine tuning. Although it would seem that Ginsparg-Wilson criterion is difficult to obtain, Hasenfratz et al. (2001) have obtained classically perfect lattice gauge theories that obey the Ginsparg-Wilson criterion to a very high accuracy. Fermions obeying the Ginsparg-Wilson criterion, in principle, have *non-local* couplings in spacetime.

Luscher (1999) further developed the Ginsparg-Wilson result by defining a modified chiral symmetry transformation for the lattice – which depends on the lattice gauge field – and that reduces to the continuum one for zero lattice spacing. Based on the modified definition of chiral transformations on a lattice, Luscher (2000) obtained a ground-breaking result in formulating an exact nonperturbative theory of lattice chiral gauge theories.

12.5 Dirac Field: Boundary Conditions

To analyze the Dirac field's state space, and in particular, to calculate the evolution kernel, the boundary conditions for the Dirac lattice action need to be ascertained, and hence the action for finite (Euclidean) lattice time is analyzed. Consider a finite open lattice of N sites in the time direction and a periodic lattice of L^3 sites for the spatial dimensions. We are essentially concerned with the coupling in the time direction. Hence, using the periodicity of the spatial lattice, define the Fourier transform in the spatial coordinates by the following

$$\sum_{\mathbf{p}} e^{i\mathbf{p}\cdot\mathbf{n}} = \prod_{i=1}^{3}\Big[\frac{1}{L}\sum_{p_i=0}^{L-1} e^{2\pi i p_i n_i/L}\Big], \quad \mathbf{p} = (p_1, p_2, p_3) \tag{12.5.1}$$

$$\psi_{(\mathbf{n},n_0)} = \sum_{\mathbf{p}} e^{i\mathbf{p}\cdot\mathbf{n}}\psi_{\mathbf{p}n_0} \quad ; \quad \bar{\psi}_{(\mathbf{n},n_0)} = \sum_{\mathbf{p}} e^{-i\mathbf{p}\cdot\mathbf{n}}\bar{\psi}_{\mathbf{p}n_0} \tag{12.5.2}$$

This gives for the Lagrangian (dropping the subscript 0 on the time coordinates so that $n_0 \to n$)

$$\mathcal{L} = -\bar{\psi}_{\mathbf{p}n}L\psi_{\mathbf{p}n} + K\big\{\bar{\psi}_{\mathbf{p}n}(1-\gamma_0)\psi_{\mathbf{p}n+1} + \bar{\psi}_{\mathbf{p}n+1}(1+\gamma_0)\psi_{\mathbf{p}n}\big\} \tag{12.5.3}$$

Using the scalar product only in the spatial indices yields

$$L = \Big(1 - 2K\sum_{i=1}^{3}\cos p_i\Big) + i2K\sum_{i=1}^{3}\gamma_i \sin p_i$$

$$= \alpha + i\gamma \cdot \beta \tag{12.5.4}$$

where, from Eq. 12.3.2

$$i\gamma \cdot \beta = i2K\sum_{i=1}^{3}\gamma_i \sin p_i = 2K\begin{bmatrix} 0 & \sigma \\ -\sigma & 0 \end{bmatrix} \quad : \quad \sigma = \sum_{i=1}^{3}\sin p_i \sigma_i \tag{12.5.5}$$

Define the upper and lower two-component spinors, denoted by subscript u and ℓ, respectively, by the following

$$\psi_{\mathbf{p}n} = \begin{pmatrix} \psi_{\mathbf{p}nu} \\ \psi_{\mathbf{p}n\ell} \end{pmatrix}, \quad \bar{\psi}_{\mathbf{p}n} = (\bar{\psi}_{\mathbf{p}nu}, \bar{\psi}_{\mathbf{p}n\ell}) \tag{12.5.6}$$

Note that from the definition of γ_0 in (2.2), $\frac{1}{2}(1 \pm \gamma_0)$ are the protection operators for the upper and lower components.

The fermionic degrees of freedom, in the momentum representation, completely factorize, with no coupling between different momenta; henceforth the momentum index is completely suppressed and restored only when necessary. Hence, the coupling in the time direction of the Dirac lattice Lagrangian is given as follows

$$\mathcal{L} = -\bar{\psi}_n L \psi_n + 2K\bar{\psi}_{n\ell}\psi_{n+1\ell} + 2K\bar{\psi}_{n+1u}\psi_{nu} \tag{12.5.7}$$

The equation above shows that the Lagrangian propagates degrees of freedom $(\bar{\psi}_{n\ell}, \psi_{nu})$ at time n to $(\bar{\psi}_{n+1u}, \psi_{n+1\ell})$ at time $n+1$.

It follows from Eq. 12.5.7 that the boundary fermionic coordinates values at time $n = 0$ is $(\bar{\psi}_{0\ell}, \psi_{0u})$, and the other remaining fermionic coordinates, namely $(\bar{\psi}_{0u}, \psi_{0\ell})$, do not couple to the next time instant at $n = 1$ and hence are decoupled from the lattice Dirac field. Similarly, the final boundary fermionic coordinates at final time $n = N$ are $(\bar{\psi}_{Nu}, \psi_{N\ell})$ and the other remaining fermionic coordinates at $n + N$, namely $(\bar{\psi}_{N\ell}, \psi_{Nu})$ do not couple to earlier times and decouple from the lattice Dirac action.[3]

Hence choose as the initial fermion field configuration the coordinate eigenstate tensor product defined on the spatial lattice

$$\prod_{\mathbf{n}} \otimes |\bar{\psi}_{(\mathbf{n},0,\ell)}, \psi_{(\mathbf{n},0,u)}\rangle \equiv |\bar{\psi}_\ell, \psi_u\rangle : \text{ initial coordinate state vector} \tag{12.5.8}$$

Similarly, choose for the final fermion field configuration the conjugate coordinate eigenstate tensor product

$$\prod_{\mathbf{n}} \otimes \langle \bar{\psi}_{(\mathbf{n},N,u)}, \psi_{(\mathbf{n},N,\ell)} | \equiv \langle \bar{\psi}_u, \psi_\ell | : \text{ final dual coordinate state vector} \tag{12.5.9}$$

The boundary values of the Dirac field are shown in Figure 12.1; for every point in space, the lattice the boundary values are as given in the Figure 12.1.

12.6   Dirac Fermionic State Space

The results of fermion calculus discussed in Chapter 6 are generalized to the case of the Dirac field. Eqs. 12.5.8 and 12.5.9 show that the coordinate basis for the state of the Dirac field factorizes into a tensor product over the lattice sites. Hence, one can analyze each lattice site one by one and compose the field's basis states using a tensor product and for this reason the index for the lattice site is suppressed.

---

[3] A derivation of the boundary conditions for the Dirac field in the continuum time theory is given in Section 12.2.

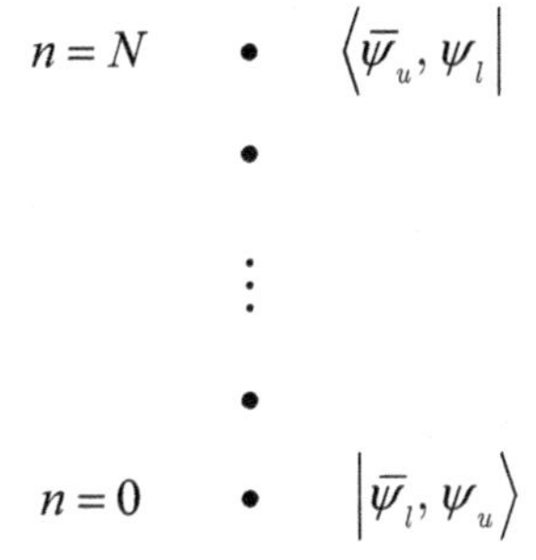

$$n = N \qquad \bullet \qquad \left\langle \bar{\psi}_u, \psi_l \right|$$

$$\bullet$$

$$\vdots$$

$$\bullet$$

$$n = 0 \qquad \bullet \qquad \left| \bar{\psi}_l, \psi_u \right\rangle$$

Fig. 12.1   The boundary values of the Dirac field; only the time lattice is displayed.

The Dirac field's coordinate as well as its dual coordinate for a single lattice site are given by anticommuting $\bar{\psi}$ and ψ four-component fermionic variables. The Dirac field's degrees of freedom

$$\psi = \begin{bmatrix} \psi_u \\ \psi_\ell \end{bmatrix} \quad ; \quad \bar{\psi} = [\bar{\psi}_u, \bar{\psi}_\ell]$$

are four-component Dirac (complex) spinors.

To define operators and state vectors, choose $\bar{\psi}_\ell, \psi_u$ as the coordinate variables and $\bar{\psi}_u, \psi_\ell$ as the conjugate variables. $\bar{\psi}_\ell, \psi_u$ is the analogue of the x coordinate and $\bar{\psi}_u, \psi_\ell$ that of the p coordinate of single-particle quantum mechanics.

The coordinate operators $\hat{\bar{\psi}}_\ell, \hat{\psi}_u$ and their eigenstates $|\bar{\psi}_\ell, \psi_u\rangle$ as well as the conjugate equations are defined as

$$\begin{bmatrix} \hat{\bar{\psi}}_\ell \\ \hat{\psi}_u \end{bmatrix} |\bar{\psi}_l, \psi_u\rangle = \begin{bmatrix} \bar{\psi}_l \\ \psi_u \end{bmatrix} |\bar{\psi}_l, \psi_u\rangle \qquad (12.6.1)$$

$$\langle \bar{\psi}_u, \psi_l| \begin{bmatrix} \hat{\bar{\psi}}_u, \ \hat{\psi}_\ell \end{bmatrix} = \langle \bar{\psi}_u, \psi_\ell| [\bar{\psi}_u, \ \psi_\ell] \qquad (12.6.2)$$

$\hat{\bar{\psi}}_\ell$ creates spin-up and spin-down particles when applied on the vacuum state and $\hat{\psi}_u$ does the same for antiparticles.

The coordinate eigenstates are $|\bar{\psi}_\ell, \psi_u\rangle$ and the dual coordinate eigenstates are $\langle \bar{\psi}_u, \psi_\ell|$. The two components of doublets are labeled by subscripts 1 and 2. Let $|f\rangle$ be an element of Hilbert state space. Then, due to the anticommuting property of the fermionic variables $\bar{\psi}_\mu, \psi_\ell$, the general coordinate Taylor expansion for the state vector is given by

$$f(\bar{\psi}_u, \psi_\ell) = \langle \bar{\psi}_u, \psi_\ell | f \rangle$$

$$= f_0 + \sum_{i=1}^{2} (f_{1i} + f_{2i}\psi_{\ell 1}\psi_{\ell 2})\bar{\psi}_{ui} + f_3\bar{\psi}_{u1}\bar{\psi}_{u2} + f_6\psi_{\ell 1}\psi_{\ell 2}$$

$$+ \sum_{i=1}^{2} (f_{4i} + f_{5i}\bar{\psi}_{u1}\bar{\psi}_{u2})\psi_{\ell i} + \sum_{ij=1}^{2} f_{7ij}\bar{\psi}_{ui}\psi_{\ell j} + f_8\bar{\psi}_{u1}\bar{\psi}_{u2}\psi_{\ell 1}\psi_{\ell 2}$$

The state vector is specified by sixteen complex coefficients $f_{\alpha;ij}$ and corresponds to the sixteen fermion states allowed for a given momentum (or at a given position) by the Pauli exclusion principle.

Conjugation is defined in Section 6.5 by the following operations:

- $\psi \to \bar{\psi}\gamma_0,\ \bar{\psi} \to \gamma_0\psi$.
- the order of the fermion variables is reversed.
- all the coefficients are complex conjugated.

Let $|f\rangle$ be a state vector and $\langle g|$ be the conjugate state vector. Then, denoting conjugation by †, yields

$$f(\bar{\psi}_u, \psi_\ell) = \langle\bar{\psi}_u, \psi_\ell|f\rangle \ ;\ g^\dagger(\bar{\psi}_\ell, \psi_u) = \langle g|\bar{\psi}_\ell, \psi_u\rangle$$

Recall conjugation is defined by

$$\psi \to \bar{\psi}\gamma_0 \ ,\ \bar{\psi} \to \gamma_0\psi$$

In particular

$$\psi_u \to \bar{\psi}_u \ ;\ \psi_\ell \to -\bar{\psi}_\ell$$

The conjugate state vector is given by

$$f^\dagger(\bar{\psi}_\ell, \psi_u) = \langle f|\bar{\psi}_\ell, \psi_u\rangle = f^*(\psi_u, -\bar{\psi}_\ell) \tag{12.6.3}$$

and with the order of the fermion variables being reversed in $f^*(\psi_u, -\bar{\psi}_\ell)$.

For a general choice of the metric, the scalar product is

$$\langle f|g\rangle = \sum_{\alpha\beta} f_\alpha^* t_{\alpha\beta} g_\beta \tag{12.6.4}$$

where α, β are the general index. Note that the metric $t_{\alpha\beta}$ is chosen to be positive definite. The scalar product can be represented by means of fermionic integration as follows

$$\langle f|g\rangle = \int d\bar{\psi}d\psi\, f^\dagger(\bar{\psi}_\ell, \psi_u)T(\bar{\psi}, \psi)g(\bar{\psi}_u, \psi_\ell) \tag{12.6.5}$$

where the metric $T(\bar{\psi}, \psi)$ is determined by the metric $t_{\alpha\beta}$.

The scalar product of the state vector with a conjugate state vector is positive definite, namely

(i) $\langle f|f\rangle \geq 0$ and $\langle f|f\rangle = 0$ iff $|f\rangle = 0$;

(ii) $\langle f|\hat{G}^\dagger|g\rangle = \langle g|\hat{G}|f\rangle^*$

The resolution of the identity operator acting on the fermionic Hilbert space is given by

$$\mathbb{I} = \int d\bar{\psi}d\psi\, |\bar{\psi}_\ell, \psi_u\rangle T(\bar{\psi}, \psi)\langle\bar{\psi}_u, \psi_\ell| \tag{12.6.6}$$

For the diagonal metric (which appears in the continuum case)

$$T(\bar{\psi}, \psi) = \exp(-\bar{\psi}\psi) = \exp\left\{ -\sum_{s=1}^{2}(\bar{\psi}_{us}\psi_{us} + \bar{\psi}_{ls}\psi_{ls}) \right\} \tag{12.6.7}$$

The metric in Eq. 12.6.7 implies $t_{\alpha\beta} = \delta_{\alpha\beta}$ in Eq. 12.6.4. The completeness equation for metric in Eq. 12.6.7 is given by

$$\mathbb{I} = \int \prod_{\alpha=1}^{4} d\bar{\psi}_\alpha d\psi_\alpha |\bar{\psi}_\ell, \psi_u) e^{-\bar{\psi}\psi} (\bar{\psi}_u, \psi_\ell| \qquad (12.6.8)$$

The metric in Eq. 12.6.7, using the completeness equation as a consistency condition, yields

$$\langle\bar{\psi}_u, \psi_\ell|\bar{\psi}_\ell, \psi_u\rangle = \exp\left\{ \sum_s (\bar{\psi}_{us}\psi_{us} + \bar{\psi}_{\ell s}\psi_{\ell s}) \right\} = \exp(\bar{\psi}\psi) \qquad (12.6.9)$$

12.7 Hilbert Space Metric and Transfer Matrix

The state space of the (four-dimensional) Dirac field is fixed by the kinetic term in the Lagrangian. The lattice (Euclidean) Dirac Lagrangian, obtained by discretizing the continuum case is given by Eq. 12.3.12

$$\mathcal{L} = -\bar{\psi}_n\psi_n + K\sum_{u=0}^{3}\{\bar{\psi}_n(1-\gamma_\mu)\psi_{n+\hat{u}} + \bar{\psi}_{n+\hat{u}}(1+\gamma_\mu)\psi_n\}$$

$$= -\bar{\psi}_n\psi_n + 2K(\bar{\psi}_{n\ell}\psi_{(n+\hat{o})\ell} + \bar{\psi}_{(n+\hat{o})u}\psi_{nu})$$

$$+ K\sum_{i=1}^{3}\{\bar{\psi}_n(1-\gamma_i)\psi_{n+\hat{i}} + \bar{\psi}_{n+\hat{i}}(1+\gamma_i)\psi_n\}$$

Hence $(\bar{\psi}_{n\ell}, \psi_{nu})$ propagates to $(\bar{\psi}_{(n+\hat{o})u}, \psi_{(n+\hat{o})\ell})$ yielding the coordinates for the Dirac Hilbert space as

$$\langle\bar{\psi}_{(n+\hat{o})u}, \psi_{(n+\hat{o})\ell}|e^{-\epsilon\hat{H}}|\bar{\psi}_{n\ell}, \psi_{nu}\rangle \qquad (12.7.1)$$

Recall, from Eqs. 12.5.8 and 12.5.9, the Dirac field's coordinate eigenstates are given by

$$\langle\bar{\psi}_u, \psi_\ell| = \bigotimes_{\vec{x}}\langle\bar{\psi}_{\vec{x}u}, \psi_{\vec{x}\ell}| \quad : \text{Coordinates of the Dirac Hilbert space}$$

$$|\bar{\psi}_\ell, \psi_u) = \bigotimes_{\vec{x}}|\bar{\psi}_{\vec{x}\ell}, \psi_{\vec{x}u}) \quad : \text{Coordinates of the dual Dirac Hilbert space}$$

A state functional of the Dirac field $|\Phi\rangle$ has the coordinate representation given by

$$\langle\bar{\psi}_u, \psi_\ell|\Phi\rangle = \Phi\left[\bar{\psi}_u, \psi_\ell\right] \qquad (12.7.2)$$

Consider a d-dimensional Euclidean spacetime lattice with lattice spacing a. Let the $(d-1)$-dimensional spatial lattice be of infinite size and the time lattice be of finite size N. The finite time Wilson lattice action is given by Eq. 12.3.12

$$S_N = -\sum_{n_0=1}^{N}\sum_{\vec{n}}\bar{\psi}_n\psi_n$$

$$+K\sum_{n_0=0}^{N}\sum_{\vec{n}}\sum_{\mu=0}^{3}\{\bar{\psi}_n(1-\gamma_\mu)\psi_{n+\mu} + \bar{\psi}_{n+\mu}(1+\gamma_\mu)\psi_n\} \qquad (12.7.3)$$

Note that in the first term in the action above, the sum over the time lattice is from $1 \leq n_0 \leq N$; in the second term of the lattice action, the sum is over $0 \leq n_0 \leq N$ and the boundary conditions at $n_0 = 0$ and $n_0 = N+1$ are correctly incorporated.

The lattice Hamiltonian H is given by the path integral as follows:

$$\lim_{N \to \infty} \langle \bar\psi_u, \psi_\ell | \exp\{-a(N+1)H\} | \bar\psi_\ell, \psi_u \rangle = \prod_{n_0=1}^{N} \prod_{\vec{n}} \int d\bar\psi_n d\psi_n \exp(S_N) \quad (12.7.4)$$

Consider the case for $N = 1$; from Eq. 12.5.7, the action is given by

$$S_1 = -\sum_{n=1}^{1} \bar\psi_n L \psi_n + 2K \sum_{n=0}^{1} \{\bar\psi_{n\ell}\psi_{n+1\ell} + \bar\psi_{n+1u}\psi_{nu}\}$$
$$= -\bar\zeta L \zeta + 2K\bar\psi_\ell \zeta_\ell + 2K\bar\zeta_u \psi_u + 2K\bar\zeta_\ell \psi_\ell + 2K\bar\psi_u \zeta_u \quad (12.7.5)$$

where

$$\psi_1 = \zeta \; ; \; \bar\psi_1 = \bar\zeta \; ; \; \bar\psi_{0\ell} = \bar\psi_\ell \; ; \; \psi_{0u} = \psi_u \; ; \; \bar\psi_{2u} = \bar\psi_u \; ; \; \psi_{2\ell} = \psi_\ell$$

Eqs. 12.5.4 and 12.5.5 yield

$$-\bar\zeta L \zeta = -\alpha \bar\zeta \zeta - 2K\bar\zeta_u \sigma \zeta_\ell + 2K\bar\zeta_\ell \sigma \zeta_u$$

Hence

$$e^{S_1} = \exp\left\{ 2K\bar\psi_u \zeta_u + 2K\bar\zeta_\ell \psi_\ell + 2K\bar\zeta_\ell \sigma \zeta_u \right\} \exp\left\{ -\alpha \bar\zeta \zeta \right\}$$
$$\times \exp\left\{ 2K\bar\psi_\ell \zeta_\ell + 2K\bar\zeta_\ell \psi_\ell - 2K\bar\zeta_u \sigma \zeta_\ell \right\} \quad (12.7.6)$$

and

$$\langle \bar\psi_u, \psi_\ell | \exp(-2aH) | \bar\psi_\ell, \psi_u \rangle = \int d\bar\zeta d\zeta \exp\{S_1(\bar\psi, \psi; \bar\zeta, \zeta)\} \quad (12.7.7)$$

Consider the following transfer matrix of the Dirac Hamiltonian

$$\langle \bar\psi_u, \psi_\ell | e^{-aH} | \bar\psi_\ell, \psi_u \rangle = \exp\left\{ 2K\sum_{\vec{n}} \bar\psi_{\vec{n}}\psi_{\vec{n}} - K\sum_{\vec{n}i}(\bar\psi_{\vec{n}}\gamma_i\psi_{\vec{n}+i} - \bar\psi_{\vec{n}+i}\gamma_i\psi_{\vec{n}}) \right\} \quad (12.7.8)$$

where $\vec{n}$ is a $(d-1)$-dimensional spatial lattice point with $i = 1, ...(d-1)$. Fourier transforming the Dirac field to momentum space yields the transfer matrix

$$\langle \bar\psi_u, \psi_\ell | e^{-aH} | \bar\psi_\ell, \psi_u \rangle = \exp\left\{ 2K\sum_{\vec{p}} \bar\psi_{\vec{p}}\psi_{\vec{p}} - 2iK\sum_{\vec{p}i} \bar\psi_{\vec{p}}\gamma_i \sin(p_i)\psi_{\vec{p}} \right\} \quad (12.7.9)$$

Recall from Eq. 12.3.2 that the gamma matrices have the following representation

$$\gamma_i = i \begin{bmatrix} 0 & -\sigma_i \\ \sigma_i & 0 \end{bmatrix} \quad \Rightarrow \quad -2iK\sum_i \gamma_i \sin(p_i) = 2K \begin{bmatrix} 0 & -\sigma \\ \sigma & 0 \end{bmatrix} \; : \; \sigma = \sum_i \sigma_i \sin(p_i)$$

For notational simplicity suppress the summation over the momentum index.

$$\psi_{\vec{p}} = \psi \; ; \; \bar\psi_{\vec{p}} = \bar\psi$$

The Dirac spinor's two component representation yields

$$\psi_{\vec{p}} = \begin{bmatrix} \psi_u \\ \psi_\ell \end{bmatrix} \quad ; \quad \bar{\psi}_{\vec{p}} = [\bar{\psi}_u, \bar{\psi}_\ell]$$

and hence

$$2K[\bar{\psi}_u, \bar{\psi}_\ell] \begin{bmatrix} 0 & -\sigma \\ \sigma & 0 \end{bmatrix} \begin{bmatrix} \psi_u \\ \psi_\ell \end{bmatrix} = 2K(-\bar{\psi}_u \sigma \psi_\ell + \bar{\psi}_\ell \sigma \psi_u)$$

The Fourier representation, using the notation of Eq. 12.5.5, yields

$$\langle \bar{\psi}_u, \psi_\ell | e^{-aH} | \bar{\psi}_\ell, \psi_u \rangle = \exp \left\{ 2K\bar{\psi}\psi + 2K\bar{\psi}_\ell \sigma \psi_u - 2K\bar{\psi}_u \sigma \psi_\ell \right\} \qquad (12.7.10)$$

Note the nearest neighbor term, containing the σ matrix, couples only the ket coordinates $\bar{\psi}_\ell, \psi_u$ and the bra dual coordinates $\bar{\psi}_u \sigma \psi_\ell$, with no coupling of the dual bra and ket coordinates. This is reflected in the fact that the boundary terms will not appear in potential terms in the action, as in the example discussed in Noteworthy 12.1. From Eqs. 12.7.5, 12.7.6, 12.7.7 and 12.7.10

$$\langle \bar{\psi}_u, \psi_\ell | \exp(-2aH) | \bar{\psi}_\ell, \psi_u \rangle = \int d\bar{\zeta} d\zeta \exp\{S_1(\bar{\psi}, \psi; \bar{\zeta}, \zeta)\} \qquad (12.7.11)$$

$$= \exp \left\{ 2K\bar{\psi}_u \sigma \psi_\ell - 2K\bar{\psi}_\ell \sigma \psi_u \right\}$$

$$\times \int d\bar{\zeta} d\zeta \langle \bar{\psi}_u, \psi_\ell | \exp(-aH) | \bar{\zeta}_\ell, \zeta_u \rangle T(\bar{\zeta}, \zeta) \langle \bar{\zeta}_u, \zeta_\ell | \exp(-aH) | \bar{\psi}_\ell, \psi_u \rangle \quad (12.7.12)$$

The Hamiltonian given by Eq. 12.7.10 is obtained by comparing Eq. 12.7.12 with Eq. 12.7.6; the comparison is sufficient to obtain the Hamiltonian.

The following terms in Eq. 12.7.6

$$-2K\bar{\zeta}_u \sigma \zeta_\ell + 2K\bar{\zeta}_\ell \sigma \zeta_u$$

are the value of the Hamiltonian for the bra and ket states $|\bar{\zeta}_\ell, \zeta_u\rangle \langle \bar{\zeta}_u, \zeta_\ell|$ that are summed over in the path integral. The terms

$$2K\bar{\psi}_\ell \sigma \psi_u - 2K\bar{\psi}_u \sigma \psi_\ell$$

are in the expression for the Hamiltonian given in Eq. 12.7.10 but do not appear in S_1, and hence need to be introduced **by hand** in Eq. 12.7.12 for it to equal Eq. 12.7.11. These terms do not appear inside the path integral in Eq. 12.7.12 as these are boundary terms and do not appear in the potential terms of the action. See discussion in Noteworthy 12.1.

It is the remarkable properties of the representation of the gamma matrices chosen for the lattice Dirac field that the gamma matrix γ_0 and γ_i ; $i = 1, 2, 3$ together 'conspire' to yield a well-defined Hamiltonian. It was crucial that all γ_i ; $i = 1, 2, 3$ be off-diagonal for the coordinates and the dual coordinates to yield the factorization of the terms in the Hamiltonian and yield the boundary terms that made the equality of Eqs. 12.7.11 and 12.7.12 possible – and which in turn yields the Hamiltonian.

In fact, if one computes $\langle \bar{\psi}_u, \psi_\ell | \exp(-aNH) | \bar{\psi}_\ell, \psi_u \rangle$, the boundary terms that do not appear in the potential terms of the action go to zero as $N \to \infty$, as is the case for the case discussed in Noteworthy 12.1.

Noteworthy 12.1: Boundary Terms in the Action

Consider a path integral for $x(t)$ with action given by

$$S = -\int_0^T dt \left(\frac{1}{2} m^2 \left(\frac{dx}{dt} \right)^2 + V(x) \right)$$

and with boundary conditions

$$x(0) = x \quad ; \quad x(T) = x'$$

On discretizing the action one obtains, for $T = N\epsilon$

$$S = -\frac{1}{2} m^2 \epsilon \sum_{n=0}^{N-1} (x_{n+1} - x_n)^2 - \epsilon \sum_{n=1}^{N-1} V(x_n)$$

with boundary conditions

$$x_0 = x \quad ; \quad x_N = x'$$

Note the boundary values occur only in the kinetic term and the potential term does not have terms $\epsilon V(x_0), \epsilon V(x_N)$ that depend on x_0, x_N; these missing boundary terms go to zero as $\epsilon \to 0$ and, like the case for the lattice Dirac Hamiltonian, need to be introduced by hand for a system with two lattice steps in time. Similarly, in deriving the Hamiltonian for the Dirac field, the boundary values do not appear in the potential of the action, as the potential has no coupling with the boundary values; these boundary terms also become zero in the limit of zero lattice spacing.

Hence, we conclude that the Hamiltonian given in Eq. 12.7.9 is the correct expression. The metric on Hilbert space is determined by the terms in Eq. 12.7.6 that do not couple to the boundary values and yield

$$T(\bar{\zeta}, \zeta) = \exp \left\{ - \alpha \bar{\zeta} \zeta \right\} \tag{12.7.13}$$

Note that the Wilson metric for the Hilbert space of lattice Dirac field, given in Eq. 12.7.15 depends on the action through the parameter K. In general, the action (Lagrangian) contains *both* the Hamiltonian and the (kinematical) structure of state space. Recall the completeness equation is given in Eq. 12.6.6 for the general lattice metric $T(\bar{\psi}, \psi)$

$$\mathbb{I} = \int d\psi d\bar{\psi} | \bar{\psi}_\ell, \psi_u \rangle T(\bar{\psi}, \psi) \langle \bar{\psi}_u, \psi_\ell |$$

From Eq. 12.7.13, writing out in full the metric gives

$$T(\bar{\psi}, \psi) = \exp \left\{ - \alpha \bar{\psi} \psi \right\} = \exp \left\{ - \sum_{\vec{p}} \left(1 - 2K \sum_i \cos(p_i) \right) \bar{\psi}_{\vec{p}} \psi_{\vec{p}} \right\} \tag{12.7.14}$$

Note that the metric given above in Eq. 12.7.14 is identical to the metric studied in Eq. 6.3.3, since the Wilson metric has factorized in momentum space.

Fourier transforming the Dirac field to the space lattice yields, from Eq. 12.7.14

$$T(\bar{\psi}, \psi) = \exp\left\{ -\sum_{\vec{n}} \bar{\psi}_{\vec{n}} \psi_{\vec{n}} + K \sum_{\vec{n}_i} \left(\bar{\psi}_{\vec{n}} \psi_{\vec{n}+i} + \bar{\psi}_{\vec{n}+i} \psi_{\vec{n}} \right) \right\} \quad (12.7.15)$$

$$= \exp\left\{ -\sum_{\vec{m}\vec{n}} \bar{\psi}_{\vec{m}} M(\vec{m}, \vec{n}) \psi_{\vec{n}} \right\}$$

The inner product of the basis states, similar to the result in Eq. 6.3.5, is given by

$$\langle \bar{\psi}_u, \psi_\ell | \bar{\psi}_\ell, \psi_u \rangle = \frac{1}{\alpha} \exp\left\{ +\alpha \bar{\psi} \psi \right\}$$

$$= \frac{1}{\det M} \exp\left\{ \sum_{\vec{n}} \bar{\psi}_{\vec{n}} \psi_{\vec{n}} - K \sum_{\vec{n}_i} \left(\bar{\psi}_{\vec{n}} \psi_{\vec{n}+i} + \bar{\psi}_{\vec{n}+i} \psi_{\vec{n}} \right) \right\} \quad (12.7.16)$$

where

$$\det M = \prod_{\vec{p}} \alpha(\vec{p})$$

For the four-dimensional continuum limit, the Hilbert space metric is given by the limit of $a \to 0$ for $d = 4$ and yields, since $2K = 1/(4 + ma)$, the following

$$T(\bar{\psi}, \psi) = \exp\left\{ -\left(\frac{1}{2K} - 3 \right) a^3 \sum_n \bar{\psi}(x) \psi(x) \right\} \to \exp\left(-\int d^3 x \, \bar{\psi}(x) \psi(x) \right)$$

Hence, for $d = 4$ the continuum metric is given by

$$T(\bar{\psi}, \psi) = \exp\left\{ -\int d^3 x \, \bar{\psi}_{\vec{x}} \psi_{\vec{x}} \right\} = \exp\left\{ -\int \frac{d^3 p}{(2\pi)^3} \bar{\psi}_{\vec{p}} \psi_{\vec{p}} \right\} \quad (12.7.17)$$

and the inner product of the basis states with the dual state is given by

$$\langle \bar{\psi}_u, \psi_\ell | \bar{\psi}_\ell, \psi_u \rangle = \exp\left\{ \int d^3 x \, \bar{\psi}_{\vec{x}} \psi_{\vec{x}} \right\} = \exp\left\{ \int \frac{d^3 p}{(2\pi)^3} \bar{\psi}_{\vec{p}} \psi_{\vec{p}} \right\} \quad (12.7.18)$$

12.8 Dirac Lattice Hamiltonian

The transfer matrix is given by the matrix elements of $\exp\{-aH\}$.

The Wilson lattice metric given in Eq. 12.7.14 yields the following coordinate inner product

$$\langle \bar{\psi}_u, \psi_\ell | \bar{\psi}_\ell, \psi_u \rangle = \exp\left\{ \sum_n \bar{\psi}_n \psi_n - K \sum_{ni} (\bar{\psi}_n \psi_{n+i} + \bar{\psi}_{n+i} \psi_n) \right\} \quad (12.8.1)$$

where a normalization constant has been dropped.

The (lattice) Hamiltonian H is defined in terms of the transfer matrix by the following

$$\langle \bar{\psi}_u, \psi_\ell | e^{-aH} | \bar{\psi}_\ell, \psi_u \rangle = e^{-aH(\bar{\psi}, \psi)} \langle \bar{\psi}_u, \psi_\ell | \bar{\psi}_\ell, \psi_u \rangle \quad (12.8.2)$$

The lattice equation for the transfer matrix e^{-aH} is given by Eq. 12.7.9

$$\langle \bar{\psi}_u, \psi_\ell | e^{-aH} | \bar{\psi}_\ell, \psi_u \rangle = \exp \left\{ 2K \sum_n \bar{\psi}_n \psi_n - K \sum_{ni} (\bar{\psi}_n \gamma_i \psi_{n+i} - \bar{\psi}_{n+i} \gamma_i \psi_n) \right\}$$

$$(12.8.3)$$

The left hand side of Eq. 12.7.9, using Eq. 6.8.4, yields

$$\lim_{a \to 0} \langle \bar{\psi}_u, \psi_\ell | e^{-aH} | \bar{\psi}_\ell, \psi_u \rangle \simeq \langle \bar{\psi}_u, \psi_\ell | \bar{\psi}_\ell, \psi_u \rangle \left\{ 1 - aH(\bar{\psi}, \psi) + O(a^2) \right\}$$

$$\simeq \frac{1}{\det M} \exp \left\{ \sum_n \bar{\psi}_n, \psi_n - K \sum_{ni} (\bar{\psi}_n \psi_{n+i} + \bar{\psi}_{n+i} \psi_n) - aH(\bar{\psi}, \psi) \right\} \quad (12.8.4)$$

Hence, from Eqs. 12.8.3 and 12.8.4, the Dirac lattice Hamiltonian is given by

$$H = -\frac{1}{a}(2K - 1) \sum_{\vec{n}} \bar{\psi}_{\vec{n}} \psi_{\vec{n}}$$

$$+ \frac{K}{a} \sum_{\vec{n}i} \left(\bar{\psi}_{\vec{n}}(1 - \gamma_i)\psi_{\vec{n}+i} - \bar{\psi}_{\vec{n}+i}(1 + \gamma_i)\psi_{\vec{n}} \right) + \frac{1}{a} \ln \det M \quad (12.8.5)$$

The results obtained in this Section are further analyzed for the eigenstates of the Dirac field in Section 13.2.

Noteworthy 12.2: Dirac Lagrangian and Metric

The metric on Hilbert space given in Eq. 12.7.15

$$T(\bar{\psi}, \psi) = \exp \left\{ -\sum_{\vec{n}} \bar{\psi}_{\vec{n}} \psi_{\vec{n}} + K \sum_{\vec{n}i} (\bar{\psi}_{\vec{n}} \psi_{\vec{n}+i} + \bar{\psi}_{\vec{n}+i} \psi_{\vec{n}}) \right\}$$

is the result of the hyper-cubic symmetric lattice Lagrangian given in Eq. 12.3.12. An alternative choice for the Dirac Hamiltonian, discussed in Section 14.8, can be obtained by choosing the canonical metric that is given by

$$T_c(\bar{\psi}, \psi) = \exp \left\{ -\sum_{\vec{n}} \bar{\psi}_{\vec{n}} \psi_{\vec{n}} \right\} \quad (12.8.6)$$

The canonical metric given in Eq. 12.8.6 yields an asymmetric action functional. Transforming from the axial gauge back to a manifestly gauge-invariant expression yields the lattice action $(n = (n_0, \vec{n}))$

$$\mathcal{S} = -\sum_n \bar{\psi}_n \psi_n + \frac{1}{2}(1 - m_0\epsilon) \sum_n \left[\bar{\psi}_{n+\hat{0}}(1 + \gamma_0) U_{n0}^\dagger \psi_n \bar{\psi}_n (1 - \gamma_0) U_{n0} \psi_{n+\hat{0}} \right]$$

$$- \epsilon \sum_{nm} \bar{\psi}_n \begin{pmatrix} 0 & h_{nm} \\ h_{nm}^\dagger & 0 \end{pmatrix} \psi_m \delta_{n_0, m_0} + \mathcal{S}_G$$

The pure gauge field action $\mathcal{S}_G$ is not affected by the metric. The lattice gauge theory Hamiltonian, from Eq. 14.8.1, is given by

$$\hat{H}(\bar{\psi}, \psi) = m_0 \sum_n \bar{\psi}_n \psi_n + \sum_{nm} \bar{\psi}_n \begin{pmatrix} 0 & h_{nm} \\ h_{nm}^\dagger & 0 \end{pmatrix} \psi_m + H_G$$

12.9 Lattice Path Integral

The probability amplitude, or the evolution kernel, is the amplitude that the field configuration $|\bar{\psi}_{n\ell}, \psi_{nu}\rangle$ to be in the conjugate configuration $\langle\bar{\psi}_{nu}, \psi_{n\ell}|$ after time T. That is, the evolution kernel $\mathcal{K}$ is defined by

$$\mathcal{K}(\bar{\psi}_u, \psi_\ell; \bar{\psi}_\ell, \psi_u; T) \equiv \langle\bar{\psi}_u, \psi_\ell|e^{-TH}|\bar{\psi}_\ell, \psi_u\rangle \qquad (12.9.1)$$

$\mathcal{K}(\bar{\psi}_u, \psi_\ell; \bar{\psi}_\ell, \psi_u; T)$ can be expressed as a Feynman path integral over the anticommuting fermion field variables, with appropriate boundary conditions on the integration variables. Let $T = Na$, and impose the boundary conditions

$$n = 0; \qquad \bar{\psi}_{0\ell} = \bar{\psi}_\ell, \qquad \psi_{0u} = \psi_u$$
$$n = N; \qquad \bar{\psi}_{Nu} = \bar{\psi}_u, \qquad \psi_{N\ell} = \psi_\ell$$

$\mathcal{K}$ is obtained by integrating over all the fermion field variables from $n = 1$ to $N - 1$, and yields

$$\mathcal{K}(\bar{\psi}_u, \psi_\ell; \bar{\psi}_\ell, \psi_u; T) = \prod_{\mathbf{p}} \prod_{n=1}^{N-1} \int d\bar{\psi}_{\mathbf{p}n} d\psi_{\mathbf{p}n} \exp(\mathcal{S})$$

where (suppressing the momentum summation)

$$\mathcal{S} = -\sum_{n=1}^{N-1} \bar{\psi}_n L \psi_n + 2K \sum_{n=0}^{N-1} \{\bar{\psi}_{n\ell}\psi_{n+1\ell} + \bar{\psi}_{n+1u}\psi_{nu}\}$$

$\mathcal{S}$ is the action for finite time Na; note that the first term in $\mathcal{S}$ has $N - 1$ terms whereas the second term coupling nearest neighbors in time have N terms. The boundary values appear in $\mathcal{S}$ only in the nearest-neighbor time coupling term. Since $\mathcal{S}$ is quadratic, the path integral can be performed for $\mathcal{K}$ by solving the classical field equation, together with the given boundary conditions.

Suppose $\bar{\xi}_n, \xi_n$ satisfy the classical field equation and the given boundary condition, that is

$$\delta\mathcal{S}(\bar{\xi},\xi)/\delta\bar{\psi}_n = 0 \quad ; \quad \delta\mathcal{S}(\bar{\xi},\xi)/\delta\psi_n = 0, \qquad (12.9.2)$$
$$\bar{\xi}_{0N} = \bar{\psi}_l \, , \, \xi_{0u} = \psi_u \, ; \quad \bar{\xi}_{Nu} = \bar{\psi}_u \, , \, \xi_{Nl} = \psi_l \qquad (12.9.3)$$

Define the new integration variables $\bar{\zeta}_n, \zeta_n$ by

$$\psi_n = \zeta_n + \xi_n, \qquad \bar{\psi}_n = \bar{\zeta}_n + \bar{\xi}_n$$
$$d\psi_n = d\zeta_n, \qquad d\bar{\psi}_n = d\bar{\zeta}_n$$

$\bar{\xi}_n, \xi_n$ satisfying the boundary conditions yields

$$\bar{\zeta}_{0l} = \bar{\zeta}_{Nu} = 0 \quad ; \quad \zeta_{0u} = \zeta_{Nl} = 0$$

Hence, the new variables $\zeta_n, \bar{\zeta}_n$ are independent of the boundary values, and form a so-called open chain boundary fermion system. Making the change of variables in $\mathcal{S}$ gives

$$\mathcal{S} = -\sum_{n=1}^{N-1} \bar{\zeta}_n L \zeta_n + K \sum_{n=1}^{N-1} \left(\bar{\zeta}_{nl}\zeta_{n+1l} + \bar{\zeta}_{n+1u}\zeta_{nu}\right) + 2K\left(\bar{\xi}_{0l}\xi_{1l} + \bar{\xi}_{Nu}\xi_{N-1u}\right)$$

The integration variables ζ_n, $\bar{\zeta}_n$ have decoupled from the classical solution ξ_n, $\bar{\xi}_n$ and yield the result

$$\mathcal{K}(\bar{\psi}_u, \psi_l, \bar{\psi}_\ell, \psi_u : T) = \langle \bar{\psi}_u, \psi_l | e^{-aNH} | \bar{\psi}_\ell, \psi_u \rangle = C(N) \exp(\mathfrak{H}(\xi, \bar{\xi})) \qquad (12.9.4)$$

where

$$\mathfrak{H}(\xi, \bar{\xi}) = 2K(\bar{\xi}_{0l}\xi_{1l} + \bar{\xi}_{Nu}\xi_{N-1u}) \qquad (12.9.5)$$

$$C(N) = \prod_{\mathbf{p}} \prod_{n=1}^{N-1} \int d\zeta_{\mathbf{p}n} d\bar{\zeta}_{\mathbf{p}n} \exp \mathcal{S}(\zeta, \bar{\zeta}) \qquad (12.9.6)$$

The normalization constant $C(T = Na)$ is evaluated in Section 12.9.1.

Note from Eq. 12.9.5 that only the boundary values of $\bar{\xi}_n$ appear in the solution. Hence only the time dependence of the classical solution for ξ_n is required; using the field equation given in Eq. 12.9.2 yields

$$\alpha \xi_{nu} = -\beta \cdot \sigma \xi_{nl} + 2K\xi_{n-1u} \qquad (12.9.7)$$

$$\alpha \xi_{nl} = -\beta \cdot \sigma \xi_{nu} + 2K\xi_{n+1l} \qquad (12.9.8)$$

$$\alpha = \left(1 - 2K \sum_i \cos p_i\right) \;\; ; \;\; \beta_i = 2K \sin p_i$$

Make the ansatz

$$\xi_{nu} = e^{\lambda n} A + e^{-\lambda n} B$$

$$\xi_{nl} = e^{\lambda n} C + e^{-\lambda n} D$$

Solving the field equations in Eqs. 12.9.7 and 12.9.8 with the boundary conditions given in Eq. 12.9.3 gives the solution [Baaquie (1983b)]

$$\begin{bmatrix} \xi_{nu} \\ \xi_{nl} \end{bmatrix} = \frac{1}{\Sigma(N)} \begin{bmatrix} \Sigma(N-n) & -\frac{\mathrm{sh}\lambda n}{\alpha} \beta \cdot \sigma \\ -\frac{\mathrm{sh}\lambda(N-n)}{\alpha} \beta \cdot \sigma & \Sigma(n) \end{bmatrix} \begin{bmatrix} \psi_u \\ \psi_l \end{bmatrix} \qquad (12.9.9)$$

with

$$2\mathrm{ch}\lambda = \frac{\alpha}{2K}\left\{1 + \left(\frac{2K}{\alpha}\right) + \frac{\beta^2}{\alpha^2}\right\}$$

$$\Sigma(n) = \mathrm{sh}\lambda n - \frac{2K}{\alpha}\mathrm{sh}\lambda(n-1)$$

There is a similar equation for $\bar{\xi}_n$.

Substituting the above solution for ξ_n in Eq. 12.9.5 gives, restoring the momentum index, the result

$$\mathfrak{H}(\bar{\psi}_{\mathbf{p}u}\psi_{\mathbf{p}l}, \bar{\psi}_{\mathbf{p}l}\psi_{\mathbf{p}u}; N)$$

$$= \sum_{\mathbf{p}} \frac{2K}{\Sigma(N)} (\bar{\psi}_{\mathbf{p}u}\bar{\psi}_{\mathbf{p}l}) \begin{bmatrix} \mathrm{sh}\lambda & -\frac{\mathrm{sh}\lambda(N-1)}{\alpha} \beta \cdot \sigma \\ \frac{\mathrm{sh}\lambda(N-1)}{\alpha} \beta \cdot \sigma & \mathrm{sh}\lambda \end{bmatrix} \begin{bmatrix} \psi_{\mathbf{p}u} \\ \psi_{\mathbf{p}l} \end{bmatrix}$$

$$= \sum_{p} \frac{2K}{\Sigma(N)} \left\{ \frac{\mathrm{sh}\lambda(N-1)}{\alpha} \bar{\psi}_{\mathbf{p}}(-i\beta \cdot \gamma)\psi_{\mathbf{p}} + \mathrm{sh}\lambda \, \bar{\psi}_{\mathbf{p}}\psi_{\mathbf{p}} \right\} \qquad (12.9.10)$$

where

$$\psi_{\mathbf{p}} = \begin{bmatrix} \psi_{\mathbf{p}u} \\ \psi_{\mathbf{p}l} \end{bmatrix}, \qquad \bar{\psi}_{\mathbf{p}} = [\bar{\psi}_{\mathbf{p}u}, \bar{\psi}_{\mathbf{p}l}]$$

The result given in Eq. 12.9.10 is exact, and can be used to study Dirac fermions on a finite lattice. For example, a direct calculation for $\mathfrak{H}$ with $N = 2$ verifies the result.

The result for $d = 2$ can be obtained from Eq. 12.9.10 by replacing $\beta \cdot \sigma$ by $2K \sin p$ and considering the ψ and $\bar{\psi}$ to be doublets.

The Dirac equation is linear and hence the continuum limit does not need renormalization; taking the continuum limit of $a \to 0$ greatly simplifies the result. Define the following dimensional quantities

$$T = Na, \qquad \mathbf{k} = \mathbf{p}a \ ; \quad t = na \ ; \quad \int_{\mathbf{k}} = \frac{1}{(2\pi)^3} \int d^3\mathbf{k} = \lim_{a \to 0} \frac{1}{a^3} \sum_{\mathbf{p}}$$

$$\omega = \sqrt{\mathbf{k}^2 + m^2}, \qquad \mu(t) = \omega \mathrm{ch}\omega t + m \mathrm{sh}\omega t$$

To take the limit $a \to 0$, the fields ξ, $\bar{\xi}$ are rescaled using Eq. 12.3.6 and only the non-vanishing terms are retained. The continuum solution of the Dirac field's classical solution is [Baaquie (1983b)]

$$\begin{bmatrix} \xi_{\mathbf{k}u}(t) \\ \xi_{\mathbf{k}l}(t) \end{bmatrix} = \frac{1}{\mu(T)} \begin{bmatrix} \mu(T-t) & -\mathrm{sh}\omega t \mathbf{k} \cdot \sigma \\ -\mathrm{sh}\omega(t-T)\mathbf{k} \cdot \sigma & \mu(t) \end{bmatrix} \begin{bmatrix} \psi_{\mathbf{k}u} \\ \psi_{\mathbf{k}l} \end{bmatrix}$$

$$[\bar{\xi}_{\mathbf{k}u}(t) \ \ \bar{\xi}_{\mathbf{k}l}(t)] = [\bar{\psi}_{\mathbf{k}u} \ \ \bar{\psi}_{\mathbf{k}l}] \begin{bmatrix} \mu(T) & -\mathrm{sh}\omega(t-T)\mathbf{k} \cdot \sigma \\ -\mathrm{sh}\omega t \mathbf{k} \cdot \sigma & \mu(T-t) \end{bmatrix} \frac{1}{\mu(T)}$$

The continuum classical solutions, yield from Eq. 12.9.5, the following

$$\mathfrak{F}(\xi, \bar{\xi}) = \int_{\mathbf{k}} \{\bar{\xi}_{\mathbf{k}u}(T)\xi_{\mathbf{k}u}(T) + \bar{\xi}_{\mathbf{k}l}(0)\xi_{\mathbf{k}l}(0)\}$$

giving the final result [Baaquie (1983b)]

$$\mathfrak{F}(\xi, \bar{\xi}) = \int_{\mathbf{k}} \left\{ \frac{\mathrm{sh}\omega T}{\mu}(-\bar{\psi}_{\mathbf{k}u}\mathbf{k} \cdot \sigma\psi_{\mathbf{k}l} + \bar{\psi}_{\mathbf{k}l}\mathbf{k} \cdot \sigma\psi_{\mathbf{k}u}) + \frac{\omega}{\mu}(\bar{\psi}_{\mathbf{k}u}\psi_{\mathbf{k}u} + \bar{\psi}_{\mathbf{k}l}\psi_{\mathbf{k}l}) \right\}$$

$$\mu = \omega \mathrm{ch}\omega T + m\mathrm{sh}\omega T \tag{12.9.11}$$

The normalization constant $C(T)$ is evaluated in Subsection 12.9.1 and yields the final result

$$\mathcal{K}(\bar{\psi}_u, \psi_l; \bar{\psi}_l, \bar{\psi}_u; \ T) = \Pi_p \left(\frac{\mu(T)}{\omega}\right)^2 \exp(\mathfrak{F}) = C(T)\exp(\mathfrak{F}) \tag{12.9.12}$$

12.9.1   *Normalization constant*

The normalization constant $C(T)$ is calculated in the continuum limit. Recall that from Eqs. 12.9.1 and 12.9.4

$$\mathcal{K} \equiv \langle \bar{\psi}_u, \psi_l | e^{-TH} | \bar{\psi}_l, \psi_u \rangle = C(T) \exp \mathfrak{F} \qquad (12.9.13)$$

To directly calculate $C(T)$, requires performing the fermion path integral given in Eq. 12.9.6 and then taking the continuum limit. The path integral can be avoided by, instead, separately evaluating the antiperiodic trace of $\mathcal{K}$ and $\exp(\mathfrak{H})$; this procedure yields $C(T)$.[4]

Using standard finite time methods [Kapusta (1993)], the left hand side of Eq. 12.9.13 yields

$$\tilde{\mathrm{Tr}} \mathcal{K} = \Pi_p \left[2\mathrm{ch}\left(\frac{1}{2}\omega T\right) \right]^4 \;\; ; \;\; \omega = \sqrt{\mathbf{p}^2 + m^2}$$

where $\tilde{\mathrm{Tr}}$ stands for the antiperiodic trace.

The antiperiodic trace of the right hand side of Eq. 12.9.13 yields

$$\tilde{\mathrm{Tr}} \exp \mathfrak{F} = \int d\bar{\psi} d\psi e^{-\bar{\psi}\psi} \exp \mathfrak{H}(\bar{\psi}_l, -\bar{\psi}_u; \bar{\psi}_u - \psi_l)$$

$$= \Pi_{\mathbf{k}} \det \left(\frac{1}{\mu} \begin{bmatrix} \mu + \omega & \mathrm{sh}\omega T \mathbf{k} \cdot \sigma \\ -\mathrm{sh}\omega T \mathbf{k} \cdot \sigma & \mu + \omega \end{bmatrix} \right) = \Pi_{\mathbf{k}} \left(\frac{4\omega}{\mu} \mathrm{ch}^2 \left(\frac{1}{2}\omega T\right) \right)^2$$

Hence, from Eq. 12.9.13 and equations above

$$C(T) = \Pi_{\mathbf{k}} \left(\frac{\mu}{\omega} \right)^2 = \Pi_{\mathbf{k}} \left(\frac{\omega \mathrm{ch}\omega T + m \mathrm{sh}\omega T}{\omega} \right)^2 \qquad (12.9.14)$$

The continuum and lattice formulation have the following similarities.

- The result for $\mathfrak{H}$ is dimensionally correct. In particular, the break-up of the spinors into upper and lower components for the lattice survives the continuum limit; that is, the initial and final continuum field configurations are given by $\{\bar{\psi}_l(\vec{x}), \psi_u(\vec{x})\}$ and $\{\bar{\psi}_u(\vec{x}), \psi_l(\vec{x})\}$, respectively, although the continuum theory does not explicitly have this to start with. It was shown in Section 12.2 that the continuum formulation determines the field coordinates based on the choice of the Dirac γ matrices.
- The solutions for the continuum $\xi(t)$ and $\bar{\xi}(t)$ can be derived directly from the continuum theory, and discussed in Chapter 13.
- The result for $\mathfrak{H}(\xi, \bar{\xi})$ given in Eq. 12.9.11 can be directly obtained from the continuum theory; one needs to use the fermionic coordinates in continuous space and the evolution kernel can be obtained from the continuum eigenstates of the Dirac field. This is discussed in Chapter 13.
- A continuum action with the correct boundary terms is written in Chapter 13 using the lattice results as a guide.

---

[4]Note that $C(T)$ could equally be obtained by taking the periodic trace of both sides of Eq. 12.9.1.

12.10 Evolution Kernel

The Hamiltonian is defined using the evolution kernel by noting that, from Eq. 12.7.18

$$-\frac{\partial}{\partial T}\mathcal{K}\Big|_{T=0} = \langle\bar{\psi}_u\psi_l|\hat{H}|\bar{\psi}_l\psi_u\rangle$$

$$= H(\bar{\psi},\psi)\langle\bar{\psi}_u\psi_l|\bar{\psi}_l\psi_u\rangle$$

$$= H(\bar{\psi},\psi)\exp\left(\int_p \bar{\psi}_p\psi_p\right)$$

and $H(\bar{\psi},\psi)$ is the Hamiltonian. Using Eq. 12.9.12 for the kernel yields

$$-\frac{\partial}{\partial T}\mathcal{K}\Big|_{T=0} = -\left[\frac{\partial C(0)}{\partial T} + \frac{\partial\mathfrak{H}(0)}{\partial T}\right]\exp\mathfrak{H}(0)$$

Dropping a constant yields the Dirac Hamiltonian

$$H(\bar{\psi},\psi) = \int_p \bar{\psi}_p(i\gamma\cdot p + m)\psi_p \qquad (12.10.1)$$

$H(\bar{\psi},\psi)$ is a functional of both the field configuration $\bar{\psi}_l(\mathbf{x})$, $\psi_u(\mathbf{x})$ as well as the conjugate coordinates $\bar{\psi}_u(\mathbf{x})$, $\psi_l(\mathbf{x})$, and is analogous to the function $H(p,x)$ defined on the phase space for a single particle.

The evolution kernel contains the complete eigenenergies and eigenfunctions of the quantum field [Baaquie (2014)]. Let E_n and $|\Phi_n\rangle$ be the eigenenergies and eigenfunctions. The completeness equation given in Eq. 12.7.17 defines the matrix elements and scalar products of the wave functions, and yields

$$\langle\Phi_m|H|\Phi_n\rangle = E_n\delta_{nm}$$

The time evolution operator can be written as

$$e^{-TH} = \sum_n e^{-TE_n}|\Phi_n\rangle\langle\Phi_n| \qquad (12.10.2)$$

where the sum is over all the eigenfunctions, and

$$\mathcal{K}(\bar{\psi}_u,\bar{\psi}_\ell,\bar{\psi}_\ell,\psi_u;\ T) \equiv \langle\bar{\psi}_u,\psi_\ell|e^{-TH}|\bar{\psi}_\ell,\psi_u\rangle$$

$$= \sum_n e^{-TE_n}\Phi_n(\bar{\psi}_u,\psi_\ell)\Phi_n^\dagger(\bar{\psi}_\ell,\psi_u)$$

$$= \prod_p\sum_{n_p} e^{-TE_n(p)}\Phi_{n_p}(\bar{\psi}_{pu},\psi_{p\ell})\Phi_{n_p}^\dagger(\bar{\psi}_{pl},\psi_{pu}) \qquad (12.10.3)$$

where the conjugate $\Phi_n^\dagger$ is defined in Eq. 12.6.3. The last equation above is due to the factorization of the kernel in momentum space. The eigenfunctions will hence be solved for each $\mathbf{p}$ separately.

To evaluate E_n and Φ_n, note that in the kernel $\mathcal{K}$, besides an overall factor of $e^{2\omega T}$ in $C(T)$, only powers of $e^{-\omega T}$ appear in $\mathfrak{H}(T)$. Hence, expanding $\mathcal{K}$ in powers of $e^{-\omega T}$ yields (for each p)

$$\mathcal{K} = e^{2\omega T}(\mathcal{K}_0 + e^{-\omega T}\mathcal{K}_1 + e^{-2\omega T}\mathcal{K}_2 + e^{-3\omega T}\mathcal{K}_3 + e^{-4\omega T}\mathcal{K}_4) \qquad (12.10.4)$$

Note that the expansion for $\mathcal{K}$ terminates due to the anticommutation of the fermions. Hence, comparing Eqs. 12.10.3 and 12.10.4 yields, after some algebra, all the eigenfunctions. The calculation is done in Section 12.11, and the results are summarized here. For each momentum p, there are five eigenenergies and sixteen eigenfunctions, which are orthonormal. The term excitation is used for a particle or antiparticle operator applied on the vacuum state and the term pair stands for applying an particle-antiparticle operator to the vacuum state.

The Dirac field has the following eigenenergies and eigenfunctions.

(1) $E_0 = -2\omega$: a unique vacuum state Ω;
(2) $E_1 = -\omega$: four eigenfunctions of a single excitation;
(3) $E_2 = 0$: six eigenfunctions, four of which are pairs and two are two excitation states;
(4) $E_3 = \omega$. four eigenfunctions which are a single excitation mixed with excitation and pair;
(5) $E_4 = 2\omega$: one eigenfunction, which is the superposition of the vacuum state, a single pair and two pairs.

These sixteen orthonormal eigenfunctions form a complete basis for the Hilbert space of states; it is a consequence of the exclusion principle that only sixteen states are allowed for each momentum $\mathbf{p}$.

For starters, the vacuum state is calculated. Let $T \to \infty$; then from Eqs. 12.9.11 and 12.9.12

$$\mathcal{K} = e^{-TE_0}\Omega(\bar{\psi}_u\psi_l)\Omega^\dagger(\bar{\psi}_l\psi_u) + O(e^{-T(E_1-E_0)})$$

$$= e^{T(2\sum_p \omega_p)}\Pi_p\left(\frac{m+\omega}{2\omega}\right)^2 \exp\left\{-\int_p \frac{1}{\omega+m}\bar{\psi}_{pu}\cdot\sigma\psi_{pl}\right\}$$

$$\times \exp\left\{\int_p \frac{1}{\omega+m}\bar{\psi}_{pl}p\cdot\sigma\psi_{pu}\right\} + O(e^{-\omega T})$$

From Eqs. 12.10.3 and 12.10.4, the vacuum energy and the vacuum eigenfunction is given by

$$E_0 = -2\sum_p \omega_p = -2\sum_p \sqrt{p^2 + m^2}$$

$$\Omega(\bar{\psi}_u\psi_l) = \langle\bar{\psi}_u\psi_l|\Omega\rangle$$

$$= \prod_p \left(\frac{m+\omega}{2\omega}\right)\exp\left(-\int_p \frac{1}{\omega+m}\bar{\psi}_{pu}p\cdot\sigma\psi_{pl}\right) \qquad (12.10.5)$$

The energy for the vacuum is a negative divergent quantity and so is the constant dropped in Eq. 12.10.1 in obtaining $H(\bar{\psi},\psi)$. Both these divergent quantities arise due to the ordering of operators chosen and can be removed by appropriately re-ordering (normal ordering) the Hamiltonian. Normal ordering is not done since firstly only energies with E_0 subtracted will enter our calculations; and secondly normal ordering the Hamiltonian would alter the Lagrangian – and which is not desirable for the path integral approach.

For the vacuum state, note that for $T \to \infty$ the coordinates $(\bar{\psi}_u, \psi_l)$ and $(\bar{\psi}_l, \psi_u)$ – referring to the later and earlier times – completely factorize in $\mathcal{K}$, and hence allow for defining the vacuum state Ω. This factorization is a reflection of the uniqueness of the vacuum, and non-factorization would be an indication of the multiplicity of vacuum states, linked with symmetry breaking.

The vacuum consists of the superposition of particle-antiparticle pairs of all momentum $\mathbf{p}$. The field in the vacuum state is correlated over spatial distance of m^{-1} and the spin orientation of the field in the vacuum state is coupled by the helicity operator $\mathbf{p} \cdot \sigma / |\mathbf{p}|$.

12.11 Energy Eigenfunctions

The eigenenergies and eigenfunctions for the free Dirac field are evaluated [Baaquie (1983b)]. The momentum index is suppressed since the field factorizes in this representation.

Recall from Eqs. 12.10.4 and 12.10.3

$$\mathcal{K} = \left(\frac{\mu(T)}{\omega}\right)^2 \exp\left\{ \frac{\mathrm{sh}\omega T}{\mu(T)}(-\bar{\psi}_u \mathbf{p} \cdot \sigma \psi_\ell + \bar{\psi}_\ell \mathbf{p} \cdot \sigma \psi_u) + \frac{\omega}{\mu(T)}(\bar{\psi}_u \psi_u + \bar{\psi}_\ell \psi_\ell) \right\}$$

$$= \sum_n e^{-TE_n} \Phi_n(\bar{\psi}_u, \psi_\ell) \Phi_n^\dagger(\bar{\psi}_\ell, \psi_u)$$

To simplify the algebra, consider the so-called *helicity* basis. Let the Pauli matrices be

$$\sigma_1 = \begin{bmatrix} 0 & 1 \\ 1 & 0 \end{bmatrix} \quad \sigma_2 = \begin{bmatrix} 0 & -i \\ i & 0 \end{bmatrix}, \quad \sigma_3 = \begin{bmatrix} 1 & 0 \\ 0 & -1 \end{bmatrix}$$

Then

$$\mathbf{p} \cdot \sigma = p \begin{bmatrix} \cos\theta & -e^{-i\phi}\sin\theta \\ e^{-i\phi}\sin\theta & -\cos\theta \end{bmatrix}$$

$$= pU\sigma_3 U^\dagger$$

where

$$\mathbf{p} = p(\cos\phi\sin\theta, \ \sin\phi\sin\theta, \ \cos\theta) \ ; \ p = |\mathbf{p}|$$

$$U = \begin{bmatrix} \cos\frac{1}{2}\theta & -e^{-i\phi}\sin\frac{1}{2}\theta \\ e^{-i\phi}\sin\frac{1}{2}\theta & \cos\frac{1}{2}\theta \end{bmatrix}, \quad UU^\dagger = 1$$

The matrix U rotates the spin of the doublets to lie parallel (or antiparallel) to their three-momentum $\mathbf{p}$, the upper component having helicity $+1$ and the lower component of the doublet having helicity -1. Choose the helicity basis given by

$$\psi \to \begin{bmatrix} U & 0 \\ 0 & U \end{bmatrix}\psi \ ; \ \bar{\psi} \to \bar{\psi} \begin{bmatrix} U^\dagger & 0 \\ 0 & U^\dagger \end{bmatrix}$$

For the doublets, use the notation

$$\psi_{u(l)} = \begin{bmatrix} \psi^1_{u(l)} \\ \psi^2_{u(l)} \end{bmatrix}, \quad \bar{\psi}_{u(l)} = (\bar{\psi}^1_{u(\ell)}, \bar{\psi}^2_{u(\ell)})$$

In the new basis the evolution kernel is given by

$$K = \left(\frac{\mu(T)}{\omega}\right)^2 \exp\left\{ -\frac{p}{\mu(T)}(\bar{\psi}_u\sigma_3\psi_l - \bar{\psi}_l\sigma_3\psi_u)\mathrm{sh}\omega T + \frac{\omega}{\mu(T)}(\bar{\psi}_u\psi_u + \bar{\psi}_l\psi_l) \right\}$$

(12.11.1)

To expand K in powers of $e^{-\omega T}$, expand $\mathrm{sh}\omega T$ and $\mu^{-1}(T)$ to $e^{-4\omega T}$ and write out the exponential term and the normalization constant to $O(e^{-4\omega T})$. This yields

$$K = e^{2\omega T}(K_0 + e^{-\omega T}K_1 + e^{-2\omega T}K_2 + e^{-3\omega T}K_3 + e^{-4\omega T}K_4)$$

Eqs. 12.10.3 and 12.11.1 – and after considerable algebra – yield the results given below.

(I) $E_0 = -2\omega$: ground state; non-degenerate

$$K_0 = \left(\frac{m+\omega}{2\omega}\right)^2 \exp\left\{ -\frac{p}{m+\omega}(\bar{\psi}_u\sigma_3\psi_l - \bar{\psi}_l\sigma_3\psi_u) \right\}$$
$$= \Omega(\bar{\psi}_u, \psi_\ell)\Omega^\dagger(\bar{\psi}_u, \psi_\ell)$$

(12.11.2)

where the vacuum state is given by

$$\Phi_0 = \Omega(\bar{\psi}_u, \psi_\ell)$$
$$= \left(\frac{m+\omega}{2\omega}\right)\exp\left\{ -\frac{p}{m+\omega}\bar{\psi}_u\sigma_3\psi_l \right\}$$

(II) $E_0 = -\omega$: first excited level; four-fold degenerate

$$K_1 = \frac{2\omega}{\omega+m}\Omega(\bar{\psi}_u, \psi_\ell)\bar{\psi}\psi\Omega^\dagger(\bar{\psi}_\ell, \psi_u) = \sum_{i=1}^{4}\Phi_i\Phi_i^\dagger$$

where

$$\Phi_1 = N\bar{\psi}_u^1\Omega(\bar{\psi}_u, \psi_\ell) \quad ; \quad \Phi_2 = N\bar{\psi}_u^2\Omega(\bar{\psi}_u, \psi_\ell)$$
$$\Phi_3 = N\psi_l^1\Omega(\bar{\psi}_u, \psi_\ell) \quad ; \quad \Phi_4 = N\psi_l^2\Omega(\bar{\psi}_u, \psi_\ell)$$
$$N = \sqrt{2\omega/(\omega+m)}$$

The first excited level consists of a single excitation on the vacuum.

(III) $E_2 = 0$: second excited level; six-fold degenerate

$$K_2 = 2\left(\frac{\omega-m}{\omega+m}\right)\Omega(\bar{\psi}_u, \psi_\ell)\left[1 + \frac{\omega}{p}(\bar{\psi}_u\sigma_3\psi_u) + \left(\frac{\omega}{p}\right)^2(\bar{\psi}\psi)^2\right]\Omega^\dagger(\bar{\psi}_u, \psi_\ell)$$
$$= \sum_{i=5}^{10}\Phi_i\Phi_i^\dagger$$

where

$$\Phi_5 = N'\frac{\omega}{p}\bar{\psi}_u\sigma_1\psi_l\Omega(\bar{\psi}_u,\psi_\ell) \quad ; \quad \Phi_6 = N'\frac{\omega}{p}\bar{\psi}_u\sigma_2\psi_2\Omega(\bar{\psi}_u,\psi_\ell)$$

$$\Phi_7 = N'\left(1+\frac{\omega}{p}\bar{\psi}_u\sigma_3\psi_\ell\right)\Omega((\bar{\psi}_u,\psi_\ell) \quad ; \quad \Phi_8 = N'\frac{\omega}{p}\bar{\psi}_u\psi_\ell\Omega(\bar{\psi}_u,\psi_\ell)$$

$$\Phi_9 = N'\sqrt{2}\frac{\omega}{p}\bar{\psi}_u^1\bar{\psi}_u^2\Omega(\bar{\psi}_u,\psi_\ell) \quad ; \quad \Phi_{10} = N'\sqrt{2}\frac{\omega}{p}\psi_\ell^1\psi_\ell^2\Omega(\bar{\psi}_u,\psi_\ell)$$

$$N' = \sqrt{2(\omega-m)/(\omega+m)}$$

The states Φ_5, Φ_6, and Φ_8 are the spin-one pair states and Φ_8 is the singlet pair state. State Φ_7 is a superposition of the vacuum state and a pair. States Φ_9 and Φ_{10} are the two excitation states.

(IV) $E_3 = \omega$: third excited level; four-fold degenerate

$$\mathcal{K}_3 = \frac{2\omega(\omega-m)}{(\omega+m)^2}\Omega(\bar{\psi}_u,\psi_\ell)\left[\bar{\psi}\psi + \frac{2\omega}{p}\bar{\psi}\psi(\bar{\psi}_u\sigma_3\psi_l - \bar{\psi}_l\sigma_3\psi_u)\right.$$

$$\left. + \left(\frac{\omega}{p}\right)^2(\bar{\psi}\psi)^3\right]\Omega^\dagger(\bar{\psi}_u,\psi_\ell) = \sum_{i=11}^{14}\Phi_i\Phi_i^\dagger$$

where

$$\Phi_{11} = N''\bar{\psi}_u^1\left(1+\frac{2\omega}{p}\bar{\psi}_u\sigma_3\psi_l\right)\Omega(\bar{\psi}_u,\psi_\ell)$$

$$\Phi_{12} = N''\bar{\psi}_u^2\left(1+\frac{2\omega}{p}\bar{\psi}_u\sigma_3\psi_l\right)\Omega(\bar{\psi}_u,\psi_\ell)$$

$$\Phi_{13} = N''\bar{\psi}_l^1\left(1+\frac{2\omega}{p}\bar{\psi}_u\sigma_3\psi_l\right)\Omega(\bar{\psi}_u,\psi_\ell)$$

$$\Phi_{14} = N''\bar{\psi}_l^2\left(1+\frac{2\omega}{p}\bar{\psi}_u\sigma_3\psi_l\right)\Omega(\bar{\psi}_u,\psi_\ell)$$

$$N'' = \sqrt{2\omega(\omega-m)}/(\omega+m).$$

These states are a superposition of a single excitation and an excitation and a pair.

(V) $E_4 = 2\omega$: fourth excited level; non-degenerate

$$\mathcal{K}_4 = \Phi_{15}\Phi_{15}^\dagger,$$

$$\Phi_{15} = \left(\frac{\omega-m}{2\omega}\right)\exp\left(\frac{p}{\omega-m}\bar{\psi}_u\sigma_3\psi_l\right)$$

The state Φ_{15} consists of a mix of no particle, one pair and two pairs. Note the similarity in form of Φ_{15} with the vacuum state Ω, the two being related by $m \to -m$.

In summary, the energy spectrum of the free fermion field, for each momentum **p**, consists of five energy levels and sixteen eigenfunctions. These eigenfunctions are orthonormal

$$\langle\Phi_n|\Phi_m\rangle = \delta_{nm}$$

and form a complete basis for the Hilbert space of states.

The eigenfunctions for the full field theory are elements of the Hilbert space formed by the tensor product over $\mathbf{p}$, and are given by

$$\Phi_n[\bar{\psi}_u, \psi_\ell] = \prod_{\mathbf{p}} \Phi_{n\mathbf{p}}(\bar{\psi}_{n\mathbf{p}}, \psi_{\ell\mathbf{p}})$$

and the eigenenergies given by

$$E_n = \sum_{\mathbf{p}} E_{n\mathbf{p}}(\mathbf{p})$$

These basis states can be used as a complete basis for doing perturbation theory, and are an alternative to the usual Fock-space approach.

12.12 Propagator

The propagator is calculated using the evolution kernel and the vacuum state. Recall that, for $T > 0$, the propagator is

$$G_{pT}^{\alpha\beta}\delta(\mathbf{p} - \mathbf{p}') = \langle\Omega|\psi_{\mathbf{p}T}^{\alpha}\bar{\psi}_{\mathbf{p}'0}^{\beta}|\Omega\rangle$$

where ψ_{pT}^{α}, $\bar{\psi}_{\mathbf{p}'0}^{\beta}$ are Heisenberg field operators.

From standard field theory the Euclidean propagator, in matrix notation, is given by [Peskin and Schroeder (1995)]

$$\begin{aligned}
G_{\mathbf{p}T} &= \int_{-\infty}^{+\infty} \frac{\mathrm{d}p_0}{2\pi} \frac{e^{ip_0 T}}{ip_0\gamma_0 + i\mathbf{p}\cdot\gamma + m} \\
&= \frac{e^{-\omega T}}{2\omega} \begin{bmatrix} m+\omega & -\mathbf{p}\cdot\sigma \\ \mathbf{p}\cdot\sigma & m-\omega \end{bmatrix}
\end{aligned} \tag{12.12.1}$$

The Heisenberg operators have the time evolution

$$\psi_{pT} = e^{TH}\psi_p e^{-TH}$$
$$\bar{\psi}_{pT} = e^{TH}\bar{\psi}_p e^{-TH}$$

with $\psi_p\bar{\psi}_p$ being the Schrödinger operators. Ignoring from now on the momentum conserving δ-functions, and using the completeness equation gives

$$G_{pT}^{\alpha\beta} = e^{TE_0}\langle\Omega|\psi_p^{\alpha}e^{-TH}\bar{\psi}_p^{\beta}|\Omega\rangle,$$

$$G_{pT}^{\alpha\beta} = e^{TE_0}\int \mathrm{d}\bar{\xi}\mathrm{d}\xi\mathrm{d}\bar{\zeta}\mathrm{d}\zeta\langle\Omega|\bar{\xi}_\ell, \xi_u\rangle e^{-\int_p \bar{\xi}\xi}\xi_p^{\alpha}\langle\bar{\xi}_u, \xi_\ell|e^{-TH}|\bar{\zeta}_\ell, \zeta_u\rangle e^{-\int_p \bar{\zeta}\zeta}\bar{\zeta}_p^{\beta}\langle\bar{\zeta}_u, \zeta_\ell|\Omega\rangle$$

$$= e^{TE_0}\int_{\bar{\xi}\xi, \bar{\zeta}\zeta}[\xi_p^{\alpha}\bar{\zeta}_p^{\beta}]\cdot\Omega^{\dagger}(\bar{\xi}_\ell, \xi_u)\mathcal{K}(\bar{\xi}_u, \xi_\ell, \bar{\zeta}_\ell, \zeta_u;\ T)\Omega(\bar{\zeta}_u, \zeta_\ell)e^{-\int_p(\bar{\xi}_p\xi_p + \bar{\zeta}_p\zeta_p)}$$

$$= N\int_{\bar{\xi}\xi, \bar{\zeta}\zeta} \xi_p^{\alpha}\bar{\zeta}_p^{\beta}\exp(\mathcal{F})$$

The exponential of the effective action $\mathcal{F}$ is equal to the product of the vacuum wave functional and its dual, the evolution kernel and the metric; Eqs. 12.10.4 and 12.10.5 yield

$$\mathcal{F} = \int_p \left\{ \frac{1}{\omega + m} \bar{\xi}_{pl} p \cdot \sigma \xi_{pu} + \frac{\mathrm{sh}\omega T}{\mu} (-\bar{\xi}_{pu} p \cdot \sigma \xi_{pl} + \bar{\zeta}_{pl} p \cdot \sigma \zeta_{pu}) \right.$$
$$\left. + \frac{\omega}{\mu} (\bar{\xi}_{pu} \zeta_{pu} + \bar{\zeta}_{pl} \xi_{pl}) - \frac{1}{\omega + m} \bar{\zeta}_{pu} p \cdot \sigma \zeta_{pl} - \bar{x} i_p \xi_p - \bar{\zeta}_p \zeta_p \right\} \qquad (12.12.2)$$

$$N = \Pi_p e^{-2\omega T} \left(\frac{\mu}{\omega} \right)^2 \left(\frac{2\omega}{\omega + m} \right)^2 \ ; \quad \mu \equiv \mu(T)$$

The generating functional for $\mathcal{F}$ is evaluated as this is useful for evaluating the n-point functions. Let h_p, $\bar{h}_p$ be the four-component fermionic sources; define

$$W(\bar{h}, h) = N \int_{\bar{\xi}\xi, \bar{\zeta}\zeta} \exp\left(\mathcal{F} + \int_p \bar{h}_p \xi_p + \int_p \bar{\zeta}_p h_p \right) \qquad (12.12.3)$$

$$= N \Pi_p \left(e^{\omega T} \frac{\omega}{\mu} \right)^2 \int_{\bar{\zeta}\zeta} \exp\left\{ -\bar{\zeta} \left(1 + \frac{1}{\omega + m} i\alpha \cdot p \right) \zeta \right\} \qquad (12.12.4)$$

$$\times \exp\left\{ e^{-\omega T} \left(\bar{h_u} + \bar{h_l} \frac{p \cdot \sigma}{\omega + m} \right) \zeta_u + \bar{\zeta} h \right\}$$

$$= \exp\left(\int_p \bar{h}_p M_p h_p \right) \qquad (12.12.5)$$

where

$$M_p = \frac{e^{-\omega T}}{2\omega} \begin{bmatrix} m + \omega & -p \cdot \sigma \\ p \cdot \sigma & m - \omega \end{bmatrix} \qquad (12.12.6)$$

Hence, the propagator is given by

$$G_{pT}^{\alpha\beta} = \frac{\delta^2}{\delta h_p^\beta \delta \bar{h}_p^\beta} W(\bar{h}, h) \Big|_{h=0, \bar{h}=0} = M_p^{\alpha\beta} \qquad (12.12.7)$$

Eqs. 12.12.1 and 12.12.6 show that equation above is the required result.

The propagator $G_{pT}^{\alpha\beta}$ can also be evaluated using the eigenfunctions. The notation of G_{pT}^{uu}, G_{pT}^{ul} is used for the 2×2 matrix entries in Eq. 12.12.6. In terms of the eigenfunctions of the Dirac field, the propagator is given by

$$G = e^{TE_0} \langle \Omega | \psi_p e^{-TH} \bar{\psi}_p | \Omega \rangle = \sum_n e^{-T(E_n - E_0)} \langle \Omega | \psi_p | \Phi_n \rangle \langle \Phi_n | \bar{\psi}_p | \Omega \rangle \qquad (12.12.8)$$

where Eq. 12.10.2 has been used to obtain the equation above. It can be shown that only the first excited level, that is, states with $E_1 = -\omega$, contributes to Eq. 12.12.8. Hence

$$G = e^{-\omega T} \sum_{i=1}^{4} \langle \Omega | \psi_p | \Phi_i \rangle \langle \Phi_i | \bar{\psi}_p | \Omega \rangle \qquad (12.12.9)$$

where the states $\Phi_1, \ldots, \Phi_4$ are given in Section 12.11. Consider evaluating $G^{\ell u}$ as an example. From Eq. 12.12.9

$$G_{\alpha\beta}^{\ell u} = e^{-\omega T} \sum_{i=1}^{4} \langle \Omega | \psi_{\ell p}^{\alpha} | \Phi_i \rangle \langle \Phi_i | \bar{\psi}_{pu}^{\beta} | \Omega \rangle$$

Using

$$\bar{\psi}_{pu}^{\beta} | \Omega \rangle = N^{-1} | \Phi_\beta \rangle, \quad \beta = 1, 2 \ ; \quad N = \sqrt{2\omega/(\omega + m)}$$

yields

$$G_{\alpha\beta}^{\ell u} = e^{-\omega T} \sum_{i=1}^{4} \langle \Omega | \psi_{p\ell}^{\alpha} | \Phi_i \rangle \langle \Phi_i | \Phi_\beta \rangle N^{-1}$$

$$= N^{-1} e^{-\omega T} \langle \Omega | \psi_{p l}^{\alpha} | \Phi_\beta \rangle = e^{-\omega T} \langle \Omega | \psi_{p\ell}^{\alpha} \bar{\psi}_{pu}^{\beta} | \Omega \rangle \tag{12.12.10}$$

where the orthonormality equation has been used. The wave functions in Section 12.11 are given in the helicity basis. To evaluate the propagator the completeness equation

$$\mathbb{I} = \int D\bar{\psi} D\psi | \bar{\psi}_\ell, \psi_u \rangle T(\bar{\psi}, \psi) . \langle \bar{\psi}_u, \psi_\ell |$$

is required.

To evaluate Eq. 12.12.10, note that the vacuum functionals generate the term given in Eq. 12.11.2; performing the fermion integration using Eq. 6.6.11 yields

$$G_{\alpha\beta}^{\ell u} = e^{-\omega T} \int D\bar{\psi} D\psi \langle \Omega | \bar{\psi}_\ell, \psi_u \rangle T(\bar{\psi}, \psi) \langle \bar{\psi}_u, \psi_\ell | \psi_{p\ell}^{\alpha} \bar{\psi}_{pu}^{\beta} | \Omega \rangle$$

$$= e^{-\omega T} \int D\bar{\psi} D\psi \psi_{\mathbf{p}\ell}^{\alpha} \bar{\psi}_{pu}^{\beta} \Omega(\bar{\psi}_u, \psi_\ell) \Omega^\dagger(\bar{\psi}_u, \psi_\ell) \exp\left\{ -\int_{\mathbf{p}} \bar{\psi}_{\mathbf{p}} \psi_{\mathbf{p}} \right\}$$

$$= \left(\frac{m+\omega}{2\omega} \right)^2 e^{-\omega T} \int D\bar{\psi} D\psi \psi_{p\ell}^{\alpha} \bar{\psi}_{pu}^{\beta}$$

$$\times \exp\left\{ -\int_{\mathbf{p}} \frac{p}{m+\omega} (\bar{\psi}_u \sigma_3 \psi_l - \bar{\psi}_l \sigma_3 \psi_u) - \int_{\mathbf{p}} \bar{\psi}_{\mathbf{p}} \psi_{\mathbf{p}} \right\}$$

$$= \frac{e^{-\omega T}}{2\omega} p(\sigma_3)_{\alpha\beta} \ ; \quad \int_{\mathbf{p}} \equiv \int \frac{d^3 p}{(2\pi)^3}$$

Transforming back to the original momentum basis yields, in matrix notation, the expected result

$$G^{\ell u} = \frac{e^{-\omega T}}{2\omega} \mathbf{p} \cdot \sigma$$

Similarly, all the other elements of G_{pT} can be calculated using the states of the first excited level.

12.13   Summary

The free Dirac field's evolution kernel was evaluated using the lattice formulation. The field coordinates $\bar{\psi}_u(x)$, $\psi_\ell(x)$ were used due to the lattice approach. In Chapter 13, the continuum Dirac theory is recast using $\bar{\psi}_u(x)$, $\psi_\ell(x)$ as the dynamical degrees of freedom.

The main difficulty with the treatment of the lattice Dirac field is the question of chiral symmetry. For the choice of the γ_μ's made, the matrix γ_5 interchanges upper and lower components. Hence chiral transformations are non-diagonal on the coordinates chosen for the state space; to describe the state space of a chiral field another set of coordinates need to be used. Other ways of defining the lattice fermions that are non-local in time cannot be used as the state space discussed in this Chapter crucially hinges on the nearest-neighbor coupling in time, and on the use of the projectors $(1 \pm \gamma_0)$ in the Lagrangian.

The calculation performed can be extended to interacting Dirac field and consists of perturbatively calculating with an open boundary interacting fermion system. The boundary values enter via the classical solution that acts as a background field for the quantum variables [Baaquie (1986)].

The evolution kernel can be used to calculate the transition amplitudes for non-local field configurations like vortices, strings, and so on. It also provides a different perspective for studying quantum field theories.

Chapter 13

Dirac Hamiltonian

The Feynman path integral is usually defined using the infinite time action, since one is normally calculating the scattering matrix for which infinite time processes are required. In effect, one evaluates the time ordered vacuum expectation values of the field operators using the infinite time path integral. However, for time independent bound states, the correct way to describe the system is to use the time independent eigenenergies and eigenfunctionals of the quantum field, and one is naturally led to the Hamiltonian formulation.

For example, in Baaquie (1983a, 2018) the eigenenergies of the (bound) eigenstates of two-dimensional Quantum Electrodynamics are evaluated, and it is found that the single fermion eigenenergies and eigenstates are eliminated in the interacting theory showing the confinement of the fundamental fermions. Since the mesons and hadrons are relativistic quark confining bound states of the interacting quark-gluon quantum field, the description of such bound states necessitates the study of the Hamiltonian approach to Quantum Chromodynamics.

The first step is to study the Hamiltonian formalism for the free quark field. In this Chapter, the Dirac Hamiltonian is extensively studied using the anticommuting fermionic variables, and in particular its close connection with path integration is shown.

The discussion in Chapter 12 on the Dirac field is based on the path integral. The eigenfunctions obtained earlier in Section 12.11, using the path integral, are re-derived using the Hamiltonian – illustrating the complimentary nature of the Hamiltonian and path integral formulations. One of the advantages of using fermion calculus is that the Hilbert space of the Dirac field can be given an explicit realization in terms of fermionic variables and many features of the Dirac field state vectors have a transparent representation.

The Dirac vacuum, which is the single most important state of the Dirac field, is derived exactly and its correlations and other properties are studied in detail. This is in contrast to the approach taken in the standard canonical approach [Peskin and Schroeder (1995)], for which the state space in general is defined in terms of the action of anticommuting creation and destruction operators. Furthermore, as shown in Section 13.8, the fermionic variables allow an explicit computation of the evolution kernel of the Dirac field, something that is not clear how to carry out in the canonical formulation.

One of the main results of this Chapter is to derive the boundary term for the finite time Dirac action – and show that the evolution kernel is given precisely by the boundary term action.

13.1   Fermionic Operators

The results obtained in Section 6.3 are applied to the Dirac field. The coordinate operators are

$$\langle \bar{\psi}_u, \psi_\ell | \begin{pmatrix} \bar{\psi}_u^{op} \\ \psi_\ell^{op} \end{pmatrix} |f\rangle = \begin{pmatrix} \bar{\psi}_u \\ \psi_\ell \end{pmatrix} f(\bar{\psi}_u, \psi_\ell) \tag{13.1.1}$$

The conjugate operators are given by the completeness equation.

$$\langle \bar{\psi}_u, \psi_\ell | \begin{pmatrix} \bar{\psi}_\ell^{op} \\ \psi_u^{op} \end{pmatrix} |f\rangle = \int d\bar{\zeta} d\zeta \begin{pmatrix} \bar{\zeta}_\ell \\ \zeta_u \end{pmatrix} \langle \bar{\psi}_u, \psi_\ell | \bar{\zeta}_\ell, \zeta_u \rangle \exp(-\bar{\zeta}\zeta) f(\bar{\zeta}_u, \zeta_\ell) \tag{13.1.2}$$

The inner product yields

$$\begin{pmatrix} \bar{\zeta}_\ell \\ \zeta_u \end{pmatrix} \langle \bar{\psi}_u, \psi_\ell | \bar{\zeta}_\ell, \zeta_u \rangle = \begin{pmatrix} -\frac{\delta}{\delta \psi_\ell} \\ \frac{\delta}{\delta \bar{\psi}_u} \end{pmatrix} \exp(\bar{\psi}_u \zeta_u + \bar{\zeta}_\ell \psi_\ell) \tag{13.1.3}$$

where the fermionic derivatives are defined as in Chapter 6. Hence

$$\langle \bar{\psi}_u, \psi_\ell | \begin{pmatrix} \bar{\psi}_\ell^{op} \\ \psi_u^{op} \end{pmatrix} |f\rangle = \begin{pmatrix} -\frac{\delta}{\delta \psi_\ell} \\ \frac{\delta}{\delta \bar{\psi}_u} \end{pmatrix} f(\bar{\psi}_u, \psi_\ell) \tag{13.1.4}$$

In summary, the following is the operator representation in the coordinate basis

$$\psi = \begin{pmatrix} \frac{\delta}{\delta \bar{\psi}_u} \\ \psi_\ell \end{pmatrix}, \bar{\psi} = \left(\bar{\psi}_u, -\frac{\delta}{\delta \psi_\ell} \right) \tag{13.1.5}$$

and yield the anticommutation equations

$$\{\bar{\psi}_\alpha, \bar{\psi}_\beta\} = \{\psi_\alpha, \psi_\beta\} = 0 \quad ; \quad \{\bar{\psi}_\alpha, \psi_\beta\} = \gamma_0^{\alpha\beta} \tag{13.1.6}$$

The anticommutators are a result of our definition of the metric, and we will see that for the lattice there is a more general anticommutator than Eq. 13.1.6. Note that in the usual scheme for quantization, ψ_α is taken as the independent degree of freedom, $\bar{\psi}_\alpha$ the conjugate variable and the anticommutation relation is postulated. In the lattice approach, the anticommutator is the result of the definition of the state space and the metric defining the inner product on this space, and can have forms more general than the canonical anticommutation relation.

Let G be an arbitrary operator. Then the matrix elements $G(\bar{\psi}, \psi)$ of the operator, as in Eq. 6.8.2, are defined by

$$\langle \bar{\psi}_u, \psi_\ell | G | \bar{\psi}_\ell, \psi_u \rangle \equiv G(\bar{\psi}, \psi) \langle \bar{\psi}_u, \psi_\ell | \bar{\psi}_\ell, \psi_u \rangle = G(\bar{\psi}, \psi) \exp(\bar{\psi}\psi)$$

The left hand side is sometimes called the kernel of the operator and the quantity $G(\bar{\psi}, \psi)$ the normal symbol. The conjugate operator G is defined by conjugating each matrix element $G(\bar{\psi}, \psi)$, and operator G is Hermitian if, for arbitrary $|f\rangle$ and $\langle g|$, we have

$$\langle f | G^\dagger | g \rangle \equiv \langle g | G | f \rangle^* = \langle f | G | g \rangle \quad : \quad \text{Hermitian} \tag{13.1.7}$$

13.2   Lattice Dirac Hamiltonian

The Hamiltonian operator for the Dirac field has been obtained in Section 12.8 starting from the finite time lattice action. To obtain the Hamiltonian from the action, one essentially has to 'divide' out the metric and the inner product of the Dirac field coordinate eigenstates from the expression for e^{-aH}. All coordinates – both lattice and continuum as well as space and momentum indicies – in this Chapter are in three dimensions as the Hamiltonian makes no reference to time. For this reason, the vector notation will not be used.

The free Dirac Hamiltonian has been derived in Section 12.8 and is given, up to a constant, by Eq. 12.8.5

$$H = -\frac{1}{a}(2K - 1) \sum_{\vec{n}} \bar{\psi}_{\vec{n}}\psi_{\vec{n}}$$
$$+ \frac{K}{a} \sum_{\vec{n}i} \left(\bar{\psi}_{\vec{n}}(1 - \gamma_i)\psi_{\vec{n}+i} - \bar{\psi}_{\vec{n}+i}(1 + \gamma_i)\psi_{\vec{n}} \right) \tag{13.2.1}$$

The lattice Hamiltonian obtained in Eq. 13.2.1 is Hermitian as defined in Eqs. 6.8.5 and 13.1.7 – and hence has only real eigenvalues. Note that the mass term for H arises partly from both the lattice time derivative term of the action and from the coordinate inner product. Similarly, the nearest neighbor coupling term in H has a contribution from the inner product.

Define the following new rescaled lattice variables

$$\zeta_n = \sqrt{\frac{1 - 2K}{a}} \; \psi_n = \sqrt{\frac{a^{d-2}}{2J}} \psi(x) \; ; \; x = na$$
$$\bar{\zeta}_n = \sqrt{\frac{1 - 2K}{a}} \; \bar{\psi}_n = \sqrt{\frac{a^{d-2}}{2J}} \bar{\psi}(x) \tag{13.2.2}$$

where

$$2K = \frac{1}{(ma + d)} \; ; \; 2J = \frac{1}{(ma + d - 1)} \tag{13.2.3}$$

Hence, the Dirac Hamiltonian is given by

$$H(\bar{\zeta}, \zeta) = \sum_n \bar{\zeta}_n \zeta_n - J \sum_{ni} \left\{ \bar{\zeta}_n(1 - \gamma_i)\zeta_{n+i} + \bar{\zeta}_{n+i}(1 + \gamma_i)\zeta_n \right\} \tag{13.2.4}$$

where J is the nearest-neighbor coupling constant given by Eq. 13.2.3. Note that to obtain H from the lattice action, the time coupling terms have been dropped and the nearest-neighbor coupling in the Lagrangian suitably altered. The metric in the rescaled variables is given by

$$T(\bar{\zeta}, \zeta) = \exp\left\{ -\frac{aJ}{K} \sum_n \bar{\zeta}_n \zeta_n + aJ \sum_{ni} (\bar{\zeta}_n \zeta_{n+i} + \bar{\zeta}_{n+i}\zeta_n) \right\}$$
$$= \exp\left\{ -\int_p \lambda_p \bar{\zeta}_p \zeta_p \right\} \tag{13.2.5}$$

where

$$\lambda_p = \frac{aJ}{K}\left(1 - 2K\sum_i \cos p_i\right) \quad ; \quad \int_p \equiv \prod_{i=1}^{3} \int_{-\pi}^{+\pi} \frac{dp_i}{2\pi} \qquad (13.2.6)$$

Note that $\vec{p}$ is the dimensionless periodic momentum.

The metric is positive definite only if

$$0 < 2K < \frac{1}{d-1}$$

This constraint on K given above is valid even in the presence of coupling to the gauge field, and discussed in Chapter 14. For the metric given in Eq. 13.2.5 the operator anticommutation equation, similar to Eq. 6.8.1, is given by

$$\{\zeta_{p\alpha}^\dagger, \zeta_{p'\beta}\} = \frac{1}{\lambda_p}\delta_{pp'}\delta_{\alpha\beta} \qquad (13.2.7)$$

The lattice Hamiltonian is given by

$$H = \int_p \bar{\zeta}_p(\alpha + i\vec{\gamma}\cdot\vec{\beta})\zeta_p \qquad (13.2.8)$$

where

$$\alpha = 1 - 2J\sum_i \cos p_i \quad ; \quad \beta_i = 2J\sin p_i \qquad (13.2.9)$$

Note Eqs. 13.2.7 and 13.2.8 completely define the operator equations for the quantum field. The problem is to diagonalize H, given the anticommutation Eq. 13.2.7. Choose the coordinate frame so that $\vec{\beta} = \beta\hat{e}_z$; then, in 2×2 block matrix notation

$$\gamma_0(\alpha + i\vec{\gamma}\cdot\vec{\beta}) = \begin{pmatrix} \alpha & \beta\sigma_3 \\ -\beta\sigma_3 & -\alpha \end{pmatrix} = WD\begin{pmatrix} 1 & 0 \\ 0 & -1 \end{pmatrix}D^\dagger \qquad (13.2.10)$$

where

$$W = \sqrt{\alpha^2 + \vec{\beta}^2} \quad ; \quad D = \begin{pmatrix} M_+ & -M_-\sigma_3 \\ M_-\sigma_3 & M_+ \end{pmatrix} \quad ; \quad DD^\dagger = 1$$

and

$$M_\pm = \sqrt{\frac{W \pm \alpha}{2W}}$$

Define the following transformation

$$\eta_p = \sqrt{\lambda_p}D^\dagger\zeta_p \ , \quad \eta_p^\dagger = \sqrt{\lambda_p}\zeta_p^\dagger D \qquad (13.2.11)$$

The factor of λ_p in the anticommutation given in Eq. 13.2.7 requires a momentum dependent rescaling of $\zeta_p^\dagger, \zeta_p$ for defining $\eta_p^\dagger, \eta_p$ in Eq. 13.2.11. From Eqs. 13.2.7 and 13.2.11

$$\{\eta_{p\alpha}^\dagger, \eta_{p'\beta}\} = \delta_{pp'}\delta_{\alpha\beta}$$

and

$$H = \int_p E_p \eta_p^\dagger \gamma_0 \eta_p \qquad (13.2.12)$$

where

$$E_p = \frac{\sqrt{\alpha^2 + \vec{\beta}^2}}{\lambda_p}$$

Recall the lattice Dirac vacuum state, from Eq. 12.10.5, is given by

$$\langle \bar{\psi}_u \psi_l | \Omega \rangle = \prod_p \left(\frac{m + \omega}{2\omega} \right) \exp \left(- \int_p \frac{1}{\omega + m} \bar{\psi}_{pu} p \cdot \sigma \psi_{pl} \right)$$

For the lattice Dirac field, the energy eigenspectrum consists of equally spaced energy levels with energy of a single excitation being E_p. The eigenstates are constructed, as in Table 13.5, by acting on the lattice vacuum state using the lattice creation and annihilation operators.

As can be seen from Eq. 13.2.12 and discussed in detail before Table 13.5, only η_ℓ and $\eta_u^\dagger$ are required for generating all the eigenfunctions. Suppressing the momentum index, from Eqs. 13.2.2 and 13.2.11

$$\eta_\ell = c \left(- M_- \sigma_3 \frac{\delta}{\delta \bar{\psi}_u} + M_+ \psi_\ell \right) \; ; \; \eta_u^\dagger = c \left(M_+ \bar{\psi}_u + M_+ \sigma_3 \frac{\delta}{\delta \psi_\ell} \right)$$

where

$$c = \sqrt{\lambda \left(\frac{1 - 2K}{a} \right)} = \sqrt{\frac{J}{K} (1 - 2K) \left(1 - 2K \sum_i \cos p_i \right)}$$

13.3 Continuum Hilbert Space

Recall from Eq. 12.7.17 that for the continuum, the metric on Hilbert space is given by

$$T(\bar{\psi}, \psi) = \exp \left(- \int d^3 \bar{\psi}(x) \psi(x) \right)$$

For the Dirac quantum field, the field coordinates are defined on the spatial three-dimensional Euclidean space, and the coordinate eigenstates as in Eqs. 12.5.8 and 12.5.9, are given by the tensor product over all three-space points $\vec{x}$

$$|\bar{\psi}_\ell, \psi_u\rangle = \prod_{\vec{x}} |\bar{\psi}_{\ell x}, \psi_{ux}\rangle \; ; \; \langle \bar{\psi}_u, \psi_\ell | = \prod_x \langle \bar{\psi}_{ux}, \psi_{\ell x}|$$

The completeness equation for the continuum quantum field is then given by

$$\mathbb{I} = \prod_x \int d\bar{\psi}_x d\psi_x |\bar{\psi}_\ell, \psi_u\rangle \exp \left(- \int d^3 x \bar{\psi}_x \psi_x \right) \langle \bar{\psi}_u, \psi_\ell |$$

The completeness equation implies, as in Eq. 12.7.18, the inner product

$$\langle \bar{\psi}_u, \psi_\ell | \bar{\psi}_\ell, \psi_u \rangle = \exp \left\{ \int d^3 x (\bar{\psi}_u \psi_u + \bar{\psi}_\ell \psi_\ell) \right\} = \exp \left(\int d^3 x \bar{\psi}_{\vec{x}} \psi_{\vec{x}} \right)$$

Note the exponent on the right hand side is dimensionless, as indeed it has to be.
   Conjugation yields

$$\psi^\dagger = \bar\psi \gamma_0 \quad \Rightarrow \quad \psi_u^\dagger = \bar\psi_u \; ; \; \psi_\ell^\dagger = -\bar\psi_\ell$$

The dual coordinates have the following realization on the Hilbert space

$$\psi_{us}\Phi\left[\bar\psi_u, \psi_\ell\right] = \frac{\delta}{\delta\bar\psi_{us}}\Phi\left[\bar\psi_u, \psi_\ell\right] \quad ; \quad s = 1,2$$

$$\bar\psi_{\ell s}\Phi\left[\bar\psi_u, \psi_\ell\right] = -\psi_{\ell s}^\dagger\Phi\left[\bar\psi_u, \psi_\ell\right] = -\frac{\delta}{\delta\psi}_{\ell s}\Phi\left[\bar\psi_u, \psi_\ell\right]$$

The creation and destruction operators have a fermionic realization in terms of the coordinates of the Dirac field. The field completely factorizes in momentum space and henceforth the momentum index $\mathbf{p}$ is dropped for notational simplicity.

13.4   Continuum Hamiltonian

The continuum limit of the lattice Dirac fermions is taken to obtain the continuum Dirac Hamiltonian. From Eqs. 13.2.8 and 13.2.9, the lattice Hamiltonian is given

$$H(\zeta, \bar\zeta) = \int_{-\pi}^{+\pi} \frac{d^3q}{(2\pi)^3}\bar\zeta_q(\alpha + i\vec\gamma \cdot \vec\beta)\zeta_q$$

   For lattice spacing a and $d = 4$, from Eq. 13.2.2, the lattice and continuum fields are

$$\zeta_n = \sqrt{\frac{a^2}{2J}}\psi(x) \; ; \quad \bar\zeta_n = \sqrt{\frac{a^2}{2J}}\bar\psi(x) \; ; \quad x = na$$

where, from Eq. 13.2.3 and for $d = 4$

$$2J = \frac{1}{ma + 3} \; ; \quad 2K = \frac{1}{ma + 4}$$

   Define continuum momentum p by

$$q = pa \quad \Rightarrow \quad -\pi \le q \le +\pi \quad \Rightarrow \quad \frac{-\pi}{a} \le p \le \frac{+\pi}{a}$$

and, for $a \to 0$

$$\int_{-\pi}^{+\pi} \frac{d^3q}{(2\pi)^3} \quad \Rightarrow \quad \int_p \equiv \int_{-\infty}^{+\infty} \frac{d^3p}{(2\pi)^3}$$

Note all momentum $p.q, \cdots$ and coordinates $x, y, \cdots$ in this Chapter are henceforth three-dimensional vectors.

   The Fourier expansion of the lattice and continuum fields are related by the following

$$\zeta_q = \sum_n e^{inq}\zeta_n = \sqrt{\frac{a^2}{2J}}\frac{1}{a^3}a^3\sum_n e^{inap}\psi(x) \to \sqrt{\frac{1}{2J}}\frac{1}{a^2}\int d^3x\, e^{ipx}\psi(x)$$

Hence

$$\zeta_q = \frac{1}{a^2}\sqrt{\frac{1}{2J}}\psi_p \;\; ; \;\; \bar{\zeta}_q = \frac{1}{a^2}\sqrt{\frac{1}{2J}}\bar{\psi}_p$$

From Eq. 13.2.3 the limit $a \to 0$ yields

$$\alpha \to \frac{1}{3}am \;\; ; \;\; \beta_i \to \frac{1}{3}ap_i$$

Hence

$$H(\zeta,\bar{\zeta}) \to H(\psi,\bar{\psi}) = \frac{1}{a^4}\frac{1}{2J}a^3\int_{-\pi/a}^{+\pi/a}\frac{d^3p}{(2\pi)^3}\bar{\psi}_p\Big(\frac{1}{3}am + i\frac{1}{3}a\vec{\gamma}\cdot\vec{p}\Big)\psi_p$$

and yields the Dirac continuum Hamiltonian

$$H(\psi,\bar{\psi}) = \int_p \bar{\psi}_p(m + i\vec{\gamma}\cdot\vec{p})\psi_p \tag{13.4.1}$$

Recall from Eq. 13.2.7

$$\{\zeta_{q\alpha}^\dagger, \zeta_{q'\beta}\} = \frac{1}{\lambda_q}(2\pi)^3\delta^{(3)}(q-q')\delta_{\alpha\beta}$$

Using the limit of $a \to 0$, Eq. 13.2.6 yields

$$\lambda_q = \frac{aJ}{K}\Big(1 - 2K\sum_i \cos q_i\Big) \to a\frac{4}{3}\cdot\frac{1}{4} = \frac{a}{3} \tag{13.4.2}$$

and hence

$$\frac{1}{a^4}\frac{1}{2J}\{\psi_{p\alpha}^\dagger, \psi_{p'\beta}\} = \frac{3}{a}\frac{1}{a^3}(2\pi)^3\delta^{(3)}(p-p')\delta_{\alpha\beta}$$

$$\Rightarrow \{\psi_{p\alpha}^\dagger, \psi_{p'\beta}\} = (2\pi)^3\delta^{(3)}(p-p')\delta_{\alpha\beta} \;\; ; \;\; \bar{\psi} = \psi^\dagger\gamma_0$$

Let

$$\psi_x = \int \frac{d^3p}{(2\pi)^3}e^{ipx}\psi_p \;\; ; \;\; \bar{\psi}_x = \int \frac{d^3p}{(2\pi)^3}e^{-ipx}\bar{\psi}_p$$

The four-component fermionic co-ordinate operators satisfy

$$\{\psi_{x\alpha}^\dagger, \psi_{x'\beta}\} = \delta^{(3)}(x-x')\delta_{\alpha\beta} \tag{13.4.3}$$

Noteworthy 13.1: Canonical quantization

For completeness, the Dirac Hamiltonian is derived using the canonical approach [Baaquie (2018)]. Recall from Eq. 12.3.3 the Euclidean Dirac Lagrangian is given by

$$\mathcal{L}(x) = -\bar{\psi}\Big(\gamma_0\frac{\partial}{\partial\tau} + \gamma_i\frac{\partial}{\partial x_i} + m\Big)\psi$$

The canonical momentum of the Dirac field Π_α, for $\dot{\psi} = \partial\psi/\partial t$, is defined by

$$\Pi_\alpha = \frac{\delta\mathcal{L}}{\delta\dot{\psi}_\alpha} = i[\bar{\psi}\gamma_0]_\alpha \tag{13.4.4}$$

where the anticommuting property of the Dirac field has been used to obtain Π_α. Define the conjugate field by

$$\psi^\dagger \equiv \bar{\psi}\gamma_0$$

The Dirac Hamiltonian density, using the anticommuting property of the Dirac field is given by

$$\mathcal{H}(\vec{x}) = \frac{\partial \psi_\alpha}{\partial t}\Pi_\alpha - \mathcal{L} = \psi^\dagger(i\gamma_0\vec{\gamma}\cdot\vec{\partial} + m\gamma_0)\psi$$
$$\Rightarrow \mathcal{H}(\vec{x}) = \bar{\psi}(i\vec{\gamma}\cdot\vec{\partial} + m)\psi$$

Hence, canonical quantization yields the Dirac Hamiltonian

$$\hat{H}(\bar{\psi}, \psi) = \int d^3x \mathcal{H}(\vec{x}) = \int d^3x \bar{\psi}(\vec{\gamma}\cdot\vec{\partial} + m)\psi \qquad (13.4.5)$$

The rules of fermion conjugation show that $H^\dagger = H$: Hermitian. To prove this recall under conjugation

$$\psi \to \bar{\psi}\gamma_0 \;\; ; \;\; \bar{\psi} \to \gamma_0\psi \;\; ; \;\; (\partial)^\dagger = -\partial \;\; ; \;\; (\vec{\gamma})^\dagger = \vec{\gamma} \;\; ; \;\; \{\gamma_0, \vec{\gamma}\} = 0$$

Hence, from Eq. 13.4.5, reversing the order of the fermion variables under conjugation, using $\{\vec{\gamma}, \gamma_0\} = 0$, yields the following

$$H^\dagger = \int d^3x \bar{\psi}\gamma_0\left((\vec{\gamma})^\dagger\cdot(\vec{\partial})^\dagger + m\right)\gamma_0\psi = \int d^3x \bar{\psi}(\vec{\gamma}\cdot\vec{\partial} + m)\psi = H$$

Note that the result obtained from the fermionic state space, path integral, Lagrangian and Hamiltonian yield the same results as obtained by canonical quantization. In particular, the path integral shows that the time derivative term in the Lagrangian yields the anticommutation equation, and which is completely determined by the metric on Hilbert space.

Recall the Euclidean gamma matrices are given by

$$\vec{\gamma} = i\begin{pmatrix} 0 & -\vec{\sigma} \\ \vec{\sigma} & 0 \end{pmatrix} \;\; ; \;\; \gamma_0 = \begin{pmatrix} 1 & 0 \\ 0 & -1 \end{pmatrix}$$

The Dirac Hamiltonian in momentum space, from Eq. 13.4.1, is given by

$$H = \int \frac{d^3p}{(2\pi)^3}\bar{\psi}_p(i\vec{p}\cdot\vec{\gamma} + m)\psi_p$$

where

$$h_p = \gamma_0(i\vec{p}\cdot\vec{\gamma} + m) = \begin{pmatrix} m & \vec{p}\cdot\vec{\sigma} \\ -\vec{p}\cdot\vec{\sigma} & -m \end{pmatrix} \qquad (13.4.6)$$

Hence, Eq. 13.4.6 yields the Dirac Hamiltonian

$$H = \int_p \psi_p^\dagger h_p \psi_p$$

In momentum space, the proof that the Hamiltonian H is Hermitian is given as follows. Under conjugation

$$\psi_p \to \psi^\dagger_{-p} \; ; \; \psi^\dagger_p \to \psi_{-p}$$

Hence, reversing the order of the fermion operators under conjugation yields

$$H^\dagger = \int_p \psi^\dagger_{-p} h^\dagger_p \psi_{-p} = \int_p \psi^\dagger_{-p} h_{-p} \psi_{-p} = \int_p \psi^\dagger_p h_p \psi_p = H \; : \; \text{Hermitian}$$

Choose the coordinate frame so that $\vec{p} = p\hat{e}_z$; then

$$h = \begin{pmatrix} m & p\sigma_3 \\ -p\sigma_3 & -m \end{pmatrix} = \omega S \begin{pmatrix} \mathbb{I} & 0 \\ 0 & -\mathbb{I} \end{pmatrix} S^\dagger \tag{13.4.7}$$

where

$$\omega = \sqrt{\vec{p}^2 + m^2} \; ; \; N_\pm = \sqrt{\frac{\omega \pm m}{2\omega}}$$

$$S = \begin{pmatrix} N_+ & -N_-\sigma_3 \\ N_-\sigma_3 & N_+ \end{pmatrix} \; ; \; SS^\dagger = \mathbb{I} \tag{13.4.8}$$

Define new operators

$$\eta_p = S^\dagger \psi_p \; ; \; \eta^\dagger_p = \psi^\dagger_p S \tag{13.4.9}$$

with

$$\{\eta^\dagger_{p\alpha}, \eta_{p'\beta}\} = \delta_{p-p'}\delta_{\alpha\beta}$$

and

$$H = \int_p \omega_p \eta^\dagger_p \gamma_0 \eta_p = \int_p \omega_p \left(\eta^\dagger_{pu}\eta_{pu} - \eta^\dagger_{p\ell}\eta_{p\ell} \right) \tag{13.4.10}$$

where recall

$$\eta = \begin{pmatrix} \eta_u \\ \eta_\ell \end{pmatrix} \; ; \; \eta^\dagger = \begin{pmatrix} \eta^\dagger_u & \eta^\dagger_\ell \end{pmatrix}$$

Writing the Hamiltonian in terms of the destruction operators on the right hand side yields, from Eq. 13.4.10

$$H = \int_p \omega_p \left\{ \eta^\dagger_{pu}\eta_{pu} + \eta_{p\ell}\eta^\dagger_{p\ell} - 2(2\pi)^3\delta^3(0) \right\} \tag{13.4.11}$$

Note by anticommuting the second term $\eta^\dagger_{p\ell}\eta_{p\ell}$ in $H-$ given in Eq. 13.4.10 – changes the sign of this term in H and yield the Hamiltonian H given in Eq. 13.4.11. Re-interpreting, in Eq. 13.4.11, $\eta^\dagger_{p\ell}$ as the antiparticle *destruction* operator and operator $\eta_{p\ell}$ as the antiparticle *creation* operator, removes the entire conundrum of negative energies that seemed to emerge for the Dirac field.

The Hamiltonian H given in Eq. 13.4.11 is now positive for all eigenstates. In particular, both particles and antiparticles always have positive energies and the vacuum energy $-2(2\pi)^3\delta^3(0)\int_p \omega_p$ is a constant energy that plays no role in the spectrum of states for the Dirac field.

Consider the commutator $[AB, C] = -\{A, C\}B + A\{B, C\}$; hence

$$[H, \eta_{qa}] = \int \omega_p(-)\{\eta^\dagger_{p'\alpha}, \eta_{q,a}\}\gamma_0^{\alpha\beta}\eta_{p\beta} \tag{13.4.12}$$

or, suppressing the momentum index

$$[H, \eta_a] = -\omega\gamma_0^{ab}\eta_b \tag{13.4.13}$$

From Eq. 13.4.13

$$[H, \eta_u] = -\omega\eta_u \tag{13.4.14}$$

For eigenstate $H|\Phi\rangle = E|\Phi\rangle$, Eq. 13.4.14 yields

$$H\eta_u|\Phi\rangle = (E - \omega)\eta_u|\Phi\rangle$$

and hence state $\eta_u|\Phi\rangle$ is a state such that one particle of energy ω has been removed; by analyzing the charge of the state $\eta_u|\Phi\rangle$ one can conclude that η_u *destroys a particle* with energy ω; similarly η_ℓ *creates* an *antiparticle* with positive energy ω. Furthermore

$$[H, \eta^\dagger_{qa}] = \omega_q\gamma_0^{a\alpha}\eta^\dagger_{q\alpha} \tag{13.4.15}$$

Hence, for $\eta^\dagger = (\eta^\dagger_u, \eta^\dagger_\ell)$: $\eta^\dagger_u$ creates a particle with positive energy ω and $\eta^\dagger_\ell$ destroys an antiparticle with energy ω.

Noteworthy 13.2: Dirac negative energy 'sea'

In Dirac's original formulation, the energy spectrum had a positive and negative branch, as shown in Figure 13.1. To remove the negative energy states, Dirac introduced the concept of vacuum state in which the 'sea' of *apparently negative* energy particle states are all occupied; the Pauli exclusion principle ensures that there is only a single particle of a given spin occupying each negative energy state.

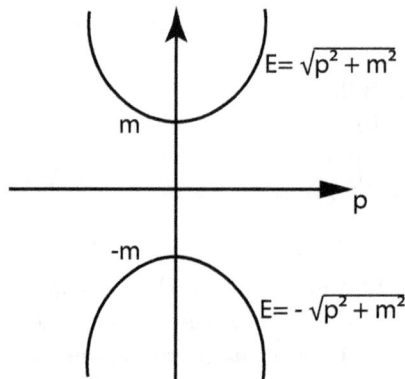

Fig. 13.1   Dirac energy spectrum with positive and negative branches.

The creation of a particle-antiparticle pair was interpreted as moving a negative energy state to the positive branch of the spectrum, with the absence – 'hole' –

left in the negative sea being identified with the antiparticle. In other words, the *absence* of a negative energy particle in the vacuum was taken to be an antiparticle. Dirac introduced these concepts to describe the antiparticle solely in terms of the concept of particles – a logical approach since the concept of the antiparticle was hitherto unknown [Baaquie (2018)].

In the field theory formulation, the Dirac quantum field consist of particles and antiparticles with *both* having positive energy – negative energy states do not exist; instead, there are two species of excitations that both have positive energy, namely the electron and positron and the vacuum, which has no particles or antiparticles. The apparently negative energy states came about because of the incorrect identification of the creation and destruction operators of the antiparticles. The operator $\eta_{p\ell}$ was incorrectly identified as destroying a negative energy state and the operator $\eta_{p\ell}^\dagger$ was incorrectly identified as creating a negative energy state. In the modern formulation, the operator $\eta_{p\ell}$ is correctly identified as *creating* a positive energy positron state and the operator $\eta_{p\ell}^\dagger$ is correctly identified as *destroying* a positive energy positron state.

Recall from Eq. 13.4.9

$$\eta = S^\dagger \psi = \begin{pmatrix} N_+ & N_-\sigma_3 \\ -N_-\sigma_3 & N_+ \end{pmatrix} \psi \ ; \quad \eta^\dagger = \psi^\dagger S = \psi^\dagger \begin{pmatrix} N_+ & -N_-\sigma_3 \\ N_-\sigma_3 & N_+ \end{pmatrix}$$

The creation and destruction operators for the Dirac electron (fermion) and positron (antifermion), suppressing the momentum index and indicating the spin index by s, s', are defined as follows:

Electron operators

$$\eta_{us}^\dagger = \psi_{us}^\dagger N_+ + N_- \sum_{s'=1}^{2} \psi_{\ell s'}^\dagger \sigma_3^{ss'} = \bar{\psi}_{us} N_+ + N_- \sum_{s'=1}^{2} \sigma_3^{ss'} \frac{\delta}{\psi_{\ell s'}}$$

: creation operator

$$\eta_{us} = N_+ \psi_{us} + N_- \sum_{s'=1}^{2} \sigma_3^{ss'} \psi_{\ell s'} = N_+ \frac{\delta}{\delta\bar{\psi}_{us}} + N_- \sum_{s'=1}^{2} \sigma_3^{ss'} \psi_{\ell s'}$$

: destruction operator $\hspace{4cm}$ (13.4.16)

Positron operators

$$\eta_{\ell s} = -N_- \sum_{s'=1}^{2} \sigma_3^{ss'} \psi_{us'} + N_+ \psi_{\ell s} = -N_- \sum_{s'=1}^{2} \sigma_3^{ss'} \frac{\delta}{\delta\bar{\psi}_{us'}} + N_+ \psi_{\ell s}$$

: creation operator

$$\eta_{\ell s}^\dagger = -N_- \sum_{s'=1}^{2} \sigma_3^{ss'} \psi_{us'}^\dagger + \psi_{\ell s}^\dagger N_+ = -N_- \sum_{s'=1}^{2} \sigma_3^{ss'} \bar{\psi}_{us'} + N_+ \frac{\delta}{\delta\psi_{\ell s}}$$

: destruction operator $\hspace{4cm}$ (13.4.17)

13.5   Dirac Field's Energy Eigenfunctionals

The energy eigenfunctionals of the Dirac field are determined using fermionic creation and annihilation operators obtained in Eqs. 13.4.16 and 13.4.17. The Dirac vacuum is defined to be a state in which there are no fermions or antifermions. Let $|\Omega\rangle$ be the Dirac vacuum state, with the coordinate realization

$$\langle \bar{\psi}_u, \psi_\ell | \Omega \rangle = \Omega \left[\bar{\psi}_u, \psi_\ell \right] \tag{13.5.1}$$

Hence, for each momentum q, the annihilation operator for fermions and antifermions, acting on the vacuum, must yield the null state. Hence, the vacuum state must satisfy the following conditions

$$\eta_{\vec{q}us}|\Omega\rangle = 0 \qquad : \text{No electrons in the vacuum state} \tag{13.5.2}$$

$$\eta^\dagger_{\vec{q}\ell s}|\Omega\rangle = 0 \qquad : \text{No positrons in the vacuum state} \tag{13.5.3}$$

Since the Dirac field is a Gaussian quantum field, assume that the vacuum state has the quadratic form given by

$$\Omega[\bar{\psi}_u, \psi_\ell] = \mathcal{N} e^{\bar{\psi}_u A \psi_\ell}$$

where the normalization is $\mathcal{N}$. Consider one of the defining equation for the vacuum state, namely Eq. 13.5.3, which yields

$$\eta^\dagger_\ell |\Omega\rangle = 0 \tag{13.5.4}$$

The coordinate realization of above equation yields, in matrix notation

$$
\begin{aligned}
\langle \bar{\psi}_u, \psi_\ell | \eta^\dagger_\ell \, | \Omega \rangle &= \mathcal{N} \left(\bar{\psi}_u N_- \sigma_3 + N_+ \frac{\delta}{\delta \psi_\ell} \right) e^{\bar{\psi}_u A \psi_\ell} \\
&= \bar{\psi}_u (N_- \sigma_3 + N_+ A) \Omega[\bar{\psi}_u, \psi_\ell] \\
&= 0 \\
\Rightarrow A &= -\frac{N_-}{N_+} \sigma_3 = -\frac{p}{m + \omega}
\end{aligned}
$$

For consistency, Eq. 13.5.2, which is another defining equation for the vacuum state, must result in the same vacuum state. Note the electron destruction operator is defined by

$$\hat{\eta}_u = N_+ \psi_u + N_- \sigma_3 \psi_\ell = N_+ \frac{\delta}{\delta \bar{\psi}_u} + N_- \sigma_3 \psi_\ell$$

and hence

$$
\begin{aligned}
\langle \bar{\psi}_u, \psi_\ell | \eta_u | \Omega \rangle &= \hat{\eta}_u \Omega[\bar{\psi}_u, \psi_\ell] \\
&= \mathcal{N} \left(N_+ \frac{\delta}{\delta \bar{\psi}_u} + N_- \sigma_3 \psi_\ell \right) e^{\bar{\psi}_u A \psi_\ell} \\
&= (N_+ A + N_- \sigma_3) \psi_\ell \Omega[\bar{\psi}_u, \psi_\ell] \\
&= 0 \\
\Rightarrow A &= -\frac{N_-}{N_+} \sigma_3 = -\frac{p}{m + \omega} \sigma_3 \qquad \text{as expected} \tag{13.5.5}
\end{aligned}
$$

Restoring the momentum index yields the following vacuum state in an arbitrary frame for momentum $\vec{p}$

$$\Omega[\bar{\psi}_u, \psi_\ell] = \prod_{\vec{p}} \left(\frac{m + \omega_p}{2\omega_p} \right) \exp \left\{ -\int \frac{d^3 p}{(2\pi)^3} \bar{\psi}_{\vec{p}u} \frac{\vec{p} \cdot \vec{\sigma}}{m + \omega} \psi_{\vec{p}\ell} \right\} : \langle \Omega | \Omega \rangle = 1 \quad (13.5.6)$$

Some of the lower energy eigenfunctions are given below. Restoring the spin index of the creation operators yields, from Eq. 13.4.16, the following electron and positron creation operators

$$\eta_{us}^\dagger = \psi_{us}^\dagger N_+ + N_- \sum_{s'=1}^{2} \psi_{\ell s'}^\dagger \sigma_3^{ss'} = \bar{\psi}_{us} N_+ + N_- \sum_{s'=1}^{2} \sigma_3^{ss'} \frac{\delta}{\psi_{\ell s'}}$$

$$\eta_{\ell s} = -N_- \sum_{s'=1}^{2} \sigma_3^{ss'} \psi_{us'} + N_+ \psi_{\ell s} = -N_- \sum_{s'=1}^{2} \sigma_3^{ss'} \frac{\delta}{\delta \bar{\psi}_{us'}} + N_+ \psi_{\ell s}$$

(I) $E_0 = -2\omega$: ground state; non-degenerate

$$\Phi_0 \equiv \Omega[\bar{\psi}_u, \psi_\ell]$$

$$= \left(\frac{m + \omega}{2\omega} \right) \exp \left\{ -\frac{p}{m + \omega} \bar{\psi}_u \sigma_3 \psi_\ell \right\}$$

$$\Rightarrow \mathbf{K}_0 = \left(\frac{m + \omega}{2\omega} \right)^2 \exp \left\{ -\frac{p}{m + \omega} (\bar{\psi}_u \sigma_3 \psi_\ell - \bar{\psi}_\ell \sigma_3 \psi_u) \right\}$$

$$= \Omega[\bar{\psi}_u, \psi_\ell] \Omega^\dagger[\bar{\psi}_\ell, \psi_u]$$

(II) $E_0 = -\omega$: first excited level; four-fold degenerate

The first excited level consists of a single excitation on the vacuum. From Table 13.5

$$\Phi_s = \langle \bar{\psi}_u, \psi_\ell | \eta_{us}^\dagger | \Omega \rangle \quad ; \quad s = 1, 2$$

$$= \left(\psi_{us}^\dagger N_+ + N_- \sum_{s'=1}^{2} \psi_{us'}^\dagger \sigma_3^{ss'} \right) \Omega[\bar{\psi}_u, \psi_\ell]$$

$$= \left(\bar{\psi}_{us} N_+ + N_- \sum_{s'=1}^{2} \sigma_3^{ss'} \frac{\delta}{\psi_{\ell s'}} \right) \Omega[\bar{\psi}_u, \psi_\ell]$$

$$= \left(\bar{\psi}_{us} N_+ + N_- \frac{p}{m + \omega} \sum_{r,s'=1}^{2} \sigma_3^{ss'} \sigma_3^{s'r} \bar{\psi}_{ur} \right) \Omega[\bar{\psi}_u, \psi_\ell]$$

$$= \left(N_+ + N_- \frac{p}{m + \omega} \right) \bar{\psi}_{us} \Omega[\bar{\psi}_u, \psi_\ell]$$

$$= \sqrt{\frac{2\omega}{\omega + m}} \bar{\psi}_{us} \Omega[\bar{\psi}_u, \psi_\ell]$$

Using the notation of Section 12.11, the application of the creation operators for the electron $\eta_{us}^\dagger$ and positron $\eta_{\ell s}$ states yields the following eigenstates:

$$\Phi_1 = \langle \bar{\psi}_u, \psi_\ell | \eta_{u1}^\dagger | \Omega \rangle = N \bar{\psi}_{u1} \Omega [\bar{\psi}_u, \psi_\ell]$$
$$\Phi_2 = \langle \bar{\psi}_u, \psi_\ell | \eta_{u2}^\dagger | \Omega \rangle = N \bar{\psi}_{u2} \Omega [\bar{\psi}_u, \psi_\ell]$$
$$\Phi_3 = \langle \bar{\psi}_u, \psi_\ell | \eta_{\ell 1} | \Omega \rangle = N \psi_{\ell 1} \Omega [\bar{\psi}_u, \psi_\ell]$$
$$\Phi_4 = \langle \bar{\psi}_u, \psi_\ell | \eta_{\ell 2} | \Omega \rangle = N \psi_{\ell 2} \Omega [\bar{\psi}_u, \psi_\ell]$$
$$N = \sqrt{\frac{2\omega}{\omega + m}}$$

Hence

$$\mathbf{K}_1 = \frac{2\omega}{\omega + m} \Omega [\bar{\psi}_u, \psi_\ell] \bar{\psi} \psi \Omega^\dagger [\bar{\psi}_\ell, \psi_u]$$
$$= \sum_{i=1}^4 \Phi_i \Phi_i^\dagger$$

All the eigenfunctions of the Dirac field can be obtained by the repeated application the electron and positron creation operators on the vacuum state $\Omega [\bar{\psi}_u, \psi_\ell]$. Using the notation of Section 12.11, all the distinct eigenstates are collected in Table 13.5. The classification of the eigenstates given in Table 13.5 shows how the eigenstates that were obtained in Section 12.11, using the path integral, can be generated from the Hamiltonian formulation using the creation and destruction operators.

Table 13.1 The Table provides a list of all the eigenfunctions of the Dirac field, together with creation operators for the fermions and antifermions required to create these states.

Energy	Eigenfunction	Degeneracy	State vector
-2ω	$|\Omega\rangle$	unique	$|\Phi_0\rangle$
$-\omega$	$(\eta_{us}^\dagger, \eta_{\ell s})|\Omega\rangle$	4-fold	$|\Phi_1\rangle, |\Phi_2\rangle, |\Phi_3\rangle, |\Phi_4\rangle$
0	$\left\{ \begin{array}{l} \eta_{us}^\dagger \eta_{\ell s'} |\Omega\rangle \\ \eta_{u1}^\dagger \eta_{u2}^\dagger |\Omega\rangle \ ; \ \eta_{\ell 1} \eta_{\ell 2} |\Omega\rangle \end{array} \right.$	6-fold	$|\Phi_5\rangle, |\Phi_6\rangle, |\Phi_7\rangle, |\Phi_8\rangle, |\Phi_9\rangle, |\Phi_{10}\rangle$
ω	$\left\{ \begin{array}{l} \eta_{u1}^\dagger \eta_{u2}^\dagger \eta_{\ell s} |\Omega\rangle \\ \eta_{\ell 1} \eta_{\ell 2} \eta_{us}^\dagger |\Omega\rangle \end{array} \right.$	4-fold	$|\Phi_{11}\rangle, |\Phi_{12}\rangle, |\Phi_{13}\rangle, |\Phi_{14}\rangle$
2ω	$\eta_{u1}^\dagger \eta_{u2}^\dagger \eta_{\ell 1} \eta_{\ell 2} |\Omega\rangle$	unique	$|\Phi_{15}\rangle$

13.6   Dirac Charge Operator

Abelian gauge invariance yields the following conserved current

$$\partial^\mu j_\mu = 0 \quad : \quad j_\mu = \bar{\psi} \gamma_\mu \psi$$

Define the charge operator $\mathcal{Q}$, which is conserved in time due to gauge invariance, by the following

$$\mathcal{Q} = \int d^3 x j_0(x) \quad \Rightarrow \quad \frac{d\mathcal{Q}}{dt} = 0$$

Consider a gauge transformation

$$\psi \to e^{ie\phi}\psi \;\; ; \;\; \bar{\psi} \to \bar{\psi}e^{-ie\phi}$$

The total charge operator $\mathcal{Q}$ can also be equivalently defined using the concept of gauge transformations. A gauge transformation on the fields state functional is generated by the charge operator $\mathcal{Q}(x)$ and given by

$$\exp\{-ie\int d^3x\phi(x)\mathcal{Q}(x)\}\Phi(\bar{\psi}_u, \psi_\ell) = \Phi(e^{-ie\phi}\bar{\psi}_u, \psi_\ell e^{ie\phi})$$

Let $\phi \simeq 0$; doing a Taylor expansion yields

$$e^{-ie\int \phi\mathcal{Q}}\Phi(\bar{\psi}_u\psi_\ell) = \Phi[e^{-ie\phi}\bar{\psi}_u, \psi_\ell e^{ie\phi}] \simeq \Phi[(1-ie\phi)\bar{\psi}_u, \psi_\ell(1+ie\phi)]$$

$$= \left(1 - ie\int \phi\bar{\psi}_u\frac{\delta}{\delta\bar{\psi}_u} + ie\int \phi\psi_\ell\frac{\delta}{\delta\psi_\ell}\right)\Phi[\bar{\psi}_u, \psi_\ell] \qquad (13.6.1)$$

For $\phi = $ constant, define the total charge operator

$$\mathcal{Q} = \int d^3x\mathcal{Q}(x)$$

Hence, Eq. 13.6.1 yields, up to a constant

$$\mathcal{Q} = e\int d^3x\left(\bar{\psi}_u\frac{\delta}{\delta\bar{\psi}_u} - \psi_\ell\frac{\delta}{\delta\psi_\ell}\right) = e\int d^3x(\bar{\psi}_u\psi_u + \psi_\ell\bar{\psi}_\ell)$$

$$= e\int d^3x\bar{\psi}\gamma_0\psi = e\int d^3x\psi^\dagger\psi = e\int_p \eta_p^\dagger\eta_p$$

$$\Rightarrow \mathcal{Q} = -|e|\int_p(\eta_{pu}^\dagger\eta_{pu} - \eta_{p\ell}\eta_{p\ell}^\dagger) \qquad (13.6.2)$$

Note the identity

$$[AB, C] = A\{B, C\} - \{A, C\}B \qquad (13.6.3)$$

yields, suppressing the momentum index, the following

$$[\mathcal{Q}, \eta_u^\dagger] = -|e|\eta_u^\dagger \quad : \quad \text{creates electron and decreases charge by } |e|$$

$$[\mathcal{Q}, \eta_\ell] = |e|\eta_\ell \quad : \quad \text{creates positron and increases charge by } |e|$$

Gauge invariance implies the important fact that

$$\mathcal{Q} = -|e|(N_{e^-} - N_{e^+}) = \text{constant} : \quad \text{conservation of total charge} \quad (13.6.4)$$

Charge conservation allows for pair creation and annihilation.

13.6.1   *Momentum and spin operators*

Momentum is a translation that shifts the underlying manifold; that is

$$e^{-i\vec{a}\vec{P}}\Phi\left[\bar{\psi}_u(x), \psi_\ell(x)\right] = \Phi\left[\bar{\psi}_u(\vec{x}+\vec{a}), \psi_\ell(\vec{x}+\vec{a})\right]$$

$$= \Phi\left[\bar{\psi}_u(\vec{x})+\vec{a}\cdot\vec{\nabla}\bar{\psi}_u, \psi_\ell(\vec{x})+\vec{a}\cdot\vec{\nabla}\psi_\ell\right]$$

$$\simeq \exp\left(\int\left\{\vec{a}\cdot\vec{\nabla}\bar{\psi}_u\frac{\delta}{\delta\bar{\psi}_u}+\vec{a}\cdot\vec{\nabla}\psi_\ell\cdot\frac{\delta}{\delta\psi_\ell}\right\}\right)\Phi\left[\bar{\psi}_u(\vec{x}), \psi_\ell(\vec{x})\right]$$

Hence, for a = constant yields

$$\Rightarrow \vec{P} = i\int d^3x\left(\vec{\nabla}\bar{\psi}_u\psi_u - \vec{\nabla}\psi_\ell\cdot\bar{\psi}_\ell\right)$$

Recall

$$\bar{\psi}_u(\vec{x}) = \int_p e^{-i\vec{p}\vec{x}}\bar{\psi}_{up} \quad ; \quad \psi_\ell(\vec{x}) = \int_p e^{-i\vec{p}\vec{x}}\psi_{\ell p}$$

Hence, up to a constant

$$\vec{P} = -i^2\int\vec{p}(\bar{\psi}_{up}\psi_{up} - \psi_{\ell p}\bar{\psi}_{\ell p})$$

$$= \int\vec{p}(\bar{\psi}_{up}\psi_{up} + \bar{\psi}_{\ell p}\psi_{\ell p}) = \int\vec{p}\bar{\psi}_p\psi_p$$

$$= \int\vec{p}\psi_p^\dagger\,\gamma_0\psi_p = \int\vec{p}\eta_p^\dagger\,\gamma_0\eta_p = \int\vec{p}(\eta_{pu}^\dagger\,\eta_{pu} - \eta_{p\ell}^\dagger\,\eta_{p\ell})$$

$$\Rightarrow \vec{P} = \int\vec{p}(\eta_{pu}^\dagger\,\eta_{pu} + \eta_{p\ell}\eta_{p\ell}^\dagger\,)$$

Hence

$$[\vec{P}, \eta_{\vec{q}u}^\dagger] = \vec{q}\eta_{\vec{q}u}^\dagger \quad : \quad \text{creates electron with positive momentum}$$

$$[\vec{P}, \eta_{\vec{q}\ell}] = -\vec{q}\eta_{\vec{q}\ell} \quad : \quad \text{creates positron with negative momentum}$$

The spin of the Dirac field is given by

$$\Sigma_3 = \frac{\hbar}{2}\int d^3x\psi^\dagger\begin{pmatrix}\sigma_3 & 0\\ 0 & \sigma_3\end{pmatrix}\psi \quad ; \quad \sigma_3 = \begin{pmatrix}1 & 0\\ 0 & -1\end{pmatrix}$$

$$= \frac{1}{2}\int_p(\eta_{pu}^\dagger\,\sigma_3\eta_{pu} - \eta_{p\ell}\sigma_3\eta_{p\ell}^\dagger\,)$$

The spin carried by fermions is given by

$$[\Sigma_3, \eta_{us}^\dagger] = \frac{\hbar}{2}\sigma_3^{ss'}\eta_{us'}^\dagger = \begin{cases}+\frac{\hbar}{2}\eta_{u1}^\dagger\\ -\frac{\hbar}{2}\eta_{u2}^\dagger\end{cases}$$

and for antifermions by

$$[\Sigma_3, \eta_{\ell s}] = -\frac{\hbar}{2}\sigma_3^{ss'}\eta_{\ell s'} = \begin{cases}-\frac{\hbar}{2}\eta_{\ell 1}\\ \frac{\hbar}{2}\eta_{\ell 2}\end{cases}$$

Recall the creation operator for particle and antiparticle $\eta^\dagger_{us}$ and $\eta_{\ell s}$ respectively, are two component spinors with $s = 1, 2$. The ($s = 1$) upper component creates a spin $1/2$ particle pointing up $|\frac{1}{2}, \frac{1}{2}\rangle$ and the ($s = 2$) lower component creates a spin $1/2$ particle pointing down $|\frac{1}{2}, -\frac{1}{2}\rangle$. Similarly for the antiparticle. Hence

$\eta^\dagger_{\vec{p}us}|\Omega\rangle = |\omega, \vec{p}, s\rangle$: fermion with energy ω, momentum $\vec{p}$ and spin s

$\eta_{\vec{p}\ell s}|\Omega\rangle = |\omega, -\vec{p}, -s\rangle$: antifermion with energy ω, momentum $-\vec{p}$ and spin $-s$

Define the electron and positron number operators

$$N^-_{\vec{p}s} = \text{number of electron in state } \vec{p}, s = \eta^\dagger_{ups}\eta_{ups}$$

$$N^\dagger_{\vec{p}s} = \text{number of positron in state } -\vec{p}, -s = \eta_{\ell ps}\eta^\dagger_{\ell ps}$$

Then, in summary

$$H = \sum_s \int_p \omega_{\vec{p}}(N^-_{ps} + N^\dagger_{ps} - 1)$$

$$Q = -|e| \sum_s \int_p (N^-_{ps} - N^\dagger_{ps} + 1)$$

$$\vec{P} = \sum_s \int_p \vec{p}(N^-_{ps} + N^\dagger_{ps})$$

$$\Sigma_3 = \int_p [(N^\dagger_{p1} - N^-_{p2}) - (N^\dagger_{p1} - N^\dagger_{p2})]$$

13.7 Finite Time Dirac Action

The evolution kernel has been studied in great detail in Chapter 12, and in Section 12.10 the lattice path integral was used to exactly evaluate the evolution kernel. A derivation of the kernel is now given that is directly based on the continuum formulation. The derivation is seen to crucially depend on the finite time Dirac action. It is shown in this Section that the finite time Dirac action needs to be supplemented with boundary terms; and, in fact, it is these boundary terms that precisely yield the evolution kernel.

A derivation of the boundary terms is given starting from the finite time lattice Dirac action. Consider an arbitrary operator G. The evolution kernel (probability amplitude) for the operator is defined by

$$K(\bar{\psi}, \psi; \theta) = \langle \bar{\psi}_u, \psi_\ell | \exp(-\theta G) | \bar{\psi}_\ell, \psi_u \rangle$$

where θ is a real variable analogous to time. K can be evaluated by using the finite time path integral. To do this, divide the interval $[0, \theta]$ into N steps where $\theta = \epsilon N$. Consider the N-fold product for $\exp(-\theta G)$ given by

$$K = \langle \bar{\psi}_u, \psi_\ell | e^{-\epsilon G} \cdots e^{-\epsilon G} | \bar{\psi}_\ell, \psi_u \rangle \tag{13.7.1}$$

From Eq. 12.6.6 the completeness equation is given by

$$\mathbb{I} = \int d\psi d\bar{\psi} |\bar{\psi}_\ell, \psi_u\rangle T(\bar{\psi}, \psi) \langle \bar{\psi}_u, \psi_\ell|$$

For the lattice Dirac field, from Eq. 12.7.15, the Wilson metric for Hilbert space is given by

$$T(\bar{\psi}, \psi) = \exp\left\{ -\sum_{\vec{n}} \bar{\psi}_{\vec{n}}\psi_{\vec{n}} + K \sum_{\vec{n}_i} \left(\bar{\psi}_{\vec{n}}\psi_{\vec{n}+i} + \bar{\psi}_{\vec{n}+i}\psi_{\vec{n}} \right) \right\}$$

Recall that the Wilson metric was introduced to have a four-dimensional cubic symmetry between the space and time for the lattice Dirac action. The term with the coefficient K in the metric reflects its dependence on the action. In the continuum limit, the Wilson metric yields the continuum Hilbert space metric. Since the lattice derivation of this Section is for the purpose of obtaining the continuum boundary terms for the action, it is simpler to start with the lattice approximation of the continuum metric given in Eq. 12.7.17. Hence, the following completeness equation is used

$$\mathbb{I} = \int d\psi d\bar{\psi} |\bar{\psi}_\ell, \psi_u\rangle e^{-\bar{\psi}\psi} \langle \bar{\psi}_u, \psi_\ell | \tag{13.7.2}$$

Introducing the continuum completeness equation $N-1$ times, given in Eq. 13.7.2, into Eq. 13.7.1 yields

$$\mathcal{K} = \prod_{n=1}^{N-1} \int d\bar{\psi}_n d\psi_n e^{-\bar{\psi}_n \psi_n} \langle \bar{\psi}_u, \psi_\ell | e^{-\epsilon G} | \bar{\psi}_{\ell N-1}, \psi_{uN-1}\rangle \cdots$$

$$\cdots \langle \bar{\psi}_{un+1}, \psi_{\ell n+1} | e^{-\epsilon G} | \bar{\psi}_{\ell n}, \psi_{un}\rangle \cdots \langle \bar{\psi}_{u1}, \psi_{\ell 1} | e^{-\epsilon G} | \bar{\psi}_\ell, \psi_u\rangle \tag{13.7.3}$$

Taking the limit of $\epsilon \to 0$ yields

$$\left\langle \bar{\psi}_{un+1}, \psi_{\ell n+1} \middle| \exp(-\epsilon G) \middle| \bar{\psi}_\ell, \psi_u \right\rangle$$

$$\simeq \left\langle \bar{\psi}_{un+1}, \psi_{\ell n+1} \middle| \bar{\psi}_\ell, \psi_u \right\rangle \left\{ 1 - \epsilon G(\bar{\psi}_{un+1}, \bar{\psi}_{\ell n}; \psi_{un}, \psi_{\ell n+1}) + O(\epsilon^2) \right\}$$

$$\simeq \exp\left\{ \bar{\psi}_{un+1}\psi_{un} + \bar{\psi}_{\ell n}\psi_{\ell n+1} - \epsilon G(\bar{\psi}_n, \psi_n) \right\} \tag{13.7.4}$$

where $G(\bar{\psi}, \psi)$ has been obtained using Eq. 12.6.8 and by dropping terms of $O(\epsilon^2)$. For simplicity, introduce the following notation

$$\bar{\psi}_{Nu} = \bar{\psi}_u , \ \psi_{N\ell} = \psi_\ell \ ; \ \bar{\psi}_{0\ell} = \bar{\psi}_\ell , \ \psi_{0u} = \psi_u$$

Hence

$$\mathcal{K} \simeq \prod_{n=1}^{N-1} \int d\bar{\psi}_n d\psi_n \exp(\mathcal{S}) \tag{13.7.5}$$

Suppressing the space lattice index $\vec{n}$, Eqs. 12.7.15, 13.7.3 and 13.7.4 yield

$$\mathcal{S} = -\sum_{n=1}^{N-1} \left\{ \bar{\psi}_n \psi_n + \epsilon G(\bar{\psi}_n, \psi_n) \right\} + \sum_{n=0}^{N-1} \left(\bar{\psi}_{un+1}\psi_{un} + \bar{\psi}_{\ell n}\psi_{\ell n+1} \right)$$

In the limit of $\epsilon \to 0$ with $\theta = \epsilon N$ fixed, the expression Eq. 13.7.5 for $\mathcal{K}$ becomes exact. To take the continuum limit, the boundary terms need to be treated very carefully. Defining the continuum derivative by

$$\partial_0 f(t) = \frac{\partial f}{\partial t} = \lim_{\epsilon \to 0} \frac{(f_{n+1} - f_n)}{\epsilon} \tag{13.7.6}$$

yields the following

$$S = -\epsilon \sum_{n=1}^{N} G(\bar{\psi}_n, \psi_n) + \bar{\psi}_{u1}\psi_{u0} + \bar{\psi}_{\ell 0}\psi_{\ell 1}$$

$$+ \sum_{n=1}^{N-1} \{(\bar{\psi}_{un+1} - \bar{\psi}_{un})\psi_{un} + \bar{\psi}_{\ell n}(\psi_{\ell n+1} - \psi_{\ell n})\}$$

$$\Rightarrow S = \epsilon \sum_{n=1}^{N-1} \{\partial_0 \bar{\psi}_{un}\psi_{un} + \bar{\psi}_{\ell n}\partial_0\psi_{\ell n} - G(\bar{\psi}_n, \psi_n)\}$$

$$+ \bar{\psi}_u(0)\psi_u(0) + \bar{\psi}_\ell(0)\psi_\ell(0) + 0(\epsilon)$$

The limit of $\epsilon \to 0$ yields the finite time Dirac action

$$S = \int_0^\theta dt\{\partial_0\bar{\psi}_u\psi_u + \bar{\psi}_\ell\partial_0\psi_\ell - G(\bar{\psi}, \psi)\} + \bar{\psi}_u(0)\psi_u(0) + \bar{\psi}_\ell(0)\psi_\ell(0)$$

$$\Rightarrow S = -\int_0^\theta dt\{\bar{\psi}\gamma_0\partial_0\psi + G(\bar{\psi}, \psi)\} + \bar{\psi}_u\psi_u(\theta) + \bar{\psi}_\ell\psi_\ell(0) \tag{13.7.7}$$

The finite time action S has a nontrivial boundary term given by

$$\bar{\psi}_u\psi_u(\theta) + \bar{\psi}_\ell\psi_\ell(0)$$

The fermionic variables $\bar{\psi}_u, \bar{\psi}_\ell$ are fixed by the boundary conditions. The terms $\psi_u(\theta), \psi_\ell(0)$ are dynamical integration variables and are coupled to the rest of the action via the time derivative term. The boundary conditions are missing in the infinite time limit; hence limiting the infinite time Dirac action to finite time – without taking the boundary terms into account – yields incorrect results.

A boundary term similar to the Dirac field occurs in Einstein's theory of general relativity. The Gibbons-Hawking-York boundary term needs to supplant the Einstein-Hilbert action for manifolds that have a boundary [York (1971); Gibbons and Hawking (1977)].

13.8 Continuum Evolution Kernel

The evolution kernel for the Dirac field is defined for Euclidean time by

$$K\left[\bar{\psi}, \psi\right] = \langle \bar{\psi}_u, \psi_\ell | e^{-\tau H_D} | \bar{\psi}_\ell, \psi_u \rangle = C(\tau)e^{\bar{\psi}\mathcal{H}\psi}$$

The eigenfunctions of the Dirac field, given in Table 13.5, yield

$$K\left[\bar{\psi}, \psi\right] = \langle \bar{\psi}_u, \psi_\ell | e^{-\tau H_D} | \bar{\psi}_\ell, \psi_u \rangle$$

$$= \sum_{n=1}^{16} e^{-\tau E_n} \Phi_n\left[\bar{\psi}_u, \psi_\ell\right] \Phi_n^\dagger\left[\bar{\psi}_\ell, \psi_u\right] \tag{13.8.1}$$

The evolution kernel is evaluated using the continuum finite time Dirac action. For this, path integral representation of the evolution kernel is required and which is given by

$$K\left[\bar{\psi}, \psi\right] = \langle \bar{\psi}_u, \psi_\ell | e^{-\tau H_D} | \bar{\psi}_\ell, \psi_u \rangle = \int D\bar{\psi} D\psi \exp\{\mathcal{S}\}$$

$$\mathcal{S} = \int_0^\tau dt \int d^3x \mathcal{L} + \mathcal{S}_B$$

The exact path integral expression for K is given by

$$K = \prod_{t=0}^{\theta} \int d\bar{\psi}(t) d\psi(t) \exp(\mathcal{S})$$

with the boundary conditions from (4.6) and (4.7) becoming

$$\bar{\psi}_u(\theta) = \bar{\psi}_u \quad , \quad \psi_\ell(\theta) = \psi_\ell$$
$$\bar{\psi}_\ell(0) = \bar{\psi}_\ell \quad , \quad \psi_u(0) = \psi_u$$

Eq. 12.6.8 yields the finite time Euclidean Dirac Lagrangian

$$\mathcal{S} = \mathcal{S}_D + \mathcal{S}_B$$
$$\mathcal{S}_D = \int_0^\tau dt \int d^3x \mathcal{L}_D \quad ; \quad \mathcal{S}_B = \int d^3x \mathcal{L}_B \qquad (13.8.2)$$
$$\mathcal{L}_D = -\bar{\psi}_D \left(\gamma_0 \partial_0 + \vec{\gamma}\vec{\partial} + m\right) \psi$$

The bulk term Lagrangian $\mathcal{L}_D$ is the one obtained in Eq. 12.3.3 for infinite spacetime case. The additional term in the action $\mathcal{S}_B$ is due to Lagrangian $\mathcal{L}_B$ that is a boundary term that exists only for finite time – being zero for infinite time. The boundary Lagrangian and action are given by

$$\mathcal{L}_B = \bar{\psi}_u(\vec{x})\psi_u(\tau, \vec{x}) + \bar{\psi}_\ell(0, \vec{x})\psi_\ell(\vec{x}) \qquad (13.8.3)$$
$$\mathcal{S}_B \equiv \int d^3x \left[\bar{\psi}_u(\vec{x})\psi_u(\tau, \vec{x}) + \bar{\psi}_\ell(0, \vec{x})\psi_\ell(\vec{x})\right]$$

Note

$$\bar{\psi}_u(\vec{x}) \quad ; \quad \psi_\ell(\vec{x})$$

are the pre-specified boundary conditions whereas

$$\bar{\psi}_\ell(0, \vec{x}) \quad ; \quad \psi_u(\tau, \vec{x})$$

are fermionic integration variables.

Consider the following observations.

- The time derivative terms in $\mathcal{L}_D$, as well as the boundary terms, are the direct result of the initial and final states as well as the metric and the scalar product for the states of the fermion field.
- If $G(\bar{\psi}, \psi)$ is the Hamiltonian, then the first term in the action, given by $\mathcal{S}_D$, is the Euclidean action for finite time θ. Hence the continuum action is not sufficient to determine the behavior for finite time since it excludes crucial boundary terms.

- The result for $\mathcal{S}$ given in Eq. 13.8.2 was derived in Baaquie (1986) using the Wilson lattice action for finite time. And, in fact, the Hamiltonian derivation given here for the finite time continuum action also has to go through an intermediate formulation using the lattice, as in Section 13.7.

In general, a large class of quadratic Lagrangians factorize in momentum space, leading to a factorization of the evolution kernel as follows

$$\mathcal{K}\left[\bar{\psi}, \psi\right] = \prod_p \mathcal{K}\left[\bar{\psi}_p, \psi_p\right] \equiv \prod_p \mathcal{K}_p$$

In the remainder of this Section, all discussions are for $\mathcal{K}_p$, with the momentum index p being restored when required.

The evolution kernel is evaluated by solving the classical field equations (with the boundary conditions) given by

$$\left.\frac{\delta \mathcal{S}_D}{\delta \psi}\right|_{\bar{\psi}=\bar{\zeta}; \ \psi=\zeta} = 0 \tag{13.8.4}$$

Note the boundary Lagrangian is not included in the variation since the variation leaves the boundary conditions unchanged. The classical Dirac action $\mathcal{S}_D$ is zero and the evolution kernel is given by the boundary term as follow

$$\mathcal{K}\left[\bar{\psi}, \psi\right] = C(\tau)e^{\bar{\psi}\mathcal{H}\psi} \tag{13.8.5}$$

$$\mathcal{H} = \int d^3x \left[\bar{\psi}_u(\vec{x})\zeta_u(\tau, \vec{x}) + \bar{\zeta}_\ell(0, \vec{x})\psi_\ell(\vec{x})\right] \tag{13.8.6}$$

and $C(\tau)$ is a normalization function.

The transition amplitude $\mathcal{K}$ is exactly evaluated for case when $G(\bar{\psi}, \psi)$ is a quadratic form. For this case, the amplitude $\mathcal{K}$ completely factorizes in momentum space, that is

$$\mathcal{K} = \prod_p \mathcal{K}_p \tag{13.8.7}$$

The momentum subscript is henceforth ignored.

13.9 Evolution Kernel: General Quadratic Case

Let

$$G(\bar{\psi}, \psi) = \bar{\psi}\gamma_0 M \psi \tag{13.9.1}$$

where M is an arbitrary 4×4 matrix. The effective finite time action is given by

$$\mathcal{S} = -\int_0^\theta dt\{\bar{\psi}\gamma_0(\partial_0 + M)\psi\} + \bar{\psi}_u\psi_u(\theta) + \bar{\psi}_\ell\psi_\ell(0)$$

Let $\bar{\zeta}, \zeta$ be the solution to the field equations, that is

$$\frac{\delta \mathcal{S}(\bar{\zeta}, \zeta)}{\delta \bar{\psi}} = \frac{\delta \mathcal{S}(\bar{\zeta}, \zeta)}{\delta \psi} = 0 \tag{13.9.2}$$

Let $\bar{\zeta}, \zeta$ satisfy the boundary conditions

$$\bar{\zeta}_u(\theta) = \bar{\psi}_u \quad , \quad \bar{\zeta}_\ell(0) = \bar{\psi}_\ell$$
$$\zeta_\ell(\theta) = \psi_\ell \quad , \quad \zeta_u(0) = \psi_u \tag{13.9.3}$$

Make the change of variables and obtain

$$\psi = \zeta + \xi \ , \quad \bar{\psi} = \bar{\zeta} + \bar{\xi}$$

with

$$\xi(0) = \xi(\theta) = 0$$
$$\bar{\xi}(0) = \bar{\xi}(\theta) = 0$$

Using the field equation given in Eq. 13.9.2 yields the following

$$\mathcal{K} = C(\theta) \exp\left\{ \bar{\psi}_u \zeta_u(\theta) + \bar{\psi}_\ell \zeta_\ell(0) \right\} \tag{13.9.4}$$

and

$$C(\theta) = \prod_{t=0}^{\theta} \int d\bar{\xi}_t d\xi_t \exp\left\{ -\int_0^\theta dt \bar{\xi} \gamma_0 (\partial_0 + M)\xi \right\} \tag{13.9.5}$$

Note that $\mathcal{K}$ is given by the boundary term, and only the time dependence of only ζ and not of $\bar{\zeta}$ is required. The coefficient term $C(\theta)$ is independent of the boundary conditions.

To solve for ζ, Eq. 13.9.2 yields the Dirac-like equation

$$\left(\frac{\partial}{\partial t} + M \right) \zeta(t) = 0$$

Introducing the constant four component spinor φ, the solution is given by

$$\zeta(t) = e^{-tM}\varphi \equiv V(t)\varphi \ : \ 0 \le t \le \theta$$

M is a 4×4 matrix; using block 2×2 notation, let

$$V(\theta) = e^{-\theta M} = \begin{pmatrix} V_{11} & V_{12} \\ V_{21} & V_{11} \end{pmatrix}$$

V_{ij} are 2×2 matrices and depend on θ. The boundary conditions given in Eq. 13.9.3 fix the spinor φ as follows.

$$\zeta_u(0) = \psi_u = \varphi_u \ ; \ \zeta(\theta) = \psi_\ell = V_{21}\psi_u + V_{22}\varphi_\ell \ \Rightarrow \ \varphi_\ell = V_{22}^{-1}(\psi_\ell - V_{21}\psi_u)$$

Hence, from above

$$\varphi = B(\theta)\psi \ \Rightarrow \ B(\theta) = \begin{pmatrix} 1 & 0 \\ -V_{22}^{-1}V_{21} & V_{22}^{-1} \end{pmatrix}$$

and hence

$$\zeta(t) = V(t)\varphi = V(t)B(\theta)\psi \ \Rightarrow \ \zeta(0) = B(\theta)\psi$$
$$\Rightarrow \ \zeta(\theta) = V(\theta)B(\theta)\psi$$

From above

$$\zeta_u(\theta) = V_{11}\psi_u + V_{12}V_{22}^{-1}(\psi_\ell - V_{21}\psi_u) \; ; \;\; \zeta_\ell(0) = \varphi_\ell = V_{22}^{-1}(\psi_\ell - V_{21}\psi_u)$$

and hence

$$\bar{\psi}_u\zeta_u(\theta) + \bar{\psi}_\ell\zeta_\ell(0) = \bar{\psi}_u\Big\{(V_{11} - V_{12}V_{22}^{-1}V_{21})\psi_u + V_{12}V_{22}^{-1}\psi_\ell\Big\} + V_{22}^{-1}(\psi_\ell - V_{21}\psi_u)$$

$$= \bar{\psi}\mathcal{H}(\theta)\psi \tag{13.9.6}$$

where

$$\mathcal{H}(\theta) = \begin{pmatrix} V_{11} - V_{12}V_{22}^{-1}V_{21} & V_{12}V_{22}^{-1} \\ -V_{22}^{-1}V_{21} & V_{22}^{-1} \end{pmatrix} \tag{13.9.7}$$

From Eq. 13.9.4, the result is given by

$$\mathcal{K} = C(\theta)\exp\{\bar{\psi}\mathcal{H}(\theta)\psi\}$$

To evaluate $C(\theta)$, one can do it indirectly by independently taking the antiperiodic (or periodic) trace in $\bar{\psi}, \psi$ of both sides of Eq. 13.9.4, as has been done in Section 12.9.1.

13.10 Evolution Kernel: Dirac Hamiltonian

Consider the free Dirac field Hamiltonian H_D with chemical potential ρQ, where Q is the charge operator. Consider the following operator

$$G = H_D - \rho Q \tag{13.10.1}$$

$$= \bar{\psi}\gamma_0\{\gamma_0(i\vec{p}\cdot\vec{\gamma} + m) - \rho\}\psi \tag{13.10.2}$$

which gives

$$M = \gamma_0(i\vec{p}\cdot\vec{\gamma} + m) - \rho = h_p - \rho \tag{13.10.3}$$

From Eq. 13.4.6

$$h_p = \gamma_0(i\vec{p}\cdot\vec{\gamma} + m) = \begin{pmatrix} m & \vec{p}\cdot\vec{\sigma} \\ -\vec{p}\cdot\vec{\sigma} & -m \end{pmatrix}$$

Choose $\vec{p} = p\hat{e}_z$; then, from Eq. 13.4.7

$$h = \begin{pmatrix} m & p\sigma_3 \\ -p\sigma_3 & -m \end{pmatrix} = \omega S \begin{pmatrix} \mathbb{I} & 0 \\ 0 & -\mathbb{I} \end{pmatrix} S^\dagger$$

where, from Eq. 13.4.8

$$\omega = \sqrt{\vec{p}^2 + m^2} \; ; \;\; N_\pm = \sqrt{\frac{\omega \pm m}{2\omega}}$$

$$S = \begin{pmatrix} N_+ & -N_-\sigma_3 \\ N_-\sigma_3 & N_+ \end{pmatrix} \; ; \;\; SS^\dagger = \mathbb{I}$$

Consider

$$V(\theta) = \exp(-\theta M) = e^{-\rho\theta}S\begin{pmatrix} e^{-\omega\theta} & 0 \\ 0 & e^{\omega\theta} \end{pmatrix} S^\dagger = \begin{pmatrix} V_{11} & V_{12} \\ V_{21} & V_{11} \end{pmatrix}$$

From Eq. 13.10.4

$$e^{\rho\theta}V_{11} = e^{-\omega\theta}N_+^2 + e^{\omega\theta}N_-^2\sigma_3^2 = e^{-\omega\theta}\frac{\omega+m}{2\omega} + e^{\omega\theta}\frac{\omega-m}{2\omega}$$

$$= \frac{1}{\omega}[\omega\cosh(\omega\theta) - m\sinh(\omega\theta)] \equiv \frac{1}{\omega}\mu(-\theta)$$

$$\rho^\theta V_{22} = e^{-\omega\theta}N_-^2\sigma_3^2 + e^{\omega\theta}N_+^2 = \frac{1}{\omega}\mu(\theta)$$

$$e^{\rho\theta}V_{12} = e^{-\omega\theta}N_+N_-\sigma_3 - e^{\omega\theta}N_+N_-\sigma_3 = -|\vec{p}\,|\sinh(\omega\theta)\sigma_3 = e^{\rho\theta}V_{21}$$

where

$$\omega = \sqrt{\vec{p}^2 + m^2} \;\; ; \;\; \mu(\theta) = \omega\cosh\omega\theta + m\sinh\omega\theta$$

Collecting the results above, and replacing $|\vec{p}\,|\sigma_3$ by $p\cdot\sigma$, yields the following:

$$V(\theta) = \exp(-\theta M)$$

$$= \frac{1}{\omega}e^{-\rho\theta}\begin{pmatrix} \mu(-\theta) & -p\cdot\sigma\sinh\omega\theta \\ -p\cdot\sigma\sinh\omega t & \mu(\theta) \end{pmatrix}$$

From Eq. 13.9.7, for the Dirac Hamiltonian the matrix elements are given by

$$\mathcal{H}_{11} = V_{11} - V_{12}V_{22}^{-1}V_{21} = \frac{1}{\omega}e^{-\rho\theta}\left(\mu(-\theta) - \frac{1}{\mu(\theta)}\vec{p}^2\sinh^2(\omega\theta)\right)$$

$$= \frac{e^{-\rho\theta}}{\omega\mu(\theta)}\left(\mu(-\theta)\mu(\theta) - \vec{p}^2\sinh^2(\omega\theta)\right) = \frac{e^{-\rho\theta}}{\omega\mu(\theta)}\left(\omega^2[\cosh^2(\omega\theta) - \sinh^2(\omega\theta)]\right)$$

$$\Rightarrow \mathcal{H}_{11} = \frac{\omega e^{-\rho\theta}}{\mu(\theta)}$$

$$\mathcal{H}_{12} = V_{12}V_{22}^{-1} = -\frac{1}{\mu(\theta)}\sinh(\omega\theta)p\cdot\sigma$$

$$\mathcal{H}_{21} = -V_{22}^{-1}V_{21} = \frac{1}{\mu(\theta)}\sinh(\omega\theta)p\cdot\sigma$$

$$\mathcal{H}_{22} = V_{22}^{-1} = \frac{\omega e^{\rho\theta}}{\mu(\theta)}$$

Collecting the results given above, and denoting momentum vector by **p**, yields

$$\mathcal{H}(\theta,\rho) = \frac{1}{\mu(\theta)}\begin{pmatrix} e^{-\rho\theta}\omega & -\mathbf{p}\cdot\sigma\sinh\omega\theta \\ \mathbf{p}\cdot\sigma\sinh\omega\theta & e^{\rho\theta}\omega \end{pmatrix}$$

The result obtained was obtained earlier in Eq. 12.9.11 using the path integral. The coefficient term $C(\theta)$ is independent of ρ and is given by

$$C(\theta) = \left(\frac{\mu(\theta)}{\omega}\right)^2$$

Note that the $\rho = 0$ case was derived in Eq. 12.9.14 using the lattice approach. The final result is given by

$$\mathcal{K} = \langle\bar{\psi}_u, \psi_\ell|\exp\{-\theta H_D - \rho Q\}|\bar{\psi}_\ell, \psi_u\rangle$$

$$= C(\theta)\exp\left\{\bar{\psi}\mathcal{H}(\theta,\rho)\psi\right\}$$

The continuum derivation clearly shows that the result has two aspects, one is solving the field equation (with the boundary conditions) and the other is using the boundary term for the action. The result for the Dirac evolution kernel using the continuum formulation yields the result obtained earlier from the lattice, verifying the correctness of the boundary term for the finite time Dirac action.

13.10.1 *Chiral charge operator*

Let G be the chiral charge operator that is given by

$$G = Q_5 = \bar{\psi}\gamma_0\gamma_5\psi$$

giving

$$V(\theta) = \exp(-\theta\gamma_5) = \begin{pmatrix} \cosh\theta & -\sinh\theta \\ -\sinh\theta & \cosh\theta \end{pmatrix}$$

and

$$H_5(\theta) = \frac{1}{\cosh\theta}\begin{pmatrix} 1 & -\sinh\theta \\ \sinh\theta & 1 \end{pmatrix}$$

$$C_5(\theta) = \cosh^2\theta$$

Hence

$$K_5 = \langle\bar{\psi}_u, \psi_\ell|\exp(-\theta Q_5)|\bar{\psi}_\ell, \psi_u\rangle$$
$$= C_5(\theta)\exp\{\bar{\psi}H_5(\theta)\psi\}$$

What is the utility of K and K_5? For example, as shown in Chapter 12, K for $\rho = 0$ contains the complete eigenfunctions of the free Dirac field. It also contains complete information on Green's function for finite time (or equivalently finite temperature). K_5 is useful for studying chiral symmetry and chiral symmetry breaking.

13.11 Summary

The Hilbert state space and operators for the Dirac field were defined using the fermionic variables that define coordinates of the state space. The continuum Dirac Hamiltonian was defined using the fermionic creation and annihilation operators. The vacuum state and all the energy eigenstates of the Dirac field were obtained.

The Hamiltonian operator is the generator of infinitesimal translations in time, and one must compose the infinitesimal transformations to achieve finite time propagation. This approach for finite time is more general than simply studying the finite time action. In fact, for the Dirac field, using the finite lattice approach, it was shown that the finite time continuum action has to be supplemented with boundary terms. It was verified in this Chapter that the boundary terms in the action yield the correct result for the evolution kernel.

PART 4

Lattice Gauge Theory Hamiltonian

Chapter 14

Lattice Gauge Theory Hamiltonian

14.1 Introduction

The lattice gauge theory Hamiltonian consists of the gauge field gauge-invariantly coupled to the Dirac field. The focus of this Chapter is to derive the Hamiltonian in the axial gauge starting from the path integral. The Hamiltonian is derived starting from the Wilson et al. (1977) lattice action.

The state space for the gauge field coupled to Wilson fermions is obtained and yields noncanonical equal-time (anti)commutation equations. The Hamiltonian is nonlocal for fermions with canonical anticommutation. The color charge operator is obtained and Gauss's law is formulated for the system. The derivation yields a boundary term in addition to the finite-time continuum action, a result derived earlier in Section 13.7 for the pure Dirac field.

Starting from a lattice gauge field Hamiltonian with a canonical metric, which yields canonical anticommutation equations, the lattice action is derived.

The results of this Chapter are largely based on Baaquie (1986). The formulation for the lattice gauge Hamiltonian, particularly those defined by Wilson et al. (1977), Drell et al. (1976) and Luscher (1999), are widely used. The earliest attempt to relate Wilson's lattice action to the Hamiltonian was made by Creutz (1977a). A ground-breaking result obtained by Luscher (2000) in defining chiral lattice fermions has been discussed in Section 12.4.

Wilson derived the matrix element of the operator $\exp(-aH)$, where H is the Hamiltonian for lattice quantum chromodynamics (QCD) and a the time lattice spacing; he also derived a noncanonical metric for the Hilbert space of the interacting theory. The Hamiltonian operator H for the free Dirac field with Wilson fermions has been derived in Section 12.8.

The Hamiltonian operator H is obtained from the transfer matrix, defined as the operator $\exp(-aH)$. The derivation consists of using the metric on the fermionic component of Hilbert space to obtain H as a differential operator with given noncanonical anticommutation relations for the field operators. The lattice gauge theory Hamiltonian is re-expressed using fermions with canonical anticommutation, and this leads to nonlocal fermion and gauge-field interactions. The quark color

271

charge operator is obtained, which is also nonlocal when expressed in terms of canonical fermions. And lastly, the lattice action is obtained starting from a lattice Hamiltonian and using canonical fermions instead of Wilson fermions.

14.2   Finite Time Action and Transfer Matrix

Consider a d-dimensional Euclidean spacetime symmetric lattice with lattice spacing a and let n denote a lattice site. Let the $(d-1)$-dimensional 'spatial' lattice be an infinite lattice and let the time lattice be open and of finite size M. Let $A_\mu^\alpha(x)$ be the continuum $SU(\mathcal{N})$ non-Abelian gauge field and $\bar\psi(x)$, $\psi(x)$ be the continuum $SU(\mathcal{N})$ quark field considered as anticommuting fermionic variables. The lattice degrees of freedom are dimensionless and, as in Chapters 7 and 13, are defined by the following (α and j are color indices)

$$g = sg_0, \quad x = na \ ; \quad B_{n\mu}^\alpha = agA_\mu^\alpha(x), \quad \mu = 0,1,2,\cdots,d-1 \ ; \quad d = 4$$

$$\psi_{nj} = \left(\frac{a^3}{2K}\right)^{1/2}\psi_j(x) \ ; \quad \bar\psi_{nj} = \left(\frac{a^3}{2K}\right)^{1/2}\bar\psi_j(x) \ ; \quad 2K = \frac{1}{4 + m_0 a} \quad (14.2.1)$$

The lattice coupling constants K and g are dimensionless and the continuum quantity sg_0 and dimensional m_0 are the bare coupling constant and bare quark mass, respectively; s is given by

$$\mathrm{Tr}(X_\alpha X_\beta) = \frac{1}{s^2}\delta_{\alpha\beta}$$

The quarks are in the fundamental representation and the gauge field in the adjoint representation of $SU(\mathcal{N})$.

Fermion calculus in the presence of the lattice gauge field is discussed in Appendix A. The transfer matrix is first evaluated, defined as the matrix elements of $\exp(-aH)$, between arbitrary initial and final field configurations.

Field configurations are defined on the $(d-1)$-dimensional spatial lattice; the coordinate eigenstates for the field are given by

$$|\bar\psi_l, \psi_\mu, U\rangle \equiv \prod_{ni}|\bar\psi_{nl}, \psi_{n\mu}, U_{ni}\rangle = \prod_{ni}|\bar\psi_{nl}\rangle \otimes |\psi_{n\mu}\rangle \otimes |U_{ni}\rangle \qquad (14.2.2)$$

such that, as expected,

$$(\hat{\bar\psi}_{nl}, \hat\psi_{n\mu}, \hat U_{ni})|\bar\psi_l, \psi_\mu, U\rangle = (\bar\psi_{nl}, \psi_{n\mu}, U_{ni})|\bar\psi_l, \psi_\mu, U\rangle \qquad (14.2.3)$$

On the left-hand side, we have Schrödinger operators denoted by a caret acting on the state, and on the right-hand side the eigenvalues; note $\bar\psi_{nl}, \psi_{n\mu}$ are anticommuting fermionic eigenvalues. The conjugate eigenstate is

$$\langle\bar\psi_u, \psi_l, U| \equiv \prod_{ni}\langle\bar\psi_{nu}, \psi_{nl}, U_{ni}|$$

and satisfies equations similar to Eq. 14.2.3. The inner product of the coordinate and conjugate eigenstates is given by

$$\langle\bar\psi_u, \psi_l, U'|\bar\psi_l, \psi_u, U\rangle = \langle\bar\psi_u, \psi_l, U|\bar\psi_l, \psi_u, U\rangle\delta(U' - U)$$

The Hilbert space for the interacting theory has a nontrivial metric $T(\bar{\psi}, \psi, U)$; the metric determines the completeness equation on Hilbert space and is given by

$$\mathbb{I} = \prod_{ni} \int d\bar{\zeta}_n d\zeta_n dV_{ni} |\bar{\zeta}_\ell, \zeta_\mu, V\rangle T(\bar{\zeta}, \zeta, V)\langle \bar{\zeta}_\mu, \zeta_\ell, V| \qquad (14.2.4)$$

where $d\bar{\zeta}$ and $d\zeta$ are fermion integrations and dV_{ni} are invariant $SU(\mathcal{N})$ integrations.

The lattice Hamiltonian H is related to the lattice action by the Feynman path integral given by

$$\langle \bar{\psi}_\mu, \psi_l, U' | \exp(-aMH) | \bar{\psi}_\ell, \psi_\mu, U \rangle = \prod_{n_0=1}^{M-1} \prod_{n,\mu} \int d\bar{\psi}_n d\psi_n dU_{n\mu} \exp(S_M) \qquad (14.2.5)$$

with boundary conditions

$$n_0 = 0 : \bar{\psi}_{(0,n)l} = \bar{\psi}_{nl}, \psi_{(0,n)u} = \psi_{nu}, U_{(0,n)i} = U_{ni}$$
$$n_0 = M : \bar{\psi}_{(M,n)u} = \bar{\psi}_{nu}, \psi_{(M,n)l} = \psi_{nl}, U_{(M,n)i} = U'_{ni} \qquad (14.2.6)$$

Recall

$$\gamma_0 = \begin{pmatrix} 1 & 0 \\ 0 & -1 \end{pmatrix}, \quad \gamma_i = i \begin{pmatrix} 0 & -\sigma_i \\ \sigma_i & 0 \end{pmatrix} ; \quad i = 1, 2, 3 \qquad (14.2.7)$$

Using the off-diagonal property of γ_i as given in Eq. 14.2.7, the Wilson action for finite time – with boundary conditions defined by Eq. 14.2.6 – is the following

$$S_M = \frac{1}{g^2} \sum_{n_0=0}^{M-1} \sum_{ni} \text{Tr}\left(U_{ni} U_{n+\hat{1}0} U_{n+\hat{0}i}^\dagger U_{n0}^\dagger + U_{n0} U_{n+\hat{0}i} U_{n+\hat{1}0}^\dagger U_{ni}^\dagger \right)$$

$$+ \frac{1}{g^2} \sum_{n_0=1}^{M-1} \sum_{nij} \text{Tr}(W_{nij})$$

$$+ K \sum_{n_0=0}^{M-1} \sum_n \sum [\bar{\psi}_n (1 - \gamma_0) U_{n0} \psi_{n+\hat{0}} + \bar{\psi}_{n+\hat{0}} (1 + \gamma_0) U_{n0}^\dagger \psi_n]$$

$$- \sum_{n_0=1}^{M-1} \sum_n \left(\bar{\psi}_n \psi_n - K \sum_i [\bar{\psi}_n (1 - \gamma_i) U_{ni} \psi_{n+\hat{1}} + \bar{\psi}_{n+\hat{1}} (1 + \gamma_i) U_{ni}^\dagger \psi_n] \right) \qquad (14.2.8)$$

where, from Eq. 7.1.2

$$W_{nij} = U_{ni} U_{n+\hat{1}j} U_{n+\hat{j}i}^\dagger U_{nj}^\dagger$$

The finite time action is exactly invariant under the local lattice gauge transformation given by

$$\psi_n \to \Phi_n \psi_n ; \quad \bar{\psi}_n \to \bar{\psi}_n \Phi_n^\dagger ; \quad U_{ni} \to U_{ni}(\Phi) \equiv \Phi_n^\dagger U_n \Phi_{n+\hat{1}} \qquad (14.2.9)$$

Note only the first and the third terms in the action Eq. 14.2.8 have $n_0 = 0$ in the time summation, and the boundary values given by Eq. 14.2.6 enter the action only through these terms coupling nearest neighbors in the time direction. Note also that at $n_0 = 0$ only $\bar{\psi}_\ell, \psi_u$ couple to later time and similarly at $n_0 = M$ only

$\bar{\psi}_u, \psi_\ell$ couple to earlier time as is required by the boundary condition given in Eq. 14.2.6. For $M = 2$, Eq. 14.2.5 yields

$$\langle \bar{\psi}_\mu, \psi_l, U' | \exp(-2aH) | \bar{\psi}_l, \psi_\mu, U \rangle$$

$$= \prod_{n,\mu} \int d\bar{\zeta}_n d\zeta_n dV_{n\mu} \exp[\mathcal{S}_2(\bar{\psi}, \psi; U, U'; \bar{\zeta}, \zeta, V)] \qquad (14.2.10)$$

However using the completeness equation Eq. 14.2.4 yields

$$\langle \bar{\psi}_\mu, \psi_\ell, U' | \exp(-2aH) | \bar{\psi}_\ell, \psi_\mu, U \rangle = \prod_{n,i} \int d\bar{\zeta}_n d\zeta_n dV_{ni}$$

$$\times \langle \bar{\psi}_\mu, \psi_l, U' | e^{-aH} | \bar{\zeta}_l, \zeta_\mu, U \rangle T(\bar{\zeta}, \zeta, V) \langle \bar{\zeta}_\mu, \zeta_l, V | e^{-aH} | \bar{\psi}_l, \psi_\mu, U \rangle \quad (14.2.11)$$

Similar to the derivation for the free Dirac field discussed in Section 13.3, by comparing Eqs. 14.2.10 and 14.2.11, it can be shown that, requiring a Hermitian Hamiltonian yields the metric

$$T(\bar{\psi}, \psi, U) = \exp\left(-\sum_n \bar{\psi}_n \psi_n + K \sum_{n,i} (\bar{\psi}_n U_{ni} \psi_{n+\hat{i}} + \bar{\psi}_{n+\hat{i}} U_{ni}^\dagger \psi_n) \right) \quad (14.2.12)$$

The non-trivial metric obtained in Eq. 14.2.12 is the gauge-invariant generalization of the metric obtained earlier in Eq. 12.7.15 for the free Dirac field. Note that the Wilson metric Eq. 14.2.12 depends on the Lagrangian through the parameter K. The norm on the Hilbert space is given by Eq. (A3) For positive norm

$$\langle \Phi | \Phi \rangle \geq 0 \quad \text{for all } |\Phi\rangle \quad \Rightarrow \quad 2K < 1/(d-1)$$

For asymptotically free theories $2K < 1/d$ so that the Wilson metric gives a positive-definite norm for quantum chromodynamics. The canonical fermion metric is obtained by setting K to zero in Eq. 14.2.12.

In anticipation of the discussion on the axial gauge in Section 14.3, consider the change of variables

$$U_{n0} = \Phi_n \Phi_{n+\hat{0}}^\dagger \quad ; \quad dU_{n0} = d\Phi_n$$

From Eq. 12.7.9, introducing a gauge-invariant coupling of the Dirac field to the gauge field yields

$$\langle \bar{\psi}_u, \psi_l, U' | \exp(-aH) | \bar{\psi}_l, \psi_u, U \rangle$$

$$= \prod_n \int d\Phi_n \exp\left[2K \sum_n (\bar{\psi}_{n\ell} \Phi_n \Phi_{n+\hat{0}}^\dagger \psi_{n\ell} + \bar{\psi}_{nu} \Phi_{n+\hat{0}} \Phi_n^\dagger \psi_{n\ell}) \right.$$

$$+ iK \sum_{n,i} (\bar{\psi}_{nu} \sigma_i U'_{ni} \psi_{n+\hat{i}\ell} - \bar{\psi}_{n\ell} \sigma_i U_{ni} \psi_{n+\hat{i}u} - \bar{\psi}_{n+\hat{i}u} \sigma_i U'^\dagger_{ni} \psi_{nl} + \bar{\psi}_{n+\hat{i}l} \sigma_i U_{ni}^\dagger \psi_{nu})$$

$$\left. + \frac{1}{g^2} \sum_{n,i} \text{Tr}(U_{ni} \Phi_{n+\hat{i}}^\dagger \Phi_{n+\hat{i}+\hat{0}} U'^\dagger_{ni} \Phi_{n+\hat{0}}^\dagger \Phi_n + h.c.) + \frac{1}{2g^2} \sum_{n,ij} \text{Tr}(W_{nij} + W'_{nij}) \right]$$

$$(14.2.13)$$

The Wilson metric $T(\bar{\psi}, \psi, U)$ plays a crucial role in ensuring that $\exp(-aH)$ is Hermitian.

Note no gauge was chosen to arrive at $\exp(-aH)$. The $\int d\Phi_n$ integrations ensure that $\exp(-aH)$ is invariant under *separate* time-independent gauge transformations for the coordinate eigenstate and the conjugate eigenstate. More precisely, in the transfer matrix given in Eq. 14.2.13, consider separate gauge-transformations on the coordinate and conjugate states by gauge transformations h_n and $\tilde{h}_n$ respectively

$$\bar{\psi}_u \to \bar{\psi}_u h^\dagger \;\; ; \;\; \psi_\ell \to h\psi_\ell \;\; ; \;\; U'_{ni} \to U'_{ni}(h) = h_n^\dagger U'_{ni} h_{n+i}$$

and

$$\bar{\psi}_\ell \to \bar{\psi}_\ell \tilde{h}^\dagger \;\; ; \;\; \psi_u \to \tilde{h}\psi_u \;\; ; \;\; U_{ni} \to U_{ni}(\tilde{h}) = \tilde{h}_n^\dagger U_{ni} \tilde{h}_{n+i}$$

Doing the change of variables in the $\int d\Phi_n$ integrations as follows

$$\Phi_n \to h_n \Phi_n \;\; ; \;\; \Phi_{n+\hat{0}} \to \tilde{h}_n \Phi_{n+\hat{0}}$$

removes the gauge-transformation h_n and $\tilde{h}_n$ from the transfer matrix given in Eq. 14.2.13 and yields the following

$$\langle \bar{\psi}_u h^\dagger, h\psi_\ell, U'(h) | \exp(-aH) | \bar{\psi}_\ell \tilde{h}^\dagger, \tilde{h}\psi_u, U(\tilde{h}) \rangle$$
$$= \langle \bar{\psi}_u, \psi_\ell, U' | \exp(-aH) | \bar{\psi}_\ell, \psi_u, U \rangle \qquad (14.2.14)$$

Hence, the transfer matrix is gauge-invariant under *separate* time-independent gauge transformations on the coordinate eigenstate and the conjugate eigenstate.

14.3 Axial Gauge: $U_{n0} = \mathbb{I}$

Consider the operator $\exp(-aH)$ acting on the state vector $|\tilde{\Psi}\rangle$

$$\langle \bar{\psi}_\mu, \psi_l, U | \exp(-aH) | \tilde{\Psi} \rangle \qquad (14.3.1)$$
$$= \prod_{n,i} \int d\zeta_n d\zeta_n dV_{ni} \langle \bar{\psi}_\mu, \psi_l, U | e^{-aH} | \bar{\zeta}_l, \zeta_\mu, V \rangle T(\bar{\zeta}, \zeta, V) \langle \bar{\zeta}_\mu, \zeta_l, V | \tilde{\Psi} \rangle$$

where $\tilde{\Psi}$ is an element of the Hilbert space. Consider the following lattice gauge transformation in Eq. 14.3.1 given by

$$\zeta_{nl} \to \Phi_n^\dagger \zeta_{nl} \;\; ; \;\; \bar{\zeta}_{n\mu} \to \zeta_{n\mu} \Phi_n \;\; ; \;\; V_{ni} \to \Phi_n^\dagger V_{ni} \Phi_{n+\hat{i}} \equiv V_{ni}(\phi) \qquad (14.3.2)$$

By a change of integration variables, and using Eq. 14.2.14, one can shift the Φ_n integrations given in Eq. 14.2.13 from $\exp(-aH)$ to the wave functional $\tilde{\Psi}$; the Φ_n integration sums $\tilde{\Psi}$ over all possible gauge transformations and in effect projects out the gauge-invariant subspace of the full Hilbert space spanned by the coordinate eigenstates given in Eq. 14.2.2. Define the gauge-invariant subspace by a collection of all Ψ such that

$$\Psi(\bar{\psi}_u, \psi_\ell, U) = \prod_n \int d\Phi_n \tilde{\Psi}(\bar{\psi}_u \Phi^\dagger, \Phi\psi_\ell, U(\Phi)) \qquad (14.3.3)$$

On this gauge-invariant subspace, the Hamiltonian is that obtained by setting

$$\Phi_n \equiv \mathbb{I}$$

in Eq. 14.2.13, and in effect is the *axial gauge*. Hence, the transfer matrix (Hamiltonian) in the axial gauge is given by

$$\langle \bar{\psi}_u, \psi_l, U' | \exp(-aH) | \bar{\psi}_l, \psi_u, U \rangle =$$

$$\exp\left\{ 2K \sum_n (\bar{\psi}_{nu}\psi_{nu} + \bar{\psi}_{nl}\psi_{nl}) + iK \sum_{n,i} (\bar{\psi}_{nu}\sigma_i U'_{ni}\psi_{n+\hat{\imath}l} \right.$$

$$-\bar{\psi}_{nl}\sigma_i U_{ni}\psi_{n+\hat{\imath}u} - \bar{\psi}_{n+\hat{\imath}u}\sigma_i U'^{\dagger}_{ni}\psi_{nl} + \bar{\psi}_{n+\hat{\imath}l}\sigma_i U^{\dagger}_{ni}\psi_{nu})$$

$$\left. +\frac{1}{g^2} \sum_{n,i} \mathrm{Tr}(U_{ni}U'^{\dagger}_{ni} + U'_{ni}U^{\dagger}_{ni}) + \frac{1}{2g^2} \sum_{n,ij} \mathrm{Tr}(W_{nij} + W'_{nij}) \right\} \qquad (14.3.4)$$

The Hamiltonian H acts on gauge-invariant wave functionals that, from Eq. 14.3.3, satisfy

$$\Psi(\bar{\psi}_u h^{\dagger}, h\psi_\ell, U(h)) = \Psi(\bar{\psi}_u, \psi_\ell, U) \qquad (14.3.5)$$

14.4   Noncanonical Fermion Anticommutation Equations

The metric $T(\bar{\psi}, \psi, U)$ determines the inner product $\langle \bar{\psi}_\mu, \psi_l, U' | \bar{\psi}_l, \psi_n, U \rangle$ as well as the fermion anticommutation equation. In the usual scheme of Hamiltonian quantization, the canonical anticommutation equations are simply postulated. In path integral quantization, however, the commutation equations of the degree of freedom with its conjugate are determined by the action, and this was shown by Feynman for the coordinate x and its conjugate p in the original paper on path integration Feynman (1948). The metric and anticommutation equations of the free Dirac field for the continuum and for the lattice have been discussed earlier in Section 12.6.

The completeness equation Eq. 14.2.4 yields an integral equation for the inner product given by

$$\langle \bar{\psi}_u, \psi_\ell, U' | \bar{\psi}_\ell, \psi_u, U \rangle =$$

$$\prod_{n,i} \int d\bar{\zeta}_n d\zeta_n dV_{ni} \langle \bar{\psi}_u, \psi_\ell, U' | \bar{\zeta}_\ell, \zeta_u, V \rangle T(\bar{\zeta}, \zeta, V) \langle \bar{\zeta}_u, \zeta, V | \bar{\psi}_\ell, \psi_u, U \rangle \qquad (14.4.1)$$

Consider the following notation for the Wilson metric

$$T(\bar{\psi}, \psi, U) = \exp\left(- \sum_{nm,jk} \bar{\zeta}_{nj} M_{nm,jk}[U] \zeta_{mk} \right) \qquad (14.4.2)$$

where, from Eq. 14.2.12, using j and k for non-Abelian indices

$$M_{nm,jk}[U] = \delta_{mn}\delta^{jk} - K \sum_j \left(\delta_{n+\hat{\imath}m} U_{ni}^{jk} + \delta_{nm+\hat{\imath}} U_{mi}^{\dagger jk} \right) \qquad (14.4.3)$$

To solve Eq. 14.4.1 note that the matrix M does not couple upper to lower Dirac components. Using this property of M, it can be shown, similar to the derivation of Eq. 12.7.16, that

$$\langle \bar{\psi}_\mu, \psi_l, U' | \bar{\psi}_l, \psi_\mu, U \rangle = \frac{1}{\det M} \exp \Big(\sum_{nm,jk} \bar{\psi}_{nj} M_{nm,jk} \psi_{mk} \Big) \prod_{n,i} \delta(U_{ni} - U'_{ni}) \quad (14.4.4)$$

To derive fermion anticommutation equations, consider the α Dirac component and j color component of the operator $\bar{\psi}_{n\ell}$ acting on Φ

$$\bar{\psi}^\alpha_{njl} \Phi(\bar{\psi}_\mu, \psi_l, U) = \langle \bar{\psi}_\mu, \psi_l, U | \bar{\psi}^\alpha_{njl} | \Phi \rangle$$

$$= \prod_{m,i} \int d\bar{\zeta}_m d\zeta_m dV_{ml} \bar{\zeta}^\alpha_{njl} \langle \bar{\psi}_\mu, \psi_l, U | \bar{\zeta}_l, \zeta_\mu, V \rangle T(\bar{\zeta}, \zeta, V) \langle \bar{\zeta}_\mu, \zeta_l, V | \Phi \rangle \quad (14.4.5)$$

where Eqs. 14.2.2 and 14.2.4 have been used to derive Eq. 14.4.5. From Eq. 14.4.4, using anticommuting fermion derivatives

$$\zeta^\alpha_{njl} \langle \bar{\psi}_\mu, \psi_\ell, U | \bar{\zeta}_\ell, \zeta_\mu, V \rangle = - \sum_{m,k} M^{-1}_{mn,kj}[U] \frac{\delta}{\delta\psi^\alpha_{mkl}} \langle \bar{\psi}_\mu, \psi_\ell, U | \bar{\zeta}_\ell, \zeta_\mu, V \rangle \quad (14.4.6)$$

Hence,

$$\bar{\psi}^\alpha_{njl} \Phi(\bar{\psi}_\mu, \psi_\ell, U) = - \sum_{m,k} M^{-1}_{mn,kj} \frac{\delta}{\delta\psi^\alpha_{mkl}} \psi(\bar{\psi}_\mu, \psi_\ell, U) \quad (14.4.7)$$

Similarly,

$$\psi^\alpha_{nju} \Phi(\bar{\psi}_u, \psi_\ell, U) = \sum_{m,k} M^{-1}_{nm,jk} \frac{\delta}{\delta\psi^\alpha_{mku}} \Phi(\bar{\psi}_\mu, \psi_\ell, U) \quad (14.4.8)$$

Using the anticommuting property of fermion variables, it follows from Eqs. 14.4.7 and 14.4.8 that the Wilson fermion equal-time anticommutator is

$$\{\bar{\psi}^\alpha_{nj}, \psi^\beta_{mk}\} = M^{-1}_{nm,kj}[U] \gamma_0^{\alpha\beta} \quad (14.4.9)$$

The noncanonical result obtained in Eq. 14.4.9 shows that the fermion equal-time anticommutator is nonlocal and depends on the gauge field U_{ni}.

The anticommutator in Eq. 14.4.9 has two possible perturbative expansions: one is as a series in powers of K; the other is as a power series in the gauge-field variable B^α_{ni} and being appropriate for $g \to 0$. These expansions for $\det M$ are discussed in Appendix B and it is shown, to one loop, that $\det M$ is free from ultraviolet mass divergences. Note for $K = 0$, the canonical anticommutation equations are recovered from Eq. 14.4.9 and is the lattice version of the canonical metric.

14.5 Lattice Gauge Theory Hamiltonian

To obtain the Hamiltonian operator from the transfer matrix $\exp(-aH)$ we use Eq. 14.4.4 to 'subtract out' the inner product from Eq. 14.3.4.

The Hamiltonian for the coupled lattice Dirac and gauge field is given by

$$H = H_G[U] + H_D[\bar{\psi}, \psi, U]$$

where $H_G[U]$ is the gauge field Hamiltonian and $H_D[\bar{\psi}, \psi, U]$ is the Hamiltonian of the Dirac field coupled with the gauge field. The Hamiltonian differential operator H, to leading order in ϵ, is obtained from the transfer matrix given in Appendix C, Eq. C3 and yields the following

$$\langle \bar{\psi}_u, \psi_\ell, U' | \exp\left(-\epsilon H\right) | \bar{\psi}_\ell, \psi_u, U \rangle \tag{14.5.1}$$

$$\simeq e^{-\epsilon \hat{H}(\bar{\psi}, \psi, U)} \langle \bar{\psi}_u, \psi_l, U | \bar{\psi}_\ell, \psi_u, U \rangle \prod_{ni} \delta(U_{ni} - U'_{ni}) \tag{14.5.2}$$

The transfer matrix is obtained from the Lagrangian $\mathcal{L}(\epsilon)$ for an asymmetric lattice given in Appendix C, Eq. C2 – with time lattice $\epsilon \to 0$ and a finite space lattice a. Note the terms independent of ϵ in the Lagrangian yield the metric and the inner product.

In the axial gauge ($U_{n0} = \mathbb{I}$), the Hamiltonian $H_D[\bar{\psi}, \psi, U]$ obtained using the asymmetric Lagrangian $\mathcal{L}(\epsilon)$ is in effect the free Dirac field Hamiltonian, as in Eq. 12.8.5, that has been rendered gauge invariant and is given by

$$H_D[\bar{\psi}, \psi, U] = -\frac{1}{a}(2K - 1)\sum_n \bar{\psi}_n \psi_n - \frac{1}{a}\ln\det M[U]$$

$$+ \frac{K}{a} \sum_{n,i} \left[\bar{\psi}_n(1 - \gamma_i)U_{ni}\psi_{n+\hat{i}} + \bar{\psi}_{n+\hat{i}} + \bar{\psi}_{n+\hat{i}}(1 + \gamma_i)U_{ni}^\dagger \psi_n \right] \tag{14.5.3}$$

where Eq. 14.4.4 has been used to obtain Eq. 14.5.3; note that, due to Eq. 14.4.4, $U_{ni} \simeq U'_{ni}$ to $O(\epsilon)$ in Eq. 14.5.1.

For the gauge field Hamiltonian, the asymmetric lattice Lagrangian in the axial gauge, given in Eq. 7.1.3, yields

$$\langle U' | e^{-\epsilon H_G} | U \rangle = \exp\left\{ \frac{1}{4g^2} \frac{a}{\epsilon} \sum_{i=1}^{3} \text{Tr}(U'_{n,i}U_{n,i}^\dagger + h.c.) + \frac{1}{4g^2}\frac{\epsilon}{a} \sum_{i\neq j=1}^{3} \text{Tr}(W'_{n,ij} + W_{n,ij}) \right\}$$

For $\epsilon \to 0$ Eq. 4.3.5 yields

$$\exp\left\{ \frac{a}{\epsilon}\frac{1}{4g^2}\sum_{ni} \text{Tr}\left(U'_{ni}U_{ni}^\dagger + U_{ni}(U')_{ni}^\dagger\right) \right\} = \exp\left(+\frac{g^2\epsilon}{a}\sum_{n,i}\nabla^2(U'_{ni}) \right)\prod_{ni}\delta(U_{ni} - U'_{ni})$$

where ∇^2 is the Laplace-Beltrami operator for $SU(\mathcal{N})$. The gauge field Hamiltonian is hence given by

$$H_G = -\frac{g^2}{a}\sum_{n,i}\nabla^2(U_{ni}) - \frac{1}{2ag^2}\sum_{n,ij}\text{Tr}(W_{nij}) \tag{14.5.4}$$

Hence, from Eqs. 14.5.3 and 14.5.4, taking ϵ to zero, the lattice gauge theory Hamiltonian in the axial gauge is given by

$$H = -\frac{g^2}{a}\sum_{n,i}\nabla^2(U_{ni}) - \frac{1}{2ag^2}\sum_{n,ij}\text{Tr}(W_{nij}) - \frac{1}{a}(2K - 1)\sum_n \bar{\psi}_n \psi_n$$

$$+ \frac{K}{a}\sum_{n,i}\left[\bar{\psi}_n(1 - \gamma_i)U_{ni}\psi_{n+\hat{i}} + \bar{\psi}_{n+\hat{i}} + \bar{\psi}_{n+\hat{i}}(1 + \gamma_i)U_{ni}^\dagger\psi_n \right]$$

$$- \frac{1}{a}\ln\det M[U] \tag{14.5.5}$$

$\hat{H}_G$ is the Hamiltonian for the lattice gauge theory with Wilson fermions. Recall from Eq. 14.4.9 that the fermion equal-time anticommutation equation is

$$\{\bar{\psi}_{nj}^{\alpha}, \psi_{mk}^{\beta}\} = M_{mn,kj}^{-1}[U]\gamma_0^{\alpha\beta} \tag{14.5.6}$$

Eqs. 14.5.5 and 14.5.6 provide a complete description for the operator formulation of the lattice theory.

The pure gauge field part of H is the Hamiltonian derived in Chapter 11. The fermion part of Eq. 14.5.5 is similar to Creutz (1977a), except for canonical anticommutation equations for the fermions instead of Eq. 14.5.6 and the use of initial and final field configurations being coherent fermion states. A generalization of coherent states, as is shown below, yield the Wilson metric. The last term in the Hamiltonian given by Eq. 14.5.5 is a reflection of the Wilson metric.

14.6   Canonical Fermions

The noncanonical anticommutation equation Eq. 14.5.6 as well as the Wilson metric given in Eq. 14.2.12 can be reduced to the canonical form by the following transformation:

$$\psi_{nj} = \sum_{m,k} M_{mn,jk}^{-1/2}\chi_{mk} \quad ; \quad \psi_{mk} = \sum_{n,j} \bar{\chi}_{nj} M_{mn,jk}^{-1/2} \tag{14.6.1}$$

Note only $\bar{\psi}_{nu}$ and $\psi_{n\ell}$ are independent variables and the transformation of $\bar{\psi}_{n\ell}$ and ψ_{nu} are fixed by Eqs. 14.4.7 and 14.4.8, respectively; Eq. 14.6.1 is to be understood in this sense. Eqs. 14.5.6 and 14.6.1 yield the canonical anticommutator

$$\{\bar{\chi}_{nj}^{\alpha}, \chi_{mk}^{\beta}\} = \delta_{nm}\delta_{jk}\gamma_0^{\alpha\beta} \tag{14.6.2}$$

For the fermion sector of the Hamiltonian, from Eq. 14.5.5

$$-aH_D = -\sum_n \bar{\chi}_n\chi_n + 2K\sum_{nm}\bar{\chi}_n M_{nm}^{-1}\chi_m$$

$$-K\sum_{pnm,j}\bar{\chi}_n M_{nm}^{-1/2}\gamma_i U_{mi} M_{m+ip}^{-1/2}\chi_p + K\sum_{pnm,j}\bar{\chi}_n M_{nm+\hat{i}}^{-1/2}\gamma_i U_{mi}^{\dagger} M_{mp}^{-1/2}\chi_p \tag{14.6.3}$$

Eqs. 14.2.12 and 14.6.1 yield the canonical fermion Hilbert space metric

$$T(\bar{\chi}, \chi) = \exp\left(-\sum_n \bar{\chi}_n\chi_n\right) \tag{14.6.4}$$

The canonical form for H_D involving $\bar{\chi}$ and χ is more suitable for perturbation theory since one can use the Fock basis for the $\bar{\chi}, \chi$ fields. The canonical form for H_D in Eq. 14.6.3 makes H_D nonlocal; noncanonical anticommutation leads to new interactions in H involving the matrix $M^{-1/2}$.

$M_{nm}^{-1/2}$ can be expanded as a power series in K and yields strings of gauge-field links of length L of the type $K^L(\prod_p U_p)$ running from $\mathbf{n}$ to $\mathbf{m}$ (see Figure 14.1). H_D has three-site nonlocal interactions between a quark at point $\mathbf{n}$, a gauge-field link between $\mathbf{m}$ to $\mathbf{m} + \hat{i}$, and an antiquark at $\mathbf{i}$ as in Figure 14.2. These nonlocal interactions also are present in the quark-color charge operator given in Eq. 14.7.1.

The relation of the coordinate eigensates $|\bar{\psi}_l, \psi_u, U\rangle$ with the fermion coherent state formalism used by Creutz (1977b) is discussed. Using the canonical fermion operators $\bar{\chi}_n, \chi_n$ given by Eq. 14.6.2 define the 'bare vacuum' state $|0\rangle$ by

$$\chi_{nu}|0\rangle = \bar{\chi}_{nl}|0\rangle = 0 \quad ; \quad \langle 0|0\rangle = 1 \tag{14.6.5}$$

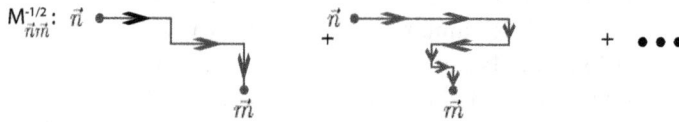

Fig. 14.1   Expansion of $M_{nm}^{1/2}$ in terms of gauge-field string operators.

The coordinate eigenstates satisfying the completeness equation Eq. 14.2.4 with the Wilson metric given in Eq. 14.2.12 can be defined (suppressing summation on all lattice and internal indices, and denoting the gauge field coordinate eigenstate by $|U\rangle$ as

$$|\bar{\psi}_l, \psi_u, U\rangle = C \exp(\bar{\psi}_l M^{1/2} \chi_l + \bar{\chi}_u M^{1/2} \psi_u)|0\rangle|U\rangle \tag{14.6.6}$$

where C is a function of $\det M[U]$. Using the rules of conjugation given in Eq. (A2), we have

$$\langle \bar{\psi}_u, \psi_l, U^\dagger | \bar{\psi}_l, \psi_u, U\rangle = \frac{\exp(\bar{\psi} M \psi)}{(\det M)} \prod_{n,i} \delta(U'_{ni} - U_{ni}) \tag{14.6.7}$$

which is simply Eq. 14.4.4. Eq. 14.6.6 is a nontrivial generalization of fermion coherent states as the gauge field is directly involved in its construction.

The limiting case of free Wilson fermions is the first step in any weak-coupling calculation. When the gauge field is set to zero, obtaining from Eqs. 14.5.5, 14.5.6 and 14.4.3, up to a constant

$$H = -\frac{1}{a}(2K-1)\sum_n \bar{\psi}_n \psi_n + \frac{K}{a}\sum_{n,i}\left[\bar{\psi}_n(1-\gamma_i)\psi_{n+\hat{i}} + \bar{\psi}_{n+\hat{i}}(1+\gamma_i)\psi_n\right] \tag{4.15a}$$

which is, as expected, the free field Dirac lattice Hamiltonian given earlier in Eq. 12.8.5.

14.7   Color Charge Operator and Gauss's Law

In the Lagrangian approach, the charge operator and Gauss's law are obtained by exploiting symmetries of the Lagrangian. In the Hamiltonian formulation, these result from transformation properties and symmetries of the wave functionals.

Consider the gauge field for $SU(\mathcal{N})$. Some Lie algebra is reviewed for the sake of ease of reading. Right and left group multiplication for group elements are given by Hermitian generators E_a^L and E_a^R, respectively, where from Eq. 2.11.8

$$[E_a^L, E_b^L] = iC^{abc}E_c^L, \quad ; \quad [E_a^R, E_b^R] = -iC^{abc}E_c^R \quad ; \quad [E_a^L, E_b^R] = 0 \tag{14.7.1}$$

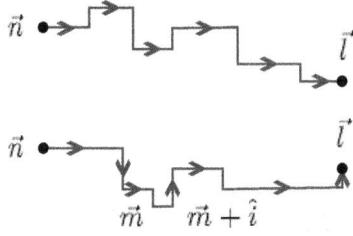

Fig. 14.2   Nonlocal interactions in the Hamiltonian for the fermion sector.

Note the minus sign in Eq. 14.7.1 for E_a^R. The operators E_a^R and E_a^L are first-order Hermitian differential operators on functions of $SU(\mathcal{N})$. For the group element U the operator equation is given by

$$E_a^R(U) = \rho_{ab}(U)E_b^L(U) \quad \Rightarrow \quad E_a^L(U) = \rho_{ab}^T(U)E_b^R(U) \tag{14.7.2}$$

where $\rho_{ab}(U)$ is the adjoint representation of U given in Eq. 2.3.4. For an arbitrary function of U, the $SU(\mathcal{N})$ Taylor theorem yields, from Eq. 2.11.11

$$f(e^{i\phi X_a}Ue^{i\sigma^a X_a}) = e^{i\phi^a E_a^R(U)}e^{i\sigma^a E_a^L(U)}f(U)$$
$$= e^{i(\phi^a E_a^R(U) + \sigma^a E_a^L(U))}f(U) \tag{14.7.3}$$

where Eq. 14.7.1 has been used to obtain Eq. 14.7.3

It has been shown in Eq. 2.11.13 that for $SU(\mathcal{N})$

$$-\nabla^2(U) = \sum_a E_a^R(U)E_a^R(U) = \sum_a E_a^L(U)E_a^L(U) \tag{14.7.4}$$

From Eq. 14.7.4, it follows that either E_a^R or E_a^L can be identified as the chromo-electric field operator for the lattice gauge field.

The analogue of Eq. 14.7.3 for the fermions is more complicated. This is discussed in Appendix A and, from Eq. (A12)

$$h(\bar{\psi}_u e^{-i\phi^a X_a}\psi_\ell) = \exp\left(-i\phi^a X_a^{jk}\bar{\psi}_{ju}^\alpha \frac{\sigma}{\delta\bar{\psi}_{ku}^\alpha} + i\phi^a X_{kl}^\alpha \frac{\sigma}{\delta\psi_{jl}^\alpha}\right)h(\bar{\psi}_u\psi_\ell)$$
$$= \exp\left(-i\phi^a \rho_n^a\right)h(\bar{\psi}_u\psi_\ell) \tag{14.7.5}$$

The charge operator ρ_n^a is the generator of gauge transformations for the lattice gauge theory and is given by

$$\rho_n^a = X_a^{jk}\left(\bar{\psi}_{nju}^\alpha \frac{\delta}{\delta\bar{\psi}_{nku}^\alpha} - \psi_{nkl}^\alpha \frac{\delta}{\delta\psi_{njl}^\alpha}\right) \tag{14.7.6}$$

Consider the (time-independent) gauge transformation $\phi_n = \exp\{i\phi_n^a X_a\}$ such that

$$U_{ni} \to \phi_n U_{ni}\phi_{n+\hat{i}}^\dagger \; ; \; \psi_n \to \phi_n\psi_n, \; \bar{\psi}_n \to \bar{\psi}_n\phi_n^\dagger \tag{14.7.7}$$

Hence, from Eqs. 14.7.3 and 14.7.5, a gauge-transformation on the wave functional is given by

$$\Psi(\bar{\psi}_u, \psi_\ell, U) \rightarrow \Psi(\bar{\psi}_u \phi^\dagger, \phi\psi_\ell, U(\phi))$$
$$= \prod_{n,i} e^{i\phi_n^a E_a^R(U_{ni})} e^{-i\phi_{n+\hat{i}}^a E_a^L(U_{ni})}$$

$$\times \prod_n \exp\left(i\phi_n^a X_a^{jk}\left(\psi_{nkl}^\alpha \frac{\partial}{\partial\psi_{njl}^\alpha} - \bar{\psi}_{nju}^\alpha \frac{\partial}{\partial\psi_{njl}^\alpha}\right)\right) \Psi(\bar{\psi}_u, \psi_\ell, U)$$

$$= \exp\left\{i\sum_{n,a}\phi_n^a\left(\sum_i[E_a^R(U_{ni}) - E_a^L(U_{n-\hat{i},i})] - \rho_n^a(\bar{\psi}, \psi)\right)\right\}\Psi(\bar{\psi}_u, \psi_\ell, U) \quad (14.7.8)$$

where ρ_n^a is the quark colour charge operator, and topologically significant surface terms in combining the chromoelectric field operators have been ignored.

The wave functionals have to be independent of ϕ_n^a in order to be gauge-invariant. Since ϕ_n^a is arbitrary in Eq. 14.7.8 gauge invariance requires the coefficient of ϕ_n^a be zero for each n, a; hence, gauge invariance requires

$$\left[\sum_i[E_a^R(U_{ni}) - E_a^L(U_{n-\hat{i},i})] - \rho_n^a\right]|\Psi\rangle = 0 \quad : \text{ Gauss's Law} \quad (14.7.9)$$

From Eq. 14.7.9, and using Eq. 14.7.2, yields

$$\sum_i[E_a^R(U_{ni}) - \rho_{ab}^T(U_{n-\hat{i},i})E_b^R(U_{n-\hat{i},i})]|\Psi\rangle = \rho_n^a|\Psi\rangle \quad (14.7.10)$$

Hence, Eqs. 14.7.8 and 14.7.9 yield the gauge-invariant subspace of lattice gauge theory given by wave functionals that satisfy

$$\Psi\left(\bar{\psi}_u\phi^\dagger, \phi\psi_\ell, U(\phi)\right) = \Psi(\bar{\psi}_u, \psi_\ell, U) \quad : \text{ Gauge-invariant}$$

It only on the gauge-invariant subspace that the Hamiltonian obtained in the axial gauge is valid.

Noteworthy 14.1: Gauss's Law: Continuum Limit

To illustrate the lattice result given in Eq. 14.7.10 consider taking the $a \rightarrow 0$ classical continuum limit; in this limit, the continuum chromoelectric field operator is the following differential operator

$$E_a^R(U_{ni}) = \left[\delta_{ab} - \frac{1}{2}C_{abc}B_{ni}^c + O(B^2)\right]\frac{\partial}{i\partial B_{ni}^b}$$
$$R_{ab}(U_{ni}) = \delta_{ab} - C_{abc}B_{ni}^c + O(B^2) \quad (14.7.11)$$

In the continuum limit, the distinction between E_a^L, E_a^R is lost. Defining the derivative by the following

$$\frac{\partial f(x)}{\partial x_i} = \lim_{a \to 0} \frac{1}{a}(f_n - f_{n-\hat{i}})$$

yields, from Eqs. 14.2.1, 14.7.10, 14.7.14 and 14.7.11 the expected continuum Gauss's law

$$\sum_i \left(\frac{\partial}{\partial x_i} E_{ai}^{cont}(x) + C^{abc} A_i^b(x) E_{ci}^{cont}(x) \right) = \bar{\psi}(x) \gamma_0 \chi_a \psi(x) \qquad (14.7.12)$$

If E_a^R is identified as the chromoelectric field operator, then Eq. 14.7.10 is Gauss's law using only E_a^R. Eq. 14.7.9 is Gauss's law for the lattice gauge theory; it is understood that this is not an operator equation and that the operator on the left-hand side of Eq. 14.7.9 is acting on wave functionals.

The derivation shows that Gauss's law is a differential statement of the wave functionals being gauge invariant and expresses local conservation of color charge at the lattice site n and for each non-Abelian index a.

It follows from Eq. 14.7.6 that

$$[\rho_n^a, \rho_m^b] = iC_{abc} \rho_n^c \delta_{n-m}$$

and hence ρ_n^a are generators of $SU(\mathcal{N})$ local gauge transformations. Inverting Eqs. 14.4.7 and 14.4.8 yields

$$\frac{\delta}{\delta \bar{\psi}_{nku}^\alpha} = \sum_{m,j} M_{nm,kj}[U] \psi_{mju}^\alpha \quad ; \quad \frac{\delta}{\delta \psi_{njl}^\alpha} = \sum_{m,k} M_{mn,kj}[U] \bar{\psi}_{mkl}^\alpha \qquad (14.7.13)$$

Combining Eqs. 14.7.6 and 14.7.13 yields, using Eq. 14.4.3 and the fact that X_α's are traceless, the following for the definition of the charge operator for the Dirac field

$$\rho_n^a = \bar{\psi}_n \gamma_0 X_a \psi_n - K \sum_i \left(\bar{\psi}_{nu} X_a U_{ni} \psi_{n+\hat{i}u} + \bar{\psi}_{nu} X_a U_{n-\hat{i},i}^\dagger \psi_{n-\hat{i}u} \right)$$

$$+ K \sum_i \left(\bar{\psi}_{n-\hat{i}\ell} U_{n-\hat{i}i} X_a \psi_{n\ell} + \bar{\psi}_{n+\hat{i}} U_{ni}^\dagger X_a \psi_{n\ell} \right) \qquad (14.7.14)$$

Note that the regulated quark charge operator ρ_n^a involves the gauge field due to the Wilson metric; for $K = 0$ the expected canonical result is recovered.

Eq. 14.7.14 together with Eqs. 14.7.9 and 14.5.6 gives a complete definition of Gauss's law.

In terms of canonical fermions $\bar{\chi}$ and χ given by Eqs. 14.6.1 and 14.6.2, the charge-density operator is given as follows. Combining Eqs. 14.7.6 and 14.7.13 yields, using Eq. 14.4.3 and the fact that X_α's are traceless, the following

$$\rho_n^a = \sum_{mp} \left(\bar{\chi}_{mu} M_{mn}^{-1/2} X_a M_{np}^{-1/2} \chi_{pu} - \bar{\chi}_{ml} M_{mn}^{-1/2} X_a M_{np}^{-1/2} \chi_{pl} \right)$$

The definition of quark charge operator involves the matrices $M^{-1/2}$ and $M^{1/2}$.

Note for the Abelian field the charge operator is in effect local and the total charge is given by

$$Q^{Abelian} = \sum_n \rho_n^{Abelian} = \sum_n \bar{\chi}_n \gamma_0 \chi_n.$$

Hence, only for the non-Abelian case does M couple to the charge operator and renders it nonlocal.

Eq. 14.7.14 is valid only classically since it has been assumed that the fields $A_\mu(x)$, $\psi(x)$, and $\bar{\psi}(x)$ are continuous and differentiable. For the quantum case, the new terms in Eq. 14.7.10 which arise due to the (lattice) cutoff all contribute to the renormalized quantum continuum limit.

14.8   Lattice Action from Lattice Hamiltonian

Suppose the starting point for the theory is taken to be the Hamiltonian. The action is derived from the Hamiltonian. Using the off-diagonal property of γ_i as given in Eq. 14.2.7, suppose the Hamiltonian, in the axial gauge, is given by

$$\hat{H}(\bar{\psi}, \psi) = m_0 \sum_n \bar{\psi}_n \psi_n + \sum_{nm} \bar{\psi}_n \begin{pmatrix} 0 & h_{nm} \\ h_{nm}^\dagger & 0 \end{pmatrix} \psi_m + H_{GF} \qquad (14.8.1)$$

with canonical anticommutation equation

$$\{\bar{\psi}_{nj}^\alpha, \psi_{mk}^\beta\} = \gamma_0^{\alpha\beta} \delta_{nm} \delta_{jk} \qquad (14.8.2)$$

The off-diagonal coupling h_{nm} can be nonlocal, and involve the gauge field for the interacting theory. The canonical metric for the anticommutation relation given in Eq. 14.8.2 is the following

$$T_c(\bar{\psi}\psi) = \exp\left(-\sum_n \bar{\psi}_n \psi_n\right) \qquad (14.8.3)$$

and with inner product given by

$$\langle \bar{\psi}_u, \psi_\ell, U' | \bar{\psi}_\ell, \psi_u, U \rangle = \exp\left(\sum_n \bar{\psi}_n \psi_n\right) \prod_{n,i} \delta(U_{ni} - U'_{ni}) \qquad (14.8.4)$$

The evolution kernel is given by

$$K = \langle \bar{\psi}_u, \psi_\ell, U' | e^{-\epsilon MH} | \bar{\psi}_\ell, \psi_u, U \rangle = \prod_{n_0=1}^{M-1} \prod_{n\mu} \int d\bar{\psi}_n d\psi_n dU_{n\mu} e^S \qquad (14.8.5)$$

where the boundary conditions for Eq. 14.8.5 are given in Eq. 14.2.6. Using the completeness equation Eq. 14.2.4 in Eq. 14.8.5 $M-1$ times, with the metric given by Eq. 14.8.3, yields

$$K = \prod_{n_0=1}^{M-1} \prod_{ni} \int d\bar{\psi}_n d\psi_n dU_{ni} e^{-\bar{\psi}_n \psi_n} \langle \bar{\psi}_{n+\hat{0}u}, \psi_{n+\hat{0}l}, U_{n+\hat{0}i} | e^{-\epsilon H} | \bar{\psi}_{nl}, \psi_{nl}, U_n \rangle \qquad (14.8.6)$$

And hence, from Eq. 14.8.4 and Eq. (A4), the fermion part, to leading order in ϵ, is given by

$$K_F \simeq \prod_n \int d\bar{\psi}_n d\psi_n dU_{ni} \exp\left\{ -\sum_n \bar{\psi}_n \psi_n + \sum_n (\bar{\psi}_{n+\hat{0}u} \psi_{nu} + \bar{\psi}_{nl} \psi_{n+\hat{0}l}) \right.$$
$$\left. - \epsilon \sum_{n_0} H_F(\bar{\psi}_{n+\hat{0}u}, \bar{\psi}_{nl}; \psi_{nu}, \psi_{n+\hat{0}l}) \right\} \qquad (14.8.7)$$

Eq. 14.8.1 yields

$$H_F(\bar{\psi}_{n+\hat{0}u}, \bar{\psi}_{nl}; \psi_{nu}, \psi_{n+\hat{0}l}) = m_0 \sum_n \left(\bar{\psi}_{n+\hat{0}u} \psi_{nu} + \bar{\psi}_{nl} \psi_{n+\hat{0}l} \right) \quad (14.8.8)$$

$$+ \sum_{nm} \left(\bar{\psi}_{n+\hat{0}u} h_{nm} \psi_{m+\hat{0}l} + \bar{\psi}_{nl} h_{nm}^\dagger \psi_{mu} \right) \quad (14.8.9)$$

From Eqs. 14.8.7 and 14.8.8, the action is given by

$$\mathcal{S}_F = - \sum_n \bar{\psi}_n \psi_n + (1 - m_0\epsilon) \sum_n \left(\bar{\psi}_{n+\hat{0}u} \psi_{nu} + \bar{\psi}_{nl} \psi_{n+\hat{0}l} \right)$$

$$+ \epsilon \sum_{nm} \bar{\psi}_n \begin{pmatrix} 0 & h_{nm} \\ h_{nm}^\dagger & 0 \end{pmatrix} \psi_m \delta_{n_0, m_0} \quad (14.8.10)$$

Transforming from the axial gauge back to a manifestly gauge-invariant expression yields the lattice action

$$\mathcal{S} = - \sum_n \bar{\psi}_n \psi_n + \frac{1}{2}(1 - m_0\epsilon) \sum_n \left[\bar{\psi}_{n+\hat{0}}(1 + \gamma_0) U_{n0}^\dagger \psi_n \bar{\psi}_n (1 - \gamma_0) U_{n0} \psi_{n+\hat{0}} \right]$$

$$- \epsilon \sum_{nm} \bar{\psi}_n \begin{pmatrix} 0 & h_{nm} \\ h_{nm}^\dagger & 0 \end{pmatrix} \psi_m \delta_{n_0, m_0} + \mathcal{S}_G \quad (14.8.11)$$

The action has Wilson projectors $1 \pm \gamma_0$ for the time coupling and is due to the fermionic coordinates of the Hilbert space. Note action $\mathcal{S}$ in Eq. 14.8.11 explicitly breaks chiral symmetry, even if $m_0 = 0$ and the Hamiltonian is chiral noninvariant. The breaking of the chiral symmetry is due to the structure of the fermionic Hilbert space, which has nonchiral invariant field coordinates, namely, $|\bar{\psi}_\ell, \psi_u\rangle$ and $\langle \bar{\psi}_u, \psi_\ell |$.

The lattice action obtained in Eq. 14.8.11 contains more information than finite time continuum action. To see this, take the limit of $M \to \infty$ and $\epsilon \to 0$ with $T = M\epsilon$ fixed, and for notational simplicity consider the continuum QCD Hamiltonian. Hence, Eq. 14.8.11 yields

$$\mathcal{S} = \int_0^T dt \int dx \mathcal{L}_{YM} + \mathcal{S}_{bd} \quad (14.8.12)$$

and

$$\mathcal{S}_{bd} = \int dx [\bar{\psi}_u(x) \psi_u(x, T) + \bar{\psi}_\ell(x) \psi_\ell(x, 0)] \quad (14.8.13)$$

where $\mathcal{L}_{YM}$ is the continuum Yang-Mills QCD Lagrangian. The first term in Eq. 14.8.12 is the finite-time action; the second term is the boundary term and is given by Eq. 14.8.13; $\bar{\psi}_u(x)$ and $\bar{\psi}_\ell(x)$ are part of the boundary conditions given in Eq. 14.2.6, whereas $\psi_u(x, T)$ and $\psi_\ell(x, 0)$ are integration variables; note boundary values $\psi_u(x)$ and $\psi_\ell(x)$ are also coupled to the action. The importance of the boundary term can be seen in the case of the free Dirac field, where

$$K(T) = \langle \bar{\psi}_u, \psi_\ell | \exp(-T H_{Dirac}) | \bar{\psi}_\ell, \psi_u \rangle$$
$$= C(T) \exp[\mathcal{S}_{bd}(\bar{\psi}, \psi, T)]$$

with $\psi(x, t)$ satisfying the classical field equations with boundary conditions given by Eq. 14.2.6, and $C(T)$ is a normalization function.

14.9 Summary

The lattice gauge theory Hamiltonian of the Dirac field coupled to the gauge field was derived using Wilson fermions. The Hamiltonian is nonlocal due to the nontrivial Wilson metric; transforming the Wilson fermions to canonical fermions makes the Hamiltonian pick up new types of nonlocal interactions.

The gauge-field color charge operator as well as the quark color charge operator were derived using the properties of the lattice gauge theory's wave functionals. The lattice quark charge operator for the non-Abelian case has anomalous pieces that depend upon the gauge field.

The Hamiltonian derived from the action is not unique. The reason being that to derive the Hamiltonian the time lattice spacing in the action has to be taken much smaller than the spatial lattice spacing, and this extension to infinitesimal time is non-unique; furthermore, a metric for Hilbert space has to be chosen. The result obtained by Creutz (1977a) reduces to the Wilson action for a symmetric spacetime lattice, but gives a simpler result for the Hamiltonian compared to the one obtained by using Wilson fermions. In particular, Creutz's Hamiltonian has a hopping parameter different from the value of K, and a different Hilbert space. Creutz's result is consistent, and can be ascribed to a different scheme for extending lattice gauge theory to infinitesimal time.

Using the results obtained in this Chapter, one can perturbatively study the lattice gauge field Hamiltonian both in the strong- and the weak-coupling sectors. To study the QCD Hamiltonian in weak coupling, a gauge has to be chosen for the lattice gauge field degrees of freedom, and is discussed in Chapter 11.

14.10 Appendix A: Fermion Calculus with Gauge Field

Let $|\Phi\rangle$ be a wave functional of the interacting quark-gauge-field system. Its coordinate representation form (2.6) is

$$\Phi(\bar{\psi}_u, \psi_\ell, U) = \langle \bar{\psi}_u, \psi_\ell, U | \Phi \rangle \qquad (A1)$$

Conjugation is defined by: (i) reverse the order of the fermion variables and complex conjugate the coefficients, (ii) $\psi \to \bar{\psi}\gamma_0$, $\bar{\psi} \to \gamma_0\psi$ and (iii) $U_{ni} \to U_{ni}^\dagger$. Hence the conjugate of Φ, denoted by $\Phi^\dagger$, is given by

$$\Phi^\dagger(\bar{\psi}_u, \psi_\ell, U) = \Phi^*(\psi_u, -\bar{\psi}_\ell, U^\dagger) = \langle \Phi | \bar{\psi}_\ell, \psi_u, U \rangle \qquad (A2)$$

The scalar product on the Hilbert space, using Eqs. (A1), (A2), and the completeness equation Eq. 14.2.4 is given by

$$\langle f | h \rangle = \int d\bar{\psi} d\psi dU \langle f | \bar{\psi}_\ell, \psi_u, U \rangle T(\bar{\psi}, \psi, U) \langle \bar{\psi}_u, \psi_\ell, U | h \rangle \qquad (A3)$$

The matrix elements of an operator G are given by

$$\langle \bar{\psi}_u, \psi_\ell, U' | G | \bar{\psi}_\ell, \psi_u, U \rangle = G(\bar{\psi}, \psi; U, U') \langle \bar{\psi}_u, \psi_\ell, U' | \bar{\psi}_\ell, \psi_u, U \rangle \qquad (A4a)$$

and Hermitian conjugation is defined as usual by

$$\langle f|G^\dagger|h\rangle = \langle h|G|f\rangle^* \qquad (A4b)$$

Note in Eq. (A4b), the metric via Eq. 14.2.4 has to be used to define the matrix element $\langle f|G|h\rangle$ and plays a central role in defining the Hermiticity of an operator. In particular, the transfer matrix $\exp(-aH)$ given in Eq. 14.2.13 is Hermitian only with the Wilson matrix given in Eq. 14.2.12. The fermonic 'Fourier transform' of $\Phi(\bar\psi_u, \psi_\ell, U)$ is defined using the Wilson metric $T(\bar\psi, \psi, U)$ given in Eq. 14.2.12, and yields

$$\Phi(\bar\psi_u, \psi_\ell, U) = \int d\bar\psi_\ell d\psi_u T(\bar\psi, \psi, U)\widetilde\Phi(\bar\psi_\ell, \psi_u, U) \qquad (A5a)$$

where $\widetilde\Phi$ denotes the Fourier transform of Φ. Inverting Eq. (A5a) using Eq. 14.4.4 yields

$$\widetilde\Phi(\bar\psi_\ell, \psi_u, U) = \int d\bar\psi_u d\phi_\ell dV \langle \bar\psi_u, \psi_\ell, V|\bar\psi_\ell, \psi_u, U\rangle \Phi(\bar\psi_u, \psi_\ell, V) \qquad (A5b)$$

The additional bosonic integration dV in Eq. (A5b) is needed for compensating for $\delta(V - U)$ that the inner product in Eq. 14.4.4. To prove Eqs. (A5a) and (A5b) requires the identity

$$\delta(\eta - \zeta) = \int d\bar\psi \exp[\bar\psi(\eta - \zeta)] \qquad (A6a)$$

$$\delta(\bar\eta - \bar\zeta) = \int d\psi \exp[(\bar\eta - \bar\zeta)\psi)] \qquad (A6b)$$

where $\eta, \zeta, \bar\eta, \bar\zeta$ are fermionic variables; the left-hand side are fermionic delta-functions that obey the usual definition given by

$$\int d\eta \delta(\eta - \zeta)f(\eta) = f(\zeta) \qquad (A6c)$$

Perform the gauge transformation

$$\psi'_\ell = e^{i\phi^a X_a}\psi_\ell, \quad ; \quad \bar\psi'_u = \bar\psi_u e^{-i\phi^a X_a} \qquad (A7)$$

Then, from Eqs. 14.2.12 and (A5a) we have in abbreviated notation

$$\Phi(\bar\psi'_u, \psi'_\ell, U)$$

$$= \int d\bar\psi_\ell d\psi_u \exp\left(-\bar\psi_u e^{-i\phi^a X_a}M\psi_u - \bar\psi_\ell M e^{i\phi^a X_a}\psi_\ell\right)\Phi(\bar\psi_\ell, \psi_u, U) \qquad (A8)$$

Let $h_u = M\psi_u$ and $\bar h_\ell = \bar\psi_\ell M$. A proof is given of the following equation.

$$\exp(-\bar\psi_u e^{-i\phi^a X_a}h_u) = \exp\left(-i\phi^a X_a^{jk}\bar\psi_{ju}\frac{\delta}{\delta\bar\psi_{ku}}\right)\exp\{-\bar\psi_u h_u\} \qquad (A9)$$

For infinitesimal ϕ^a the right-hand side of Eq. (A9) is

$$\left(1 - i\phi^a X_a^{jk}\bar\psi_{ju}\frac{\delta}{\delta\bar\psi_{ku}} + O(\phi^2)\right)e^{-\bar\psi_u h_u} = (1 + i\phi^a\bar\psi_u X_a h_u)e^{-\bar\psi_u h_u}) \qquad (A10)$$

$$\simeq \exp[-\bar\psi_u e^{i\phi^a X_a}h_u + O(\phi^2)] \qquad (A11)$$

Iterating Eq. (A10) yields Eq. (A9) for finite ϕ^a. Similarly

$$\Phi(\bar{\psi}_u e^{-i\phi^a X_a}, e^{i\phi^a X_a}\psi_\ell, U)$$

$$= \exp\left(i\phi^a X_a^{jk}\left(-\bar{\psi}_{ju}\frac{\delta}{\delta\bar{\psi}_{ku}} + \psi_{kl}\frac{\delta}{\delta\psi_{jl}}\right)\right)\Phi(\bar{\psi}_u, \psi_\ell, U) \qquad (A12)$$

where Eq. (A12) has been obtained by using the commutativity of the exponents in Eq. (A12).

14.11   Appendix B: Matrix M

The matrix M that appears in the Hamiltonian and in the noncanonical commutation equations for the Wilson fermions is analyzed. From Eqs. 14.4.2 and 14.4.3, using $\bar{\psi}, \psi$ that carry only color charge – since M carries no Dirac indices – let

$$\mathcal{S}_M(\bar{\psi}, \psi, U) = -\sum_{nm,jk} \bar{\psi}_{nj} M_{nm,jk} \psi_{mk} \qquad (B1)$$

and the matrix M is given by

$$M_{nm,jk} = \delta_{nm}\delta^{jk} - K\sum_i(\delta_{n+\hat{i},m}U_{ni}^{jk} + \delta_{n,m+\hat{i}}U_{mi}^{\dagger jk}) \qquad (B2)$$

There are two expansions for M, namely, as a power series in K and in B_{ni}^a, where $U_{ni} = \exp(iB_{ni}^a X_a)$; the expansion in B_{ni}^a is in effect an expansion as a power series in the coupling constant g.

(a) Weak-coupling expansion. Consider the following

$$U_{ni} = 1 + iB_{ni}^a X_a + \frac{i^2}{2}(B_{ni}^a X_a)^2 + O(B^3) \qquad (B3)$$

Define

$$\lambda_p = 1 - 2K\sum_i \cos p_i \quad ; \quad \sum_a X_a^2 = c_2\mathbb{I} \quad ; \quad \int_p \equiv \int_{-\pi}^{+\pi}\frac{d^3p}{(2\pi)^3} \qquad (B4)$$

Eqs. (B1) and (B3) yield (see Figure 14.3)

$$\det M = \prod_n \int d\bar{\psi}_n\psi_n \exp[\mathcal{S}_M(\bar{\psi}, \psi, B)] \qquad (B5)$$

$$= \exp\left(\frac{1}{2}c_2KJ\sum_{nia}(B_{ni}^a)^2 - \frac{1}{2}c_2K^2\sum_{nm,ija}B_{ni}^aB_{mj}^a\int_q e^{iq(n-m)}\Gamma_{ij}(q) + O(B^3)\right)$$

det M: (a)   +   (b)

Fig. 14.3   Feynman diagrams for expansion of det$M$ in powers of the gauge field.

where

$$J = \int_p \left(\frac{e^{ip_i} + e^{-ip_i}}{\lambda_p} \right) \; ; \; \Gamma_{ij}(q) = \int_p \frac{(e^{i(p+q)_i} - e^{-ip_j})(e^{i(p+q)_j} - e^{-ip_i})}{\lambda_p \lambda_{p+q}} \quad (B6)$$

The first term in Eq. (B5) comes from Figure 14.3(a) and the second term from Figure 14.3(b). Note for $d = 4$, J is linearly divergent, and so is $\Gamma_{ij}(q)$, and would give a divergent mass term for the gauge field. However, these two divergences cancel exactly (due to lattice gauge invariance) using the numerical identity in d dimensions [Baaquie (1977b)]

$$\delta_{ij} J = K \Gamma_{ij}(q = 0) \quad (B7)$$

Hence, from Eq. (B5)

$$\det M = \exp\left[-\frac{\kappa^2}{2} c_2 \sum_{nmija} B^a_{ni} B^a_{mj} \int_q e^{iq \cdot (n-m)} [\Gamma_{ij}(q) - \Gamma_{ij}(0)] + O(B^3) \right] \quad (B8)$$

The ultraviolet divergences of M have to be studied using the weak-coupling expansion [Baaquie (1977b)].

(b) Strong-coupling expansion. For the case of $g \gg 1$, the theory is expanded as a power series in K and $1/g^2$. To leading order in K, Eqs. (B1) and (B5) yield

$$\det M \simeq \exp\left(\frac{\kappa^4}{4} \sum_{n,ij} \mathrm{Tr}(U_{ni} U_{n+\hat{i}j} U^\dagger_{n+\hat{j}i} U^\dagger_{nj} + U_{nj} U_{n+\hat{j}i} U^\dagger_{n+\hat{i}j} U^\dagger_{ni}) + O(K^6) \right) \quad (B9)$$

14.12 Appendix C: Lagrangian for an Asymmetric Lattice

If one uses a symmetric spacetime lattice, and take the logarithm of the transfer matrix $\exp(-aH)$ to obtain the Hamiltonian, the result cannot be obtained in closed form and is not very useful.

Consider instead an asymmetric lattice with the time lattice spacing being ϵ and the spatial lattice spacing being a; the limit of $\epsilon \to 0$ is then taken [Baaquie (1986)]; this limit yields a closed form Hamiltonian. For the asymmetric lattice, the relations given in Eq. 14.2.1 are all valid except that

$$B^a_{n0} = \epsilon g A^a_0(x), \quad x = (n_0 \epsilon, \mathbf{n}a) \quad (C1)$$

An asymmetric lattice Lagrangian is considered such that

- the expected classical continuum limit is recovered
- the Wilson Lagrangian is recovered, for $\epsilon = a$ symmetric lattice, and
- the metric and the inner product given by Eqs. 14.2.12 and 14.4.4, respectively, are incorporated in the Lagrangian so as to obtain a Hermitian Hamiltonian.

These considerations yield the following asymmetric lattice Lagrangian density

$$\mathcal{L}(\epsilon) = \frac{1}{2}\left(\frac{\epsilon}{a}(2K-1)+1\right)\left[\bar{\psi}_{n+\hat{0}}U_{n0}^{\dagger}\psi_n + \bar{\psi}_n(1-\gamma_0)U_{n0}\psi_{n+\hat{0}}\right]$$

$$+\frac{1}{2}K\left(\frac{\epsilon}{a}-1\right)\sum_i\left[\bar{\psi}_{n+\hat{0}}(1+\gamma_0)U_{n+\hat{0}i}U_{n+\hat{i}0}^{\dagger}\psi_{n+\hat{i}}\right.$$

$$+\bar{\psi}_{n+\hat{i}}U_{n+\hat{i}0}U_{n+\hat{0}i}^{\dagger}(1-\gamma_0)\psi_{n+\hat{0}} + \bar{\psi}_n(1-\gamma_0)U_{ni}U_{n+\hat{i}0}\psi_{n+\hat{i}+\hat{0}}$$

$$\left.+\bar{\psi}_{n+\hat{i}+\hat{0}}(1+\gamma_0)U_{n+\hat{i}0}^{\dagger}U_{ni}^{\dagger}\psi_n\right] - \frac{\epsilon}{a}K\sum_i\left(\bar{\psi}_n\gamma_i U_{ni}\psi_{n+\hat{i}} - \bar{\psi}_{n+\hat{i}}\gamma_i U_{ni}^{\dagger}\psi_n\right)$$

$$-\bar{\psi}_n\psi_n + K\sum_i\left(\bar{\psi}_n U_{ni}\psi_{n+\hat{i}} + \bar{\psi}_{n+\hat{i}}U_{ni}^{\dagger}\psi_n\right)$$

$$+\frac{a}{\epsilon}\frac{1}{g^2}\sum_i\text{Tr}\left(U_{ni}U_{n+\hat{i}0}U_{n+\hat{0}i}^{\dagger}U_{n0}^{\dagger} + U_{n0}U_{n+\hat{0}i}U_{n+\hat{i}0}^{\dagger}U_{ni}^{\dagger}\right)$$

$$+\frac{\epsilon}{a}\frac{1}{g^2}\sum_{ij}\text{Tr}\left(W_{nij}\right) + \left(\frac{\epsilon}{a}-1\right)\ln\det M[U] \qquad (C2)$$

It is verified in Appendix D that this Lagrangian, because of some nontrivial cancellations, has the expected classical continuum limit.

The asymmetric Lagrangian is much more complicated than the asymmetric action used by Creutz (1977a), and this is due to the use of the Wilson metric. Note for $\epsilon = a$ the symmetric Wilson action is recovered.

There are two new types of term for the asymmetric lattice: namely, the second term and the last term in Eq. (C2). The second term arises from the inner product given by Eq. 14.4.4 taken between states which are nearest neighbors in time. The last term is also due to the inner product – which includes the determinant of the matrix M – and is discussed in Appendix B; the $\det M$ term is local in time but nonlocal in space, and is a quantum effect that vanishes in the classical limit as $\hbar$.

From the Lagrangian given by Eq. (C2), the transfer matrix (and subsequently the action) – as in Eq. 14.3.4 – is obtained by repeating Eqs. 14.2.10 to 14.2.13 and in the **axial gauge**

$$\langle\bar{\psi}_u,\psi_\ell,U'|\exp(-\epsilon H)|\bar{\psi}_\ell,\psi_u,U\rangle =$$

$$\exp\left(\left(\left(\frac{\epsilon}{a}(2K-1)+1\right)\right)\sum_n\left(\bar{\psi}_{nu}\psi_{nu}+\bar{\psi}_{nl}\psi_{nl}\right)\right.$$

$$+K\left(\frac{\epsilon}{a}-1\right)\sum_{n,i}\left(\bar{\psi}_{nu}U_{ni}'\psi_{n+\hat{i}u} + \bar{\psi}_{n+\hat{i}l}U_{ni}'^{\dagger}\psi_{nl} + \bar{\psi}_{ni}U_{ni}\psi_{n+\hat{i}l} + \bar{\psi}_{n+\hat{i}u}U_{ni}^{\dagger}\psi_{nu}\right)$$

$$+i\frac{a}{\epsilon}K\sum_{n,i}\left(\bar{\psi}_{nu}\sigma_i U_{ni}'\psi_{n+\hat{i}\ell} - \bar{\psi}_{n+\hat{i}\ell}\sigma_i U_{ni}'^{\dagger}\sigma_i\psi_{nu} + \bar{\psi}_{n\ell}\sigma_i U_{ni}\psi_{n+\hat{i}u} - \bar{\psi}_{n+\hat{i}u}\sigma_i U_{ni}^{\dagger}\sigma_i\psi_{n\ell}\right)$$

$$+\frac{a}{\epsilon}\frac{1}{g^2}\sum_{n,i}\text{tr}\left(U_{ni}U_{ni}'^{\dagger}U_{ni}'U_{ni}^{\dagger}\right) + \frac{\epsilon}{a}\frac{1}{2g^2}\sum_{n,ij}\text{tr}\left(W_{nij}+W_{nij}'\right)$$

$$\frac{2^{d/2}}{2}\left(\frac{\epsilon}{a}-1\right)\left(\ln\det M[U]+\ln\det M[U']\right) \qquad (C3)$$

14.13 Appendix D: Classical Continuum Limit

By the classical continuum limit is means the limit of $\hbar \to 0$ and $a \to 0$ for the lattice theory, with the field values and their derivatives being continuous and differentiable. The classical continuum limit for the symmetric lattice has been derived by Wilson. We essentially redo this calculation for the asymmetric lattice theory given by Eq. (C2).

Recall from Eqs. 14.2.1 and (C1) we have for spatial lattice spacing a and time lattice spacing ϵ the following:

$$x = (n_0\epsilon, na) \;\; ; \;\; B_{n_0}^\alpha = \epsilon g A_0^\alpha(x) \;\; ; \;\; B_{nl}^\alpha = ag A_i^\alpha(x)$$

$$\psi_n = \left(\frac{a^{d-1}}{2K}\right)^{1/2} \psi(x) \;\; ; \;\; \bar{\psi}_n = \left(\frac{a^{d-1}}{2K}\right)^{1/2} \bar{\psi}(x)$$

$$2K = 1/(d + m_0 a) \;\; ; \;\; g = g_0 s a^{(d-4)/2} \tag{D1}$$

The limit of $a \to 0, \epsilon \to 0$ is taken for the action defined on an infinite spacetime lattice

$$\mathcal{S} = \sum_n \mathcal{L} \tag{D2}$$

where $\mathcal{L}$ is given by Eq. (C2).

The $\ln \det M$ term in Eq. (C2) goes as $\hbar$ in the action (D2) and vanishes in the classical limit; hence it will be dropped.

In the classical limit, $A_\mu^\alpha(na)$ and similarly for $\psi(na + \hat{\mu}a)$ and so on, are expanded in a Taylor series; in particular, for the pure gauge-field part of action, it can be readily shown, using Eq. (D1) that

$$\mathcal{S}_{GF} = \frac{1}{g^2}\frac{a}{\epsilon}\sum_{n,i} \text{Tr}(U_{ni}U_{n+\hat{1}0}U_{n+\hat{0}i}^\dagger + h.c.) + \frac{\epsilon}{a}\frac{1}{g^2}\sum_{n,ij}\sum_{n,ij} \text{Tr}(W_{nij})$$

$$\simeq -\frac{1}{4}\epsilon a^{d-1}\sum_n \left(\sum_{ia}[F_{0i}^a(x)]^2 + \sum_{ija}[F_{ij}^a(x)]^2\right) \tag{D3}$$

where

$$F_{\mu v}^a(x) = \partial_\mu A_v^a + sg_0 C_{abc} A_\mu^b A_v^c \tag{D4}$$

Using

$$\epsilon a^{d-1}\sum_n \to \int d^d x \tag{D5}$$

yields the continuum Yang-Mills action from Eq. (D3).

The spacetime asymmetry is more nontrivial for the fermion sector. Taylor expanding the fields using Eq. (D1) yields

$$\psi_{n+\hat{0}} = \left(\frac{a^{d-1}}{2K}\right)^{1/2}[\psi(x) + \epsilon\partial_0\psi(x) + O(\epsilon^2)]$$

$$\psi_{n+\hat{i}} = \left(\frac{a^{d-1}}{2K}\right)^{1/2}[\psi(x) + a\partial_i\psi(x) + O(a^2)]$$

$$U_{n+\hat{0}i} \simeq 1 + iga[A_i^\alpha(x) + a\partial_0 A_i^\alpha(x) + ...]$$

$$U_{n+\hat{i}0} \simeq 1 + ig\epsilon[A_0^\alpha(x) + \epsilon\partial_i A_0^\alpha(x) + ...] \qquad (D6)$$

On carrying out the Taylor expansion of the action, we keep terms only of $O(\epsilon a^{d-1})$ and discard all higher order terms. All terms of lower order must cancel (as, in fact, they do) to have a finite classical limit. The fermion part of the action, after considerable simplifications, using $A_\mu \equiv A_\mu^\alpha X_\alpha$, is given by

$$\mathcal{S}_D \simeq -\epsilon a^{d-1}\sum_n \left\{\frac{1}{a}\left(\frac{1}{2K} - d\right)\bar{\psi}(x)\psi(x) + \left(1 + \left(\frac{\epsilon}{a} - 1\right)\left(d - \frac{1}{2K}\right)\right)\right.$$

$$\left. \times \bar{\psi}(x)\gamma_0\left[\partial_0 + isg_0 A_0(x)\right]\psi(x) + \bar{\psi}(x)\gamma_i\left[\partial_i + isg_0 A_i(x)\right]\psi(x)\right\}$$

$$+ O(\epsilon^2 a^{d-1}, \epsilon a^d) \qquad (D7)$$

From Eq. (D1)

$$1/(2K) - d = m_0 a \qquad (D8)$$

Hence, the coefficient of the time-derivative term in Eq. (D7) becomes, from Eq. (D8)

$$1 - m_0 a(\tfrac{\epsilon}{a} - 1) = 1 + O(a) \qquad (D9)$$

and consequently the time and space asymmetry vanishes in the classical continuum limit. From Eqs. (D7), (D8) and (D3), (D5)

$$\mathcal{S} = \mathcal{S}_D + \mathcal{S}_{GF} \qquad (D10)$$

$$= -\int d^dx\left\{m_0\bar{\psi}(x)\psi(x) + \bar{\psi}(x)\gamma_\mu\left[\partial_\mu + isg_0 A_\mu(x)\right]\psi(x)\right\} - \frac{1}{4}\int d^dx[F_{\mu\nu}^a]^2$$

which is the color gauge theory with bare coupling constant sg_0 and bare quark mass m_0.

Bibliography

Baaquie, Belal E. (1976). *The lattice gauge theory: Quantum field theory of gauge fields on a lattice spacetime*, Ph.D. Thesis, Cornell University.

Baaquie, Belal E. (1977a). $(2 + 1)$-dimensional Abelian lattice gauge theory, *Phys. Rev. D* **16**, 3040–3046.

Baaquie, Belal E. (1977b). Gauge fixing and mass renormalization in the lattice gauge theory, *Phys. Rev. D* **16**, 2612.

Baaquie, Belal E. (1977c). The Migdal approximate recursion equation, SLAC-PUB-1964.

Baaquie, Belal E. (1983a). Energy eigenvalues and string tension in the Schwinger model, *Phys. Rev. D* **27**, 962–968.

Baaquie, Belal E. (1983b). Evolution kernel for the Dirac field, *Nucl. Phys. B* **214**, 90.

Baaquie, Belal E. (1985). Gauge fixing the SU(N) lattice-gauge-field Hamiltonian, *Phys. Rev. D* **32**, 2774–2779.

Baaquie, Belal E. (1986). An exact realization of the Kac-Moody algebra, *Phys. Lett. B* **177**, 310–312.

Baaquie, Belal E. (1986). Lattice gauge theory: Hamiltonian, Wilson fermions, and action, *Phys. Rev. D* **33**, 2367–2379.

Baaquie, Belal E. (1988). The Kac-Moody and Virasoro commutation equations, *Nucl. Phys. B* **295**, 188–198.

Baaquie, Belal E. (1991a). Functional differential realization of the Kac-Moody algebra and group cohomology, *Mod. Phys. Lett. A* **6**, 2881–2885.

Baaquie, Belal E. (1991b). Kac-Moody gauge fields and matter fields in d dimensions, *Phys. Lett. B*, **271** 343–346.

Baaquie, Belal E. (1992). Functional differential realization of the SU(2) Kac-Moody algebra, *Int. J. Mod. Phys. A* **07**, 2509–2513.

Baaquie, Belal E. (2000). (Supersymmetric) Kac-Moody gauge fields in 3+1 dimensions, *Phys. Rev. D* **61**, 085009.

Baaquie, Belal E. (2013). *The Theoretical Foundations of Quantum Mechanics* (Springer).

Baaquie, Belal E. (2014). *Path Integrals and Hamiltonians: Principles and Methods*

(Cambridge University Press).

Baaquie, Belal E., and Parwani, Rajesh R. (1996). Asymptotically free $\hat{U}(1)$ Kac-Moody gauge fields in $3 + 1$ dimensions, *Phys. Rev. D* **54**, 5259–5273.

Baaquie, Belal E. and Willeboordse, Frederick H. (2015). *Exploring the Invisible Universe: From Black Holes to Superstrings* (World Scientific Publishing).

Baaquie, Belal E., Huang, Kerson, Peskin, Michael E, and Phua, K. K. (2015). *Ken Wilson Memorial Volume Renormalization, Lattice Gauge Theory, the Operator Product Expansion and Quantum Fields* (World Scientific Publishing).

Baaquie, Belal E. (2018). *Quantum Field Theory for Economics and Finance* (Cambridge University Press).

Chandrasekharan, S., and Wiese, U-J. (2004). An introduction to chiral symmetry on the lattice, *Progr. Part. Nucl. Phys.* **53**, 373–418.

Christ, N. H., and Lee, T. D. (1980). Operator ordering and Feynman rules in gauge theories, *Phys. Rev. D* **22**, 939–958.

Creutz, M. (1983). *Quarks, Gluons and Lattices*, 1st edn. (Cambridge University Press).

Creutz, Michael. (1977a). Gauge fixing, the transfer matrix, and confinement on a lattice, *Phys. Rev. D* **15**, 1128–1136.

Creutz, Michael. (1977b). Gauge fixing, the transfer matrix, and confinement on a lattice, *Phys. Rev. D* **15**, 1128.

Cronstrom, C., and Mickelsson, J. (1983). On Topological boundary characteristics in nonabelian gauge theory, *J. Math. Phys.* **24**, 2528 [Erratum: *ibid.* **27**, 419 (1986)].

Cutler, Roger, and Sivers, Dennis. (1978). Quantum-chromodynamic gluon contributions to large-p_T reactions, *Phys. Rev. D* **17**, 196–211.

Das, A. (2006). *Field Theory: A path integral approach* (World Scientific Publishing).

Dirac, P. A. M. (1999). *The Principles of Quantum Mechanics*, 4th edn. (Oxford University Press).

Drell, Sidney D., Weinstein, Marvin, and Yankielowicz, Shimon. (1976). Strong-coupling field theories. II. Fermions and gauge fields on a lattice, *Phys. Rev. D* **14**, 1627–1647.

Fadin, V. S., and Fiore, R. (2005). Nonforward NLO Balitsky-Fadin-Kuraev-Lipatov kernel, *Phys. Rev. D* **72**, 014018.

Feynman, R. P. (1948). Space-time approach to non-relativistic quantum mechanics, *Rev. Mod. Phys.* **20**, 367–387.

Feynman, R. P., and Hibbs, A. R. (1965). *Quantum Mechanics and Path Integrals* (McGraw-Hill).

Francesco, P., Mathieu, P., and Senechal, D. (2012). *Conformal Field Theory*, 1st edn. (Springer). '

Gepner, D., and Witten, E. (1986). String theory on group manifolds, *Nucl. Phys. B* **278**, 493 – 549.

Gibbons, G. W., and Hawking, S. W. (1977). Cosmological event horizons, thermodynamics, and particle creation, *Phys. Rev. D* **15**, 2738–2751.

Ginsparg, Paul H., and Wilson, Kenneth G. (1982). A remnant of chiral symmetry on the lattice, *Phys. Rev. D* **25**, 2649–2657.

Goddard, P, and Olive, D. (1988). *Kac-Moody and Virasoro Algebras* (World Scientific Publishing).

Gradshteyn, I. S., and Ryzhik, I. M. (1980). *Table of Integrals, Series and Products* (Academic Press).

Gross, D. J., and Witten, E. (1980). Possible third-order phase transition in the large-N lattice gauge theory, *Phys. Rev. D* **21**, 446–453.

Haber, H. E. 2017. Useful relations among the generators in the defining and adjoint representations of SU(N), https://arxiv.org/abs/1912.13302

Hasenfratz, P., Hauswirth, S., Holland, K., Jörg, T., Niedermayer, F., and Wenger, U. (2001). The construction of generalized Dirac operators on the lattice, *Int. J. Mod. Phys. C* **12**, 691–707.

Kadanoff, Leo P. (1977). The application of renormalization group techniques to quarks and strings, *Rev. Mod. Phys.* **49**, 267–296.

Kapusta, J. I. (1993). *Finite Temperature Field Theory* (Cambridge University Press).

Kardar, M. (2007). *Statistical Theory of Fields* (Cambridge University Press).

Knizhnik, V. G., and Zamolodchikov, A. B. (1984). Current algebra and Wess-Zumino model in two dimensions, *Nucl. Phys. B* **247**, 83–103.

Kogut, J, and Susskind, L. (1975). Hamiltonian formulation of Wilson's lattice gauge theories. *Phys. Rev. D* **11**, 395–408.

Luscher, Martin. (1999). Abelian chiral gauge theories on the lattice with exact gauge invariance. *Nucl. Phys. B* **549**, 295–334.

Luscher, Martin. 2000. Weyl fermions on the lattice and the non-abelian gauge anomaly. *Nucl. Phys. B* **568**, 162 – 179.

Malik, R. P. (2001). Dual BRST symmetry for QED, *Mod. Phys. Lett. A* **16**, 477–488.

Marinov, M. S., and Terentyev, M. V. (1979). Dynamics on the group manifold and path integral, *Fortschritte der Phys.* **27**, 511–545.

Menotti, P., and Onofri, E. (1981). The action of SU(N) lattice gauge theory in terms of the heat kernel on the group manifold, *Nucl. Phys. B* **190**, 288 – 300.

Mickelsson, Jouko. (1985). Chiral anomalies in even and odd dimensions. *Commun. Math. Phys.* **97**, 361–370.

Migdal, Alexander A. (1975). Gauge transitions in gauge and spin lattice systems, *Sov. Phys. JETP* **42**, 743 [*Zh. Eksp. Teor. Fiz.* **69**, 1457 (1975)].

Montvay, I., and Munster, G. (1994). *Quantum Fields on a Lattice*, 1st edn. (Cambridge University Press).

Peskin, M. E., and Schroeder, D. V. (1995). *An Introduction to Quantum Field Theory*, 1st edn. (Addison Wesley).

Polyakov, A. M. (1986). Fine structure of strings, *Nucl. Phys. B* **268**, 406.

Pressley, A., and Segal, G. (1988). *Loop Groups*, Oxford Mathematical Monographs (Oxford University Press).

Roth, H. J. (1997). *Lattice Gauge Theories: An Introduction*, 1st edn. (World Scientific).

Ryder, L. H. (2001). *Quantum Field Theory* (Cambridge University Press).

Schwinger, Julian. (1962). Gauge invariance and mass, *Phys. Rev.* **125**, 397–398.

Sugawara, Hirotaka. (1968). A field theory of currents, *Phys. Rev.* **170**, 1659–1662.

Tirrito, E., Rizzi, M., Sierra, G., Lewenstein, M., and Bermudez, A. (2019). Renormalization group flows for Wilson-Hubbard matter and the topological Hamiltonian, *Phys. Rev. B* **99**, 125106.

Varadarajan, V. S. (1974). *Lie Groups, Lie Algebras and their Representations* (Prentice-Hall).

Wadia, Spenta R. (1980). N = ∞ phase transition in a class of exactly soluble model lattice gauge theories, *Phys. Lett. B* **93**, 403 – 410.

Weinberg, S. (2010). *The Theory of Quantum Fields, Volumes I, II, III*, 1st edn. (Cambridge Univesity Press).

Wilson, K. G., and Kogut, John B. (1974). The Renormalization group and the epsilon expansion, *Phys. Rep.* **12**, 75–200.

Wilson, Kenneth G. (1974). Confinement of quarks, *Phys. Rev. D* **10**, 2445–2459.

Wilson, Kenneth G. (1983). The renormalization group and critical phenomena, *Rev. Mod. Phys.* **55**, 583–600.

Wilson, K. G., Lévy, M., and Mitter, P. (1977). *New Developments in Quantum Field Theory and Statistical Mechanics, Cargèse 1976* (Springer).

Witten, Edward. (1983). Global aspects of current algebra, *Nucl. Phys. B* **223**, 422–432.

Witten, Edward. (1989). Quantum field theory and the Jones polynomial, *Commun. Math. Phys.* **121**, 351–399.

York, James W. (1971). Gravitational degrees of freedom and the initial-value problem, *Phys. Rev. Lett.* **26**, 1656–1658.

Zelobenko, D. P. (1973). *Compact Lie Groups and Their Representations* (American Mathematical Society).

Zinn-Justin, J. (1993). *Quantum Field Theory and Critical Phenomenon* (Oxford University Press).

Index

Abelian charge conjugation
  ghost field, 204
antifermion, 94
  conjugation, 94
  Hilbert space, 95
  state space, 94

block gauge field, 161
block-spin action
  Migdal model, 161
    four-fold convolution, 163
block-spin construction
  Migdal model, 160
block-spin renormalization, 157
BRST
  Abelian charge, 202
  Abelian state space, 204
  charge operator
    Abelian, 205
  cohomology, 201
BRST symmetry
  lattice gauge field, 126
BRST transformation
  lattice gauge field, 129

Campbell-Baker-Hausdorff formula, 17
canonical quantization, 249
charge operator
  non-Abelian, 190
chiral Lagrangian, 38
chiral symmetry
  lattice, 216
chromoelectric operator
  Coulomb gauge, 194
  left-invariant, 189

  right-invariant, 189
  transverse, 194
classical solutions, 59
color charge operator, 280
  quark, 190
completeness equation
  fermion, 92
  fermion-antifermion, 96
Coulomb gauge
  non-Abelian, 191
Coulomb lattice potential
  non-Abelian, 198
covariant quatization
  Abelian gauge field, 201

Dirac
  continuum Hilbert space, 247
  fermionic state space, 220
  field coordinates, 212
  finite time action, 259, 261
  Hamiltonian, 243
  lattice, 211
  lattice equation, 213
  lattice Hamiltonian, 227
  lattice Lagrangian, 215
Dirac field
  boundary conditions, 219
  eigenfunctions, 234, 235
  evolution kernel, 233
  Hilbert space, 223
  Wilson metric, 223
Dirac Hamiltonian
  canonical metric, 284
Dirac-Feynman formula
  gauge field Hamiltonian, 188

ghost Hamiltonian, 201

electric field
  non-Abelian, 189
electron, 253
evolution kernel, 211
  $SU(\mathcal{N})$, 56
  $U(1)$ and $SU(2)$, 57
  Dirac field, 233

Faddeev-Popov
  Abelian, 122
  ghost action, 123, 126
  lattice gauge field, 120
fermion, 87
  antifermion
      eigenstates, 103
  antiperiodic trace, 232
  calculus, 87
  canonical, 279
  canonical metric
      Dirac Hamiltonian, 284
  complex, 96
      Gaussian, 98
  Gaussian integration, 96
  generation function, 98
  Hamiltonian, 103
  Hilbert space, 90
  integration, 89
  Lagrangian, 105
  lattice
      noncanonical, 276
  noncanonical anticommutation, 277
  normal ordering, 100
  path integral, 100
  real, 96
  variables, 88
fermion-antifermion
  conjugation, 95
functional differentiation, 31
  chain rule, 32

gamma matrices
  block representation, 212
  Euclidean, 213
gauge field
  generalities, 113
gauge invariant
  non-Abelian state space, 196
gauge-fixing

Abelian, 121
axial gauge
    lattice gauge theory, 275
BRST Abelian Hamiltonian, 198
d=3 Abelian, 133
Gupta-Bleuler, 198
lattice domains, 141
lattice gauge field, 118
non-Abelian, 191
non-Abelian Hamiltonian, 197
Gauss's law, 194, 196
  color charge, 190
  non-Abelian, 190, 196
Gaussian integration
  fermions, 96
Gaussian quantum field, 131
ghost field
  Abelian Hamiltonian, 199
  Abelian state space, 199
Ginsparg-Wilson lattice fermions, 217
Ginsparg-Wilson relation, 219
Gupta-Bleuler condition
  Abelian gauge field, 206

Hamiltonian
  $SU(\mathcal{N})$ Laplace-Beltrami operator, 56
  asymmetric Lagrangian, 188
  axial gauge, 188
  covariant gauge-fixing, 198
  fermionic, 103
  lattice
      Wilson action, 271
  lattice gauge field, 188
  lattice gauge theory, 278
  non-Abelian gauge-fixed, 197
Hilbert space
  Dirac field, 223
  fermionic, 90
  metric
      lattice gauge theory, 273

Kac-Moody
  $SU(2)$ algebra, 44
  $SU(2)$ central extension, 44
  $SU(\mathcal{N})$ central extension, 35
Kac-Moody algebra
  commutation equations, 46
Kac-Moody algebras, 30
Kac-Moody Casimir
  regularization, 50

Kac-Moody generators, 33
  two cocycle, 36

Lagrangian
  fermionic, 105
  lattice Dirac, 215
  lattice gauge field, 115
  Yang-Mills, 114
lattice fermions
  chiral symmetry, 216, 217
  Luscher, 219
lattice gauge field, 113
  Abelian
    mean field, 139
    Wilson loop, 137
  axial gauge, 188
  block-spin renormalization, 157
  BRST symmetry, 126
  Faddeev-Popov counter-term, 120
  field tensor, 115
  gauge transformation, 115
  gauge-fixed path integral, 119
  gauge-fixing, 118
  Hamiltonian, 188
  mass renormalization, 143
  partition function, 116
  two-dimensional, 158
  weak coupling, 116
lattice gauge theory
  Hamiltonian, 271, 278
  transfer matrix, 272
Lie group
  $SU(\mathcal{N})$, 9
  $\mathcal{SU}(2)$ generators, 23
  adjoint representation, 11
  axioms, 9
  Haar measure, 15
  Lie algebras, 10
  metric tensor, 13
  Ricci tensor, 14
loop group
  $\mathcal{SU}(2)$, 43

mass divergence
  four sources, 147
  ghost field, 148
  lattice gauge field, 149
mass renormalization
  lattice gauge field, 143
  one-loop, 153

self-energy, 144
Slavnov-Taylor identity, 154
  zero, 153
  zero mode, 146
  zero mode effective action, 146
Migdal model
  approximation
    $d = 3$, 167
    $d > 2$, 168
    planar loops, 166
  block-spin construction, 160
  confinement
    asymptotic freedom, 184
  four-fold convolution, 163
  generalization to $d > 2$, 164
Migdal model $SU(2)$
  numerical recursion equation, 176
  recursion equation
    numerical algorithm, 178
    numerical results, 182, 183
    weak coupling, 185
  weak coupling, 170
  weak coupling recursion, 172, 173
    β-function, 175
    $d = 4 + \epsilon$, 174
Migdal recursion equation
  $d > 2$ dimensions, 165

operators
  fermionic, 102
overcomplete basis
  fermion, 92

path integral
  fermionic, 100
pfaffian, 97
photon
  state space, 204
point-split regularization, 49
positron, 253

quantization
  Abelian Faddeev-Popov, 198
quantum field
  discrete Gaussian, 134
quark, 211
  color charge operator, 190

renormalization group
  fixed points, 165

renormalization group transformation
  Migdal model, 159

short distance expansion, 49
spinor, 211
state space
  non-Abelian Jacobian, 197
Sugawara Hamiltonian, 40
$SU(2)$
  measure, 16
    invariance, 17
  Migdal model
    strong coupling, 169
    strong coupling fixed point, 170
    strong coupling recursion, 170
    weak coupling, 170
    weak coupling fixed point, 170
    weak coupling recursion, 172
$SU(3)$
  Casimir operator, 81
  character functions, 81, 84
  evolution kernel, 83
  invariant measure, 85
  irreducible representations, 82
$SU(\mathcal{N})$
  path integrals, 55, 72
  differential generators, 24
    left-invariant, 24
    right-invariant, 24
  fundamental representation, 19
  irreducible representations, 18
  Laplace-Beltrami operator, 25, 197
  path integrals
    continuum limit, 74
    zero mass renormalization, 78

transfer matrix
  lattice gauge theory, 272

transition amplitude
  fermionic, 107

vielbein, 12
Virasoro algebra
  central charge, 52
  commutation equations, 49, 52

Wilson action
  Hamiltonian, 271
  lattice gauge field, 115
  modified Abelian, 132
Wilson lattice fermions, 212, 213, 216, 217, 271
Wilson loop, 113
  Abelian, 137
  topology, 113
Wilson metric, 228, 275
  Dirac field, 223
  lattice Dirac field, 226
  lattice gauge theory, 273
Wilson plaquette, 115
Wilson renormalization, 157
  Migdal model, 182
winding number
  classical paths, 61
  classical solutions, 59
Witten, 113
WZW action, 42
WZW Hamiltonian, 40, 41
WZW Lagrangian, 40, 42

Yang-Mills
  action, 114
  field tensor, 114
  gauge transformation, 114
  Lagrangian, 114

www.ingramcontent.com/pod-product-compliance
Lightning Source LLC
Chambersburg PA
CBHW081529190326
41458CB00015B/5503